D0935506

GENETIC DATA ANALYSIS

GENETIC DATA ANALYSIS

Methods for Discrete Population Genetic Data

Bruce S. Weir
North Carolina State University

Sinauer Associates, Inc. Publishers
Sunderland, Massachusetts

GENETIC DATA ANALYSIS

Sunderland, MA 01375 USA.

Library of Congress Cataloging-in-Publication Data

Weir, B. S. (Bruce S.), 1943-
Genetic data analysis: methods for discrete population genetic data / Bruce S. Weir.
p. cm.
Includes bibliographical references.
ISBN 0-87893-871-0
ISBN 0-87893-872-9 (pbk.)
1. Genetics – Statistical methods. 2. Quantitative genetics. 3. Gene frequency – Statistical methods. I. Title.
QH438.4.S73W45 1990 89-26173
575.1′072–dc20 CIP

Printed in U.S.A.

4 3 2 1

To Beth

Contents

Preface

This book describes methods for the statistical analysis of discrete population genetic data. Such data are generally counts, whether of genotypes or of base types, and I have omitted all mention of quantitative genetic data that result from measurement of genetic traits such as height or yield.

Population geneticists are now facing the exciting task of interpreting new kinds of genetic data and it seems timely to review some of the statistical techniques appropriate for restriction fragment length polymorphisms and DNA sequences. I hope that this book shows, however, that even new data can be handled with the techniques developed in the early studies of R.A. Fisher, J.B.S. Haldane and Sewall Wright and extended by many workers since then. It is my hope that this treatment will bridge the gap between the excellent treatment, 20 years ago, of Regina Elandt-Johnson in her *Probability Models and Statistical Methods in Genetics* and the current literature in scientific journals. I have tried to give a wide coverage, and have referred where appropriate to more specialized books. For example, the reader may want to read Jurg Ott's *Analysis of Human Genetic Linkage* for details on estimating linkage, especially in humans, and Masatoshi Nei's *Molecular Evolutionary Genetics* for a comprehensive treatment of molecular genetic data. This book is intended to be a guide to methods for traditional analyses of Hardy-Weinberg and linkage disequilibrium and to methods for characterizing population structure and estimating genetic distance. It also introduces the reader to some of the many methods of constructing evolutionary trees and to the difficult problems of aligning and comparing sequences.

Writing about analyses of molecular genetic data is like aiming at a moving target. Increasingly sophisticated methods are being developed rapidly, and indeed will be needed as data accumulate from the Human Genome Project. I will be grateful to receive suggestions of topics to be included in future editions, and to learn of any errors or obscurities in this edition. That any deficiencies are not more numerous is due in large part to the very fine reviewing provided by many people. It is a great pleasure to acknowledge my indebtedness to Bill Hill (Edinburgh), Wyatt Anderson, Jonathan Arnold, Jamie Cuticchia and Yunxin Fu (Athens, Georgia), Joe Felsenstein (Seattle), Mike Clegg (Riverside), and Chris Basten, Ken Dodds, Rebecca Doerge, Spencer Muse, and Jun Zhu (Raleigh). I also received advice on specific topics from Walter Fitch (Irvine), Anthony Edwards (Cambridge) and John Monahan (Raleigh). The book grew out of a set of notes I prepared for a course I have taught in Raleigh. Several students from North Carolina State and Duke Universities have helped clarify the approach I have used.

I am grateful to the many individuals and publishers who have given me permission to include unpublished data and copyrighted material.

It has been a pleasure to work with Sinauer Associates in the production of this book and a special acknowledgement is due Andy Sinauer. It was at his instigation that I embarked upon the task of learning LaTeX. He has been very patient with the delays caused by my slow learning. The result is a book that could be modestly priced and that escaped one link in the chain that can introduce typographical errors. My efforts in using this typesetting package were greatly aided by help from Chris Basten and Hal Caswell. Chris Basten helped particularly with some of the figures.

It is with some trepidation that I have included the source code listings for some computer programs. By doing so I reveal my naïvité as a programmer and risk the possibility that the user will be given wrong results. I have been encouraged to include them, however, because of the continuing requests for copies of my linkage disequilibrium and F-statistics programs. I would appreciate any comments on the usefulness of their inclusion in the book. Although the programs are not particularly efficient, I think they are correct. A user should check them against the sample input files I have also listed. They are all written in FORTRAN/77 and have run on microcomputers. I cannot accept responsibility for any analyses that may be based on these programs.

It will be quite obvious to anyone who opens these covers how much I have been influenced by Clark Cockerham's ideas. Clark has been a guide to me for 25 years. I could not have been more fortunate.

Finally, I thank Beth, Claudia and Henry for their support and their patience.

Raleigh, North Carolina
March, 1990

GENETIC DATA ANALYSIS

Chapter 1

Introduction

THE NATURE OF GENETIC DATA

This book treats data collected for discrete characters that can range from morphological traits such as flower color to individual base types in a DNA sequence. Whatever the nature of the data, the analyses to be presented are for observations on discrete heritable units that can be characterized as MENDELIZING UNITS. In other words, observations are made on units that are transmitted from parent to offspring, and whose transmission is a random process. At one time such units would have been called genes, but it is no longer desirable to restrict attention to these DNA regions that encode for proteins. Statistical theories can be developed for units that are indeed genes or that depend on physical properties of gene products, or that may have nothing to do with coding regions.

Before embarking on this study it is interesting to consider just what kinds of discrete data are used in population genetics. Several different kinds of data are shown in this chapter.

EXAMPLES OF GENETIC DATA

Phenotypic Data

The best known phenotypic data are those given by Mendel for the garden pea *Pisum sativum.* Mendel observed seven characters in a series of crosses and gave the results, from the offspring of the resulting hybrid plants, shown in Table 1.1. These data allowed Mendel to postulate a 3:1 ratio for dominant and recessive phenotypes, although there has been discussion that they fit this rule rather too well. This argument will be pursued later in the chapter.

Table 1.1 Mendel's results for seven dominant characters in *Pisum sativum.* Source: Mendel (1866).

	Character	Dominant Form		Recessive Form	
		Seed characters			
A	Seed shape	5474	Round	1850	Wrinkled
B	Cotyledon color	6022	Yellow	2001	Green
		Plant characters			
C	Seed coat color	705	Grey-brown	224	White
D	Pod shape	882	Simply inflated	299	Constricted
E	Unripe pod color	428	Green	152	Yellow
F	Flower position	651	Axial	207	Terminal
G	Stem length	787	Long	277	Short

Codominant Phenotypic Data

As Fisher (1936) pointed out in discussing Mendel's work, a problem with dominant characters where heterozygotes and dominant homozygotes cannot be distinguished is that parental types cannot be determined with certainty. This issue does not arise with codominant traits such as some blood groups, and a particularly complete data set, published by Race et al. (1949), is shown in Table 1.2.

The MN blood groups are recognized by agglutination with anti-M or anti-N serum and can be regarded as being controlled by a single locus with two alleles, M and N. The three genotypes can all be distinguished. The S blood groups are detected with anti-S serum and are controlled by a locus with alleles S and s. The two genotypes SS and Ss both agglutinate with anti-S serum and could not be differentiated by Race et al. The M and S loci are closely linked. In Table 1.2, the notation *MN.S* indicates that that the S gene may be located on either the M or N bearing chromosomes, while *MSNs*, for example, is used if it is clear that M and S are on one chromosome, and N, s are on the other. Two children linked with an "=" are monozygotic twins, while those linked with a "–" are dizygotic twins.

Allozyme Data

A major impact on population genetics was provided by development of the technique for electrophoretic detection of charge differences for variants among soluble proteins. Protein material from an individual is placed at one end of a slab of gel and allowed to migrate under the action of an electric field along the gel. Migration rates are determined by the size, shape, and charge of the protein. Material from several individuals can be accommodated in distinct lanes on a single gel. A comprehensive discussion of the implications of the finding of large amounts of electrophoretic variation was given by Lewontin (1974). The different allelic forms of a protein detected in this way are called ALLOZYMES. Observations could now be made at the level of gene products rather than on the phenotype of an individual. Although only about 30% of proteins are soluble, and although only about 25% of the differences in DNA sequences for the structural genes encoding these proteins lead to charge differences detectable by electrophoresis, the technique remains the easiest and cheapest way of collecting data on many individuals and several loci. Surveys of 10 or so loci in several hundred individuals are often reported.

The electrophoretic banding patterns used by Clegg and Allard (1973) to infer genotypes at six loci in wild oats, *Avena fatua*, are shown in Figure 1.1.

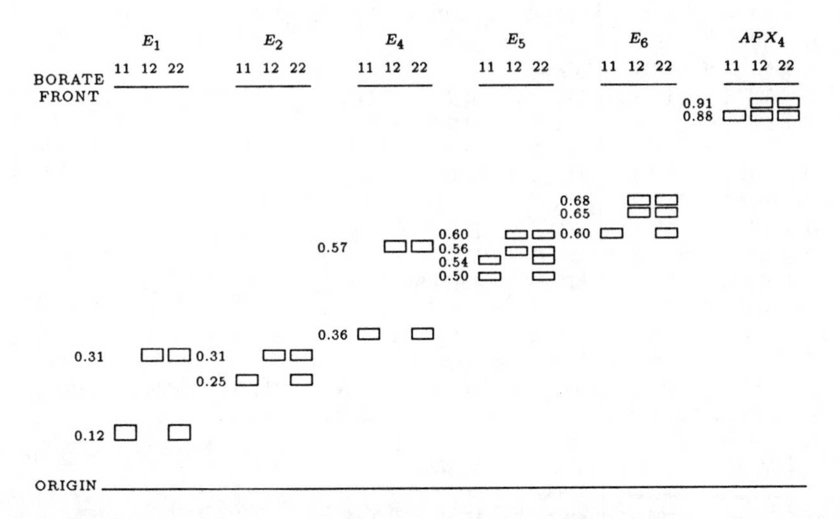

Figure 1.1 Electrophoretic banding pattern for six loci in wild oats. Source: Clegg and Allard (1973).

Table 1.2 MNS blood group types for families reported by Race et al. (1949).

	Parents		Children					
	Father	Mother	1	2	3	4	5	6
1	MSMs	MsMs	MsMs	MSMs	MSMs			
2	MM.S	MsMs	MSMs	MSMs				
3	MM.S	MM.S	MM.S	MM.S	MM.S			
4	MM.S	MM.S	MM.S					
5	MM.S	MM.S	MM.S					
6	MM.S	MM.S	MM.S	MM.S	MM.S			
7	MsMs	MsNs	MsNs	MsNs				
8	MsNs	MsMs	MsMs					
9	MsNs	MsMs	MsNs					
10	MsMs	MsNs	MsMs=	=MsMs				
11	MsMs	MsNs	MsNs	MsMs	MsNs	MsMs		
12	MsNs	MSMs	MSNs–	–MsMs				
13	MSMs	MsNs	MsNs	MSNs	MsMs			
14	MsNs	MSMs	MsMs	MSNs	MSMs			
15	MM.S	MsNs	MSMs	MSMs	MSNs	MSNs	MSNs	MSNs
16	MsNs	MM.S	MSMs	MSMs=	=MSMs			
17	MM.S	MsNs	MSMs	MSMs				
18	MsNs	MM.S	MSMs=	=MSMs				
19	MsNs	MM.S	MSNs					
20	MsNs	MM.S	MSMs	MSMs				
21	MsMs	MSNs	MSMs	MsNs	MSMs	MsNs		
22	MsMs	MSNs	MSMs	MSMs	MSMs–	–MSMs		
23	MSNs	MSMs	MM.S	MsNs				
24	MSNs	MSMs	MM.S	MsNs				
25	MsNS	MSMs	MsMs					
26	MM.S	MN.S	MM.S	MN.S	MN.S			
27	MM.S	MN.S	MM.S	MN.S				
28	MM.S	MN.S	MM.S	MM.S				
29	MN.S	MM.S	MN.S	MM.S				
30	MM.S	MN.S	MN.S					
31	MM.S	MN.S	MM.S	MN.S				
32	MM.S	MN.S	MN.S	MM.S				
33	MN.S	MM.S	MN.S–	–MM.S				
34	MM.S	MN.S	MM.S	MM.S				
35	MM.S	MN.S	MM.S	MM.S	MN.S			
36	NsNs	MsMs	MsNs	MsNs				
37	NsNs	MsMs	MsNs	MsNs	MsNs	MsNs	MsNs	MsNs
38	NsNs	MsMs	MsNs	MsNs	MsNs			
39	MsMs	NsNs	MsNs					
40	MSMs	NsNs	MsNs	MsNs				
41	MSMs	NsNs	MSNs	MsNs				
42	NsNs	MSMs	MSNs	MsNs	MSNs			
43	NsNs	MM.S	MSNs					
44	NN.S	MsMs	MsNS					
45	NsNs	MSMs	MsNs	MN.S				
46	MM.S	NN.S	MN.S					

(Continued)

Table 1.2 Continued

	Parents		Children					
	Father	Mother	1	2	3	4	5	6
47	NN.S	MM.S	MN.S=	=MN.S				
48	MsNs	MsNs	MsNs	MsNs				
49	MsNs	MsNs	MsNs	MsMs	NsNs			
50	MsNs	MsNs	NsNs					
51	MsNs	MsNs	MsNs	NsNs				
52	MSNs	MsNs	MSNs	NsNs	MsNs			
53	MsNs	MSNs	NsNs	MSNs				
54	MsNs	MSNs	MSMs	MSNs	NsNs			
55	MSNs	MsNs	NsNs	MSNs	MSMs			
56	MsNs	MSNs	MSMs	MsNs	MSNs			
57	MSNs	MsNs	NsNs	MSNs	MSNs			
58	MSNs	MsNs	NsNs					
59	MsNs	MSNs	MSNs	MsNs	MsNs	MSMs	NsNs	
60	MsNs	MsNS	MsMs					
61	MsNs	MN.S	MN.S–	–MsNs	MN.S			
62	MsNs	MN.S	NSNs					
63	MsNs	MN.S	MSMs	MN.S				
64	MN.S	MsNs	MN.S					
65	MN.S	MsNs	MN.S					
66	MSNs	MSNs	NsNs	MSNs				
67	MSNs	MSNs	MSMS	MSNs	NsNs	MSMS	MsNs	
68	MSNs	MSNs	MSMS	NsNs				
69	MN.S	MN.S	MN.S	NSNs	MsNs	NSNs		
70	MN.S	MN.S	MN.S	MM.S				
71	MN.S	MN.S	MN.S	MN.S				
72	MN.S	MN.S	MN.S	MN.S				
73	NsNs	MsNs	NsNs					
74	MSNs	NsNs	MSNs	NsNs				
75	NsNs	MSNs	MSNs	NsNs				
76	NsNs	MSNs	NsNs	NsNs				
77	MSNs	NsNs	NsNs					
78	NsNs	MSNs	MSNs	NsNs	NsNs			
79	MsNS	NsNs	MsNs	MsNs	NSNs			
80	NsNs	MN.S	NSNs					
81	MN.S	NsNs	NSNs	NSNs				
82	NsNs	MN.S	MSNs	MSNs	MSNs			
83	NSNs	MSNs	MN.S	NsNs				
84	NN.S	MN.S	MN.S	NN.S	MN.S	NN.S		
85	MN.S	NN.S	MN.S	MN.S				
86	MsNs	NSNs	MsNs	NsNs	NSNs			
87	NN.S	MsNs	MsNS					
88	NN.S	MsNs	MsNS					
89	NsNs	NsNs	NsNs	NsNs	NsNs			
90	NsNs	NSNs	NsNs					
91	NSNs	NsNs	NSNs	NsNs	NsNs			
92	NSNs	NsNs	NsNs					
93	NN.S	NsNs	NSNs	NSNs	NSNs			

All six loci in this study had two alleles, although any allele may produce one or two bands on the gel. It can be seen from Figure 1.1 that one of the bands for the E_1 locus migrates the same distance as one from the E_2 locus, so that all three genotypes can be scored at either locus only if the other

Table 1.3 Multilocus allozyme genotypes in a sample of *Aedes aegypti*. Source: J. Powell (personal communication).

Indiv.	*Sex*	*Pgi*	*Pgd*	*Gpd*	*Idh1*	*Idh2*	*Pgm*	*Mdh*	*Hk1*	*Hk2*	*Hk3*	*Hk4*
1	1	11	11	11	12	11	34	12	11	11	11	11
2	1	11	11	11	12	11	22	12	11	11	11	11
3	1	11	11	11	12	11	23	11	11	11	11	11
4	1	11	11	11	11	11	11	11	11	11	11	11
5	1	11	11	11	12	11	34	12	11	11	11	
6	1	11	11	11	11	11	22	11	11	11	11	
7	1	11	11	11	11	11	11	11	11	11	11	
8	1	11	11	11	12	11	24	12	11	11	11	
9	1	11	11	11	11	12	11	11	11	11	11	
10	1	11	11	11	11	11	12	11	11	11	11	11
11	1	11	11	11	12	11	24	11	11	11	11	11
12	1	11	11	11	12	11	24	12	11	11	11	11
13	1	11	11	11	12	11	34	11	11	11	11	11
14	1	11	11	11	11	11	11	11	11	11	11	
15	1	11	11	11	12	11	22	12	11	11	11	
16	1	11	11	11	12	11	34	11	11	11	11	11
17	1	11	11	11	12	11	23	12	11	11	11	11
18	1	11	11	11	12	11	24	11	11	11	11	11
19	1	11	11	11	12	11	22	11	11	11	11	11
20	1	11	11	11	12	11	34	11	11	11	11	11
21	1	11	11	11	11	12	11	11	11	11	11	11
22	1	11	11	11	12	11	23	12	11	11	11	11
23	1	11	11	11	12	11	34	11	11	11	11	11
24	2	11	11	11	11	11	11	11	11	11	11	11
25	2	11	11	11	11	11	34	11	11	11	11	11
26	2	11	11	11	11	11	24	12	11	11	11	11
27	2	11	11	11	11	11	11	11	11	11	11	11
28	2	11	11	11	11	11	34	12	11	11	11	11
29	2	11	11	11	11	11	24	12	11	11	11	11
30	2	11	11	11	11	11	24	11	11	11	11	11
31	2	11	11	11	11	11	23	11	11	11	11	11
32	2	11	11	11	11	11	22	11	11	11	11	11
33	2	11	11	11	11	11	23	11	11	11	11	11
34	2	11	11	11	11	11	11	11	11	11	11	11
35	2	11	11	11	11	11	34	12	11	11	11	11
36	2	11	11	11	11	11	24	11	11	11	11	11
37	2	11	11	11	11	11	34	11	11	11	11	11
38	2	11	11	11	11	11	23	12	11	11	11	11
39	2	11	11	11	11	11	11	11	11	11	11	11
40	2	11	11	11	11	11	23	11	11	11	11	11

locus is homozygous for the band that does not overlap.

The genetic basis of the banding patterns should be determined before commencing any genetic analysis of electrophoretic data. This is best accomplished by making crosses between individuals with different patterns and checking the segregation ratios in the offspring. Such a procedure was followed by Clegg and Allard, but often it may not be practical. In those cases it is common to argue that banding patterns do indeed correspond to distinct loci by analogy to the situation for the same enzymes in species where genetic tests have been made.

Data from electrophoretic studies are recorded as multilocus genotypes for each of the individuals scored. With horizontal slicing of gels (i.e., slicing along planes parallel to the direction to the lines of movement of the proteins) and application of appropriate stains to each slice, many enzymes can be scored. Genotypes at 11 loci in a sample of the yellow fever mosquito, *Aedes aegypti*, collected in Ghana and made available by J. Powell are shown in Table 1.3. Different alleles at a locus are represented by different numbers, and the pairs of numbers give the genotypes. There is no way to determine from the electrophoretic gel which of a pair of genes is paternal and which is maternal. This can cause problems when analyses are performed on two or more loci together. Another problem is that not every locus is scored in every individual. Such a lack of balance is a feature of almost all population genetic data sets. Finally, it can be seen in this table that several of the loci have only one allelic type, and so show sample MONOMORPHISM.

Protein Sequence Data

Electrophoresis works with charge differences among different forms of a protein, but a more complete picture of proteins was provided when it became possible to determine the sequence of amino acids constituting the protein. The variation revealed by this technology allowed more detailed studies of the evolutionary relationships between different species. Figure 1.2 displays a collection of protein sequences for fish protamines (Hunt and Dayhoff 1982). Protamines, or sperm histones, are polypeptides associated with nuclear DNA that replace histones in maturing sperm of vertebrates. The sequences in Figure 1.2 have been aligned to give lowest percentage differences and employ a standard one-letter code for amino acids, as shown in Table 1.4.

Protein sequences are now collected into databases so that they are widely accessible. One such database is that maintained by the National Biomedical Research Foundation in Washington, D.C.

```
 1  Tuna Y2       PRRRR--QASRPVRRRRRYRRSTAARRRRRVVRRRR
 2  Tuna Z2       PRRRR--RSSRPVRRRRRYRRSTAARRRRRVVRRRR
 3  Tuna Z1       PRRRR--RSSRPVRRRRRYRRSTAARRRRRVVRRRR
 4  Salmon AII    PRRRRRRSSSRPIRRRR-YRRAS--RRRRRGGRRRR
 5  Trout IB      PRRRRRRSSSRPIRRRR-PRRVS--RRRRRGGRRRR
 6  Salmon AI     PRRRR--SSSRPVRRRRRPR-VSR-RRRRRGGRRRR
 7  Trout IA      PRRRR--SSSRPVRRRRRPRRVSR-RRRRRGGRRRR
 8  Trout II      PRRRR--SSSRPVRRRR-ARRVSR-RRRRRGGRRRR
 9  Herring YII   PRRR-TRRASRPVRRRR-PRRVS--RRRR--ARRRR
10  Herring Z     ARRRRSRRASRPVRRRR-PRRVS--RRRR--ARRRR
11  Herring YI    ARRRRS--SSRPIRRRR-PRRRTT-RRRR-AGRRRR
12  Sturgeon B    ARRRRR--SSRPQRRRRR-RRHG--RRRR--GRR--
13  Sturgeon A    ARRRRRHASTKLKRRRRR-RRHQ--KK----SHK--
```

Figure 1.2 Alignment of fish protamine sequences. Source: Hunt and Dayhoff (1982).

Restriction Fragment Data

Information on the DNA sequence itself grew with the use of RESTRICTION ENDONUCLEASES, a class of enzymes that cleaves DNA molecules at specific recognition sites. Use of such enzymes, coupled with the ability to identify particular regions of the genome with labeled probes, made it possible to score for the presence or absence of several restriction sites in an individual. Unlike observations on proteins or phenotypes, restriction sites can be associated with noncoding as well as coding regions of the genome, although

Table 1.4 One-letter and three-letter codes for the 20 amino acids.

	Amino Acid	Codes			Amino Acid	Codes	
1	Alanine	Ala	A	11	Leucine	Leu	L
2	Arginine	Arg	R	12	Lysine	Lys	K
3	Asparagine	Asn	N	13	Methionine	Met	M
4	Aspartic acid	Asp	D	14	Phenylalanine	Phe	F
5	Cysteine	Cys	C	15	Proline	Pro	P
6	Glutamine	Gln	Q	16	Serine	Ser	S
7	Glutamic acid	Glu	E	17	Threonine	Thr	T
8	Glycine	Gly	G	18	Tryptophan	Trp	W
9	Histidine	His	H	19	Tyrosine	Tyr	Y
10	Isoleucine	Ile	I	20	Valine	Val	V

Table 1.5 Some common restriction enzymes and their recognition sequences.

Enzyme	Recognition Sequence
*Bam*HI	GGATCC CCTAGG
*Eco*RI	GAATTC CTTAAG
*Hind*III	AAGCTT TTCGAG
*Taq*I	TCGA AGCT
*Xho*I	CTCGAG GAGCTC

they still meet the criterion of being Mendelizing units. When specific DNA segments are probed and digested with a restriction enzyme, the resulting set of fragments indicates the presence or absence of recognition sites for the enzyme. Variation in fragment sizes across a sample can also be due to sequence insertions or deletions unrelated to the enzyme. A list of some common restriction enzymes is given in Table 1.5, along with their double-stranded recognition sequences.

Variation at a set of four restriction sites, five insertions or deletions, as well as the two *Adh* allozymes is shown for a series of lines of the fruit fly *Drosophila melanogaster* in Table 1.6. These data were obtained by Langley et al. (1982) from stocks of *D. melanogaster* made homozygous for independent second chromosomes. A region of about 12,000 nucleotides containing the alcohol dehydrogenase (*Adh*) locus was studied. Apart from the four variable sites, another 20 restriction sites were found to be present in all lines. In Table 1.6, a "+" indicates the presence and a "–" indicates the absence of a restriction site or an insertion or deletion. The two *Adh* allozymes are written as F or S.

The relative positions (about an origin taken at a *Bam*HI site in the *Adh* coding region) of the four restriction sites are given in Table 1.6, illustrating one of the novel features of such data. Not only is information about genetic variation made available at several locations in the genome,

Table 1.6 Restriction map variants among *D. melanogaster* lines. Source: Langley et al. (1982).

		Restriction Sites				Insertions/Deletions*				
		*Bam*HI	*Hind*III	*Hind*III	*Xho*I	Δa	Δb	Δc	Δd	Δf
Line	*Adh*	−7.1†	−3.0	+2.7	+1.2	20nt	550nt	900nt	180nt	30nt
R1	*S*	+	−	−	+	−	−	−	+	−
R2	*S*	+	−	−	+	−	−	−	+	−
M1	*S*	+	−	−	+	−	−	−	−	−
R3	*S*	+	−	−	+	−	−	−	−	−
N1	*S*	+	−	−	−	−	−	−	−	+
N2	*S*	+	−	−	−	−	−	−	−	+
R4	*S*	+	−	−	−	−	−	−	−	−
R5	*S*	+	−	−	−	−	−	−	−	−
R6	*S*	+	−	+	−	−	−	−	−	−
K1	*S*	+	−	+	−	−	−	+	−	−
K2	*S*	+	−	+	+	−	−	−	−	−
K3	*F*	−	−	−	+	−	−	−	−	−
K4	*F*	−	−	−	+	−	−	−	−	−
R8	*F*	−	+	−	+	−	−	−	−	−
M2	*F*	−	+	−	+	−	−	−	−	−
R9	*F*	−	−	−	+	+	−	−	−	−
R10	*F*	−	−	−	+	+	−	−	−	−

* a, d, f are deletions; b, c are insertions. Estimated sizes are given in numbers of nucleotides.

† These figures (thousands of nucleotides indicate the positions of the restriction sites.

but also the physical distance between these locations can generally be determined. Several databases have been established that include information on variable restriction sites, primarily in humans. They include the Howard Hughes Medical Institute Human Gene Mapping Library in New Haven, CT, and that maintained at the Centre d'Etude du Polymorphisme in Paris.

DNA Sequence Data

The ultimate step in obtaining genetic data is the determination of the sequence of bases in the DNA molecule. There are four kinds of base, `A`,`T`,`C`, and `G` as shown in Table 1.7 (`U` replaces `T` in RNA molecules), and DNA sequences can be found for coding as well as noncoding regions. Such data

Table 1.7 The four nitrogenous bases and their types. The same two purines, A and G, are found in DNA and RNA. The two pyrimidines in DNA are C and T; in RNA U replaces T.

Type	Base
Purine	Adenine
	Guanine
Pyrimidine	Cytosine
	Thymine
	Uracil

have been collected with different objectives in mind. Constructions of phylogenies for a set of species have used a single sequence from each of the species, while questions concerning variation within species require several sequences from a species.

Mitochondrial DNA sequences for the *tRNA his* gene are shown in Figure 1.3. These sequences, along with other mitochondrial sequences (Brown et al. 1982), have been used to construct evolutionary histories of man and apes. Information on the sequences of the *Adh* region in *D. melanogaster* collected from isochromosomal lines derived from 11 populations by Kreitman (1983) is shown in Figure 1.4. Among the 2,721 nucleotides sequenced in the 11 lines, 43 were found to be variable, and it is the bases at these polymorphic sites that are displayed in the figure. The reference sequence shown at the head of the table consists of the sequence of most common bases among the six lines that had the S allozyme for *Adh*.

As is the case with restriction site data, DNA sequence data contain information about the relative physical locations of the observational units in the genome. Many of the properties of the proteins encoded by DNA sequences

```
1   GTAAATATAG TTTAACCAAA ACATCAGATT GTGAATCTGA CAACAGAGGC TTACGACCCC TTATTTACC
2   GTAAATATAG TTTAACCAAA ACATCAGATT GTGAATCTGA CAACAGAGGC TCACGACCCC TTATTTACC
3   GTAAATATAG TTTAACCAAA ACATCAGATT GTGAATCTGA TAACAGAGGC TCACAACCCC TTATTTACC
4   GTAAATATAG TTTAACCAAA ACATTAGATT GTGAATCTAA TAATAGGGCC CCACAACCCC TTATTTACC
5   GTAAACATAG TTTAATCAAA ACATTAGATT GTGAATCTAA CAATAGAGGC TCGAAACCTC TTGCTTACC
```

Figure 1.3 Five mitochondrial DNA sequences. 1: Human; 2: chimpanzee; 3: gorilla; 4: orangutan; 5: gibbon. Source: Brown et al. (1982).

Reference:	CCGCAATATGGGCGCTACCCCCGGAATCTCCACTAGACAGCCT
Line	
Wa-S	........AT........TT.ACA.TAAC..............
Fl-1S	..C...............TT.ACA.TAAC..............
Af-S	...................................A....T.A
Fr-S	................GT.................A..TA...
Fl-2S	...AG...A.TC..AGGT..................C......
Ja-S	..C............G..............T.T.CAC....T.
Fl-F	..C............G.............GTCTCC.C......
Fr-F	TGCAG...A.TCG..G.............GTCTCC.CG.....
Wa-F	TGCAG...A.TCG..G.............GTCTCC.CG.....
Af-F	TGCAG...A.TCG..G.............GTCTCC.CG.....
Ja-F	TGCAGGGGA....T.G....A...G....GTCTCC.C......

Figure 1.4 Variable nucleotides among 11 *Drosophila Adh* sequences. Source: Kreitman (1983).

depend on the two- and three-dimensional structure of the molecules, and not just on the one-dimensional amino acid sequence. Advances in inferring such structures from nucleotide sequences (e.g., Zuker 1989) are beyond the scope of this book.

DNA sequences are collected into databases in the same way as are protein sequences. The principal databases are GenBank in the United States and the EMBL databank in the Federal Republic of Germany.

GENETIC AND STATISTICAL SAMPLING

One of the themes that will recur in this book is that the analysis of genetic data must be based on some theory or model. The various types of sampling used to generate the data must be kept in mind, for these will affect the properties of estimates and will determine the scope of the inferences that may be drawn from the data. Frequent mention will be made of the distinction between statistical and genetic sampling, and the distinction can be seen most easily from the idealized situation in Figure 1.5.

The data to be analyzed will be for a sample of n individuals taken from a population of size N. Statistical sampling refers to this choice of n individuals from N. Another sample, also of size n, from the same population may give different values of quantities such as gene frequency simply because individuals with different genotypes may be sampled. Most analyses will be based on multinomial sampling, in which every member of the population

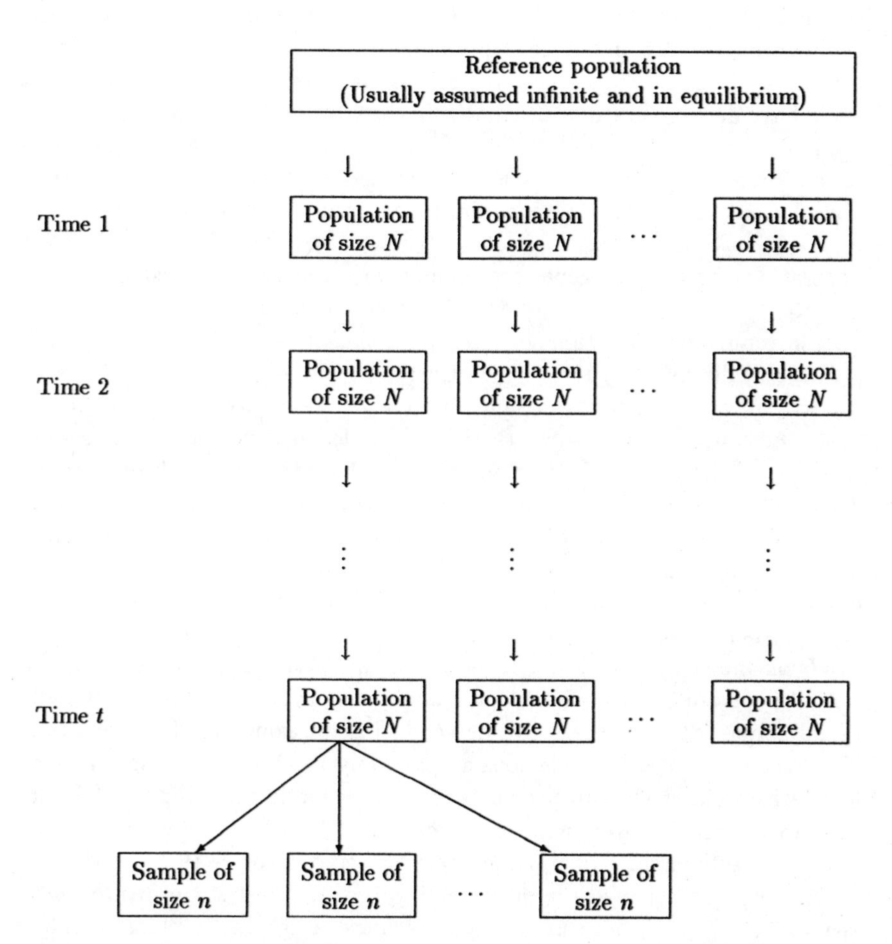

Figure 1.5 Representation of the two processes of genetic and statistical sampling.

is equally likely to be chosen at any stage to be a member of the sample. Even though repeated samples are not generally taken, there are statistical theories that predict how much variation would be likely between such samples and this variation is taken into account in making inferences about the population on the basis of sample data.

For genetic data there is another source of variation to be considered, and this arises from the sampling inherent in the transmission of genetic material from parent to offspring. The population of size N from which a sample is drawn can be regarded as just one of the many replicate populations that could have descended from the same reference, or founder, population. Even if all factors such as population size, mating structure, selection, and mutation forces were kept exactly the same, these replicate populations would differ because different genes may be transmitted at each locus between generations for the different replicates. There is variation between replicate populations because of genetic sampling, and theory is needed to accommodate this variation. A vivid experimental demonstration of the effects of drift in replicate populations of *Drosophila melanogaster* was given by Buri (1956) and discussed in the text by Hartl and Clark (1989).

Just as variation between samples from the same population cannot be measured unless there is more than one sample, so variation between replicate populations cannot be measured from data within one such population. This could pose a problem since the replicate populations are often conceptual only, but some idea of the variation can be measured from different loci. Providing the loci can be regarded as being nearly independent; their differences in pedigree mimic the differences between populations.

Whether attention is paid to both statistical and genetic sampling depends on the goals of a study. If there is interest just in the present population, it is possible to ignore the genetic sampling that has made this population different from all others that descend from the same founder population. If, on the other hand, conclusions are to apply to all populations that may have arisen under similar circumstances, the genetic sampling must be incorporated. Similar arguments are made in distinguishing between fixed and random effects in statistical models (e.g., Steel and Torrie 1980, p. 149).

Another situation arises in predicting properties of some future population. The sample size necessary to achieve a certain level of sampling variance in genetic parameter estimates may need to be determined, but the composition of the population, let alone the sample, at that future time is unknown. Both genetic and statistical sampling variation must be included in this situation.

NOTATION AND TERMINOLOGY

The distinction between population size N and sample size n has been made already. It is convenient to speak loosely of a genetic locus, **A**, either with two alleles A, a or with a series of alleles A_u. The symbol **A** will be used whether structural loci or restriction sites, or maybe even individual nucleotides, are meant.

An allele A has a population frequency p_A, which will generally mean the average frequency over all replicate populations maintained under the same conditions. If there are no disturbing forces this frequency does not change over time, and is the frequency of that allele in the reference population. The frequency in one particular replicate does change over time because of genetic sampling, and genetic drift causes each replicate population eventually to become fixed for one allele.

Frequencies of combinations of alleles at two or more loci will also be written as lower case p's, as in p_{AB} for the combination of alleles A and B, but upper case P's will be used for genotypic frequencies. At a single locus, a notation such as P_{AA} is sufficient, but for two loci it may be necessary to indicate which pairs of genes were received from the same parent. Individuals receiving the pair AB from one parent and ab from the other have a frequency written as P_{ab}^{AB}, while the other double heterozygotes have frequency P_{aB}^{Ab}. If there is no need to keep track of gametic associations, all alleles will be subscripted, as in P_{AaBb}.

In a sample, counts of genes or genotypes will be written as lower case n's with appropriate subscripts, so that for a locus $\mathbf{A}$ with alleles A and a the sample size is the sum of counts in the three genotypic classes and the allelic counts follow from the genotypic counts

$$\begin{aligned} n &= n_{AA} + n_{Aa} + n_{aa} \\ n_A &= 2n_{AA} + n_{Aa} \end{aligned}$$

Sample frequencies will be indicated by tildes, and follow from counts as

$$\begin{aligned} \tilde{p}_A &= \frac{1}{2n} n_A \\ \tilde{P}_{AA} &= \frac{1}{n} n_{AA} \end{aligned}$$

while estimates will be indicated by carets. Generally, the estimates of gene and genotypic frequencies are just the observed values

$$\hat{p}_A = \tilde{p}_A$$

but this will not be the case for estimates of squares and products of frequencies.

Population frequencies are the basic genetic quantities of interest, and care will be taken to distinguish between unobservable PARAMETERS and the STATISTICS used as their estimates. Much effort will be directed to finding high quality estimates and it is therefore necessary to know their sampling properties. There is a particular interest in the means and variances of estimators, and $\mathcal{E}$ will be used to indicate the mean or average value of a statistic over all possible samples. In statistical language, a mean of a variable is the EXPECTED VALUE of the variable. Unless otherwise stated, these expectations will be taken over all samples from a population, and over all populations. It will be shown that observed gene frequencies are UNBIASED ESTIMATES of population gene frequencies because

$$\mathcal{E}(\tilde{p}_A) = p_A$$

MENDEL AND FISHER

Whereas the work of Mendel forms the basis of genetics, the analysis of his data provides one the classic examples in genetic data analysis. Mendel presented data on seven characters in his 1866 paper. Although there is no question about the fundamental model of discrete genes controlling these characters, there has been criticism of the data. It appeared to Fisher (1936) that the data were so close to the values that Mendel expected under his theory that there must have been some manipulation, or omission, of data. Since Fisher's paper there has been an extensive literature of support for either Fisher or Mendel, but until recently little of substance has been added to the debate. Careful reviews were given by Piegorsch (1983, 1986), and the following discussion is based on Fisher (1936) and Edwards (1986). The discussion is included here because of its historic interest, and because it introduces the chi-square distribution.

Mendel crossed inbred pea lines that differed in easily scored characters, each of which was controlled by a single gene with two alleles, one dominant to the other. He introduced the notation A for the dominant and a for the recessive alleles. The crossed, or F_1, populations were heterozygous Aa and showed only the dominant character. When such heterozygotes are allowed to self, the offspring are expected to be AA, Aa, and aa in the ratio 1:2:1, or to show dominants $AA + Aa$ and recessives aa in the ratio 3:1. The data in Table 1.1 certainly are consistent with this 3:1 ratio. The characters are labeled $A - G$ in Table 1.1 to maintain consistency with Edwards' paper. The figures in Table 1.1 for the plant characters $C - G$ are repeated in rows 23 – 27 of Table 1.8.

Table 1.8 Normal and chi-square statistics for Mendel's segregations.

	Character	Segregation			Total	X	X^2
		Expected	Observed				
			F_2, seed characters				
1	*A*	3:1	5138	1749	6887	−0.7583	0.58
2	*B*	3:1	5667	1878	7545	+0.2193	0.05
			F_2, plant variability				
3	*A*	3:1	45	12	57	+0.6882	0.47
4	*A*	3:1	27	8	35	+0.2928	0.09
5	*A*	3:1	24	7	31	+0.3111	0.10
6	*A*	3:1	19	10	29	−1.1793	1.39
7	*A*	3:1	32	11	43	−0.0880	0.01
8	*A*	3:1	26	6	32	+0.8165	0.67
9	*A*	3:1	88	24	112	+0.8729	0.76
10	*A*	3:1	22	10	32	−0.8165	0.67
11	*A*	3:1	28	6	34	+0.9901	0.98
12	*A*	3:1	25	7	32	+0.4082	0.17
13	*B*	3:1	25	11	36	−0.7698	0.59
14	*B*	3:1	32	7	39	+1.0170	1.03
15	*B*	3:1	14	5	19	−0.1325	0.02
16	*B*	3:1	70	27	97	−0.6448	0.42
17	*B*	3:1	24	13	37	−1.4237	2.03
18	*B*	3:1	20	6	26	+0.2265	0.05
19	*B*	3:1	32	13	45	−0.6025	0.36
20	*B*	3:1	44	9	53	+1.3482	1.82
21	*B*	3:1	50	14	64	+0.5774	0.33
22	*B*	3:1	44	18	62	−0.7332	0.54
			F_2, plant characters				
23	*C*	3:1	705	224	929	+0.6251	0.39
24	*D*	3:1	882	299	1181	−0.2520	0.06
25	*E*	3:1	428	152	580	−0.6712	0.45
26	*F*	3:1	651	207	858	+0.5913	0.35
27	*G*	3:1	787	277	1064	−0.7788	0.61
			F_3, seed characters				
28	*A*	2:1	372	193	565	−0.4165	0.17
29	*B*	2:1	353	166	519	+0.6518	0.42

(Continued)

Table 1.8 Continued

	Character	Segregation Expected	Observed		Total	X	X^2
		F_3, plant characters					
30	C	0.63:0.37	64	36	100	+0.2252	0.05
31	D	0.63:0.37	71	29	100	+1.6743	2.80
32	E	0.63:0.37	60	40	100	−0.6029	0.36
33	F	0.63:0.37	67	33	100	+0.8462	0.72
34	G	0.63:0.37	72	28	100	+1.8813	3.54
35	E	0.63:0.37	65	35	100	+0.4322	0.19
		F_2, bifactorial experiment					
36	A	3:1	423	133	556	+0.5876	0.35
37	B among 'A'	3:1	315	108	423	−0.2526	0.06
38	B among 'a'	3:1	101	32	133	+0.2503	0.06
		F_3, bifactorial experiment					
39	A among 'AB'	2:1	198	103	301	−0.3261	0.11
40	A among 'Ab	2:1	67	35	102	−0.2100	0.04
41	B among 'aB'	2:1	68	28	96	+0.8660	0.75
42	B among Aa'B'	2:1	138	60	198	+0.9045	0.82
43	B among AA'B'	2:1	65	38	103	−0.7664	0.59
		F_2, trifactorial, seed characters					
44	A	3:1	480	159	639	+0.0685	0.00
45	B among 'A'	3:1	367	113	480	+0.7379	0.54
46	B among 'a'	3:1	122	37	159	+0.5037	0.25
		F_3, trifactorial, seed characters					
47	A among 'AB'	2:1	245	122	367	+0.0369	0.00
48	A among 'Ab'	2:1	76	37	113	+0.1330	0.02
49	B among 'aB'	2:1	79	43	122	−0.4481	0.20
50	B among Aa'B'	2:1	175	70	245	+1.5811	2.50
51	B among AA'B'	2:1	78	44	122	−0.6402	0.41

(Continued)

Table 1.8 Continued

	Character	Segregation Expected	Observed		Total	X	X^2
	F_2, trifactorial, plant character						
52	C among $AaBb$	3:1	127	48	175	−0.7419	0.55
53	C among $AaBB$	3:1	52	18	70	−0.1380	0.02
54	C among $AABb$	3:1	60	18	78	+0.3922	0.15
55	C among $AABB$	3:1	30	14	44	−1.0445	1.09
56	C among $Aabb$	3:1	60	16	76	+0.7947	0.63
57	C among $AAbb$	3:1	26	11	37	−0.6644	0.44
58	C among $aaBb$	3:1	55	24	79	−1.1043	1.22
59	C among $aaBB$	3:1	33	10	43	+0.2641	0.07
60	C among $aabb$	3:1	30	7	37	+0.8542	0.73
	F_3, trifactorial, plant character						
61	C among $AaBb$	0.63:0.37	78	49	127	−0.3488	0.12
62	C among $AaBB$	0.63:0.37	38	14	52	+1.5174	2.30
63	C among $AABb$	0.63:0.37	45	15	60	+1.9384	3.76
64	C among $AABB$	0.63:0.37	22	8	30	+1.1816	1.40
65	C among $Aabb$	0.63:0.37	40	20	60	+0.6020	0.36
66	C among $AAbb$	0.63:0.37	17	9	26	+0.2610	0.07
67	C among $aaBb$	0.63:0.37	36	19	55	+0.3903	0.15
68	C among $aaBB$	0.63:0.37	25	8	33	+1.5276	2.33
69	C among $aabb$	0.63:0.37	20	10	30	+0.4257	0.18
	Gametic ratios, first experiment, seed characters						
70	A	1:1	43	47	90	−0.4216	0.18
71	B among AA	1:1	20	23	43	−0.4575	0.21
72	B among Aa	1:1	25	22	47	+0.4376	0.19
	Gametic ratios, second experiment, seed characters						
73	A	1:1	57	53	110	+0.3814	0.15
74	B among Aa	1:1	31	26	57	+0.6623	0.44
75	B among aa	1:1	27	26	53	+0.1374	0.02
	Gametic ratios, third experiment, seed characters						
76	A	1:1	44	43	87	+0.1072	0.01
77	B among AA	1:1	25	19	44	+0.9045	0.82
78	B among Aa	1:1	22	21	43	+0.1525	0.02

(Continued)

Table 1.8 Continued

	Character	Segregation Expected	Observed		Total	X	X^2
	Gametic ratios, fourth experiment, seed characters						
79	A	1:1	49	49	98	0.0000	0.00
80	B among Aa	1:1	24	25	49	−0.1429	0.02
81	B among aa	1:1	22	27	49	−0.7143	0.51
	Gametic ratios, plant characters						
82	G	1:1	87	79	166	+0.6209	0.39
83	C among Gg	1:1	47	40	87	+0.7505	0.56
84	C among gg	1:1	38	41	79	−0.3375	0.11

* Source: Edwards (1986).

To illustrate the variability between plants, Mendel reported the counts for the two seed characters in the F_2 generation observed on each of 10 F_1 plants, and these counts are displayed in Table 1.9. For Edwards' analysis, the totals from Table 1.9 are subtracted from the entries in Table 1.1 to avoid using the same data twice. This gives 5138:1749 for seed shape and 5667:1878 for cotyledon color (rows 1 and 2 in Table 1.8). The figures in Table 1.9 are repeated in rows 3 through 22 of Table 1.8.

Individuals in the F_2 generation were selfed to reveal the presence of two genotypes in the dominant class. (The recessive class always gave recessive offspring on selfing.) Under Mendel's model, one-third of the dominant F_2's are homozygous and so should give only dominant F_3 offspring, whereas two-thirds are heterozygous and so should give dominant and recessive F_3's in the ratio 3:1. The expected ratio of heterozygotes to homozygotes among the dominant F_2's is therefore 2:1 and, once again, Mendel's data were consistent with this ratio.

A problem with this second set of experiments is that a heterozygous plant could, by chance, give a series of offspring that did not include any recessives and so could, incorrectly, be included among the dominant homozygous class. For the five plant characters, Mendel cultivated 10 seeds from each of 100 dominant F_2 plants. If all 10 offspring from a single plant were dominant, he concluded that the F_2 parent belonged to the homozygous

Table 1.9 Mendel's results for two characters on 10 individual F_1 plants. Source: Mendel (1866).

Plant	Seed Form		Cotyledon Color	
	Round	Wrinkled	Yellow	Green
1	45	12	25	11
2	27	8	32	7
3	24	7	14	5
4	19	10	70	27
5	32	11	24	13
6	26	6	20	6
7	88	24	32	13
8	22	10	44	9
9	28	6	50	14
10	25	7	44	18

class. Otherwise it was classified as a heterozygote. With each heterozygous F_2 plant having a 0.75 chance of giving a dominant selfed offspring, there is a $(0.75)^{10} = 0.0563$ chance of 10 offspring all being dominant and the F_2 being misclassified as homozygous. Instead of the probability being 2/3 that an F_2 is heterozygous, the correct probability is 2/3 times the chance of correct classification – i.e., 0.6291. The true expected ratio of homozygotes and heterozygotes among the dominant F_2's is therefore 0.3709:0.6291. Fisher pointed out that Mendel's data were much closer to the 1:2 ratio than to this corrected value, giving rise to the suggestion that some data manipulation had taken place, and that Mendel was demonstrating rather than establishing his theory. The problem does not arise with the two seed characters since they could be scored directly on the F_2 plants, and many seeds could be observed per plant. Mendel's data for the plant characters are shown in rows 30 – 35 of Table 1.8, and for the two seed characters in rows 28 and 29.

With the single-gene inheritance of dominant characters established, Mendel turned to demonstrating the independent inheritance of two or three characters. In the bifactorial experiments the two seed characters were used. Letting A, a denote the alleles for round and wrinkled seeds and B, b the alleles for yellow and green cotyledons, the cross $AABB \times aabb$ gave all $AaBb$ round yellow seeds, which, when selfed, gave all four possible classes of offspring. Edwards divided the data into three comparisons for the 3:1 expected ratios for each character separately: round versus wrinkled, yellow

Table 1.10 Mendel's F_3 bifactorial data for seed shape and cotyledon color.

		Seed Shape			
		AA	Aa	aa	Total
Cotyledon color	BB	38	60	28	126
	Bb	65	138	68	271
	bb	35	67	30	132
	Total	138	265	126	529

versus green among round seeds, and yellow versus green among wrinkled seeds (see Table 1.8, rows 36, 37, and 38). Within each dominant class, the expected ratio of 2:1 for heterozygote versus homozygote was looked for by planting the F_2 seeds, and the resulting F_3 data are shown in Table 1.10. When the wrinkled yellow seeds were planted, the F_3's were all wrinkled but were expected to segregate 2:1 for yellow and green versus yellow only (row 41 of Table 1.8). For the offspring of round green seeds there is constant color but 2:1 expected segregation for round and wrinkled versus round only seed shape (row 40). The double dominant class gave all four classes on planting, and Edwards formulated three comparisons with 2:1 expected ratios: round and wrinkled versus round only (row 39), yellow and green versus green only (row 42), and yellow and green versus yellow only among round seeds (row 43). All 30 wrinkled green F_2's gave only this kind of offspring.

Seed coat color, with alleles C, c, was added to seed shape and cotyledon color to give a trifactorial experiment. The triple homozygotes $AABBCC$ and $aabbcc$ were crossed and selfed, leading to the counts set out in Table 1.11 in the format given by Fisher. Edwards detailed the 26 comparisons possible in this data set, as shown in Table 1.8 (rows 44 –69). In that table, 'A' denotes $AA + Aa$, and 'a' denotes aa. Taken together, the bifactorial and trifactorial data lend support to independent assortment of characters.

The final set of experiments reported in 1866 by Mendel concerned gametic ratios, designed to demonstrate that egg and pollen cells of F_1 hybrids had alleles AB, Ab, aB, or ab at loci **A** and **B**. Five experiments were conducted by crossing two double homozygotes, $AABB$ and $aabb$, and then fertilizing the hybrids with the pollen of either double homozygote or fertilizing either double homozygote with pollen from the hybrids. In the first of these four arrangements, AB pollen applied to $AaBb$ plants is expected

Table 1.11 Mendel's F_2 trifactorial data for seed shape (A), cotyledon color (B), and seed coat color (C).

	AA	Aa	aa	Total
			CC	
BB	8	14	8	30
Bb	15	49	19	83
bb	9	20	10	39
			Cc	
BB	22	38	25	85
Bb	45	78	36	159
bb	17	40	20	77
			cc	
BB	14	18	10	42
Bb	18	48	24	90
bb	11	16	7	34
			Total	
BB	44	70	43	157
Bb	78	175	79	332
bb	37	76	37	150
Total	159	321	159	639

to produce equal numbers of $AABB$, $AaBB$, $AABb$, $AaBb$ F_2's, all with the dominant appearance for both characters. Selfing this collection, to distinguish the homozygous and heterozygous genotypes within dominant classes as in previous experiments, should reveal this equality of four classes. In Edwards' scheme, the appropriate comparisons are $AA + Aa$ versus aa, $BB + Bb$ versus bb among $AA + Aa$ plants and $BB + Bb$ versus bb among aa plants, and the set of 15 such comparisons is also shown in Table 1.8 (rows 70 – 84).

The X values shown in Table 1.8 are the square roots of the CHI-SQUARE goodness-of-fit statistics. If a category has an observed count of o, but a count e is expected under some hypothesis, then the goodness-of-fit test statistic is formed by summing the quantity $(o - e)^2/e$ over categories

$$X^2 = \sum_{\text{categories}} \frac{(o-e)^2}{e}$$

If the hypothesis is true, it is not likely that the *o* and *e* values in any category will be exactly equal, but they should not be too far apart. How far apart they may be and still have the data support the hypothesis is found from properties of the chi-square distribution. In particular, for two categories, there is only a 5% chance that X^2 will exceed 3.84 if the hypothesis is true (see the column headed 0.05 for one degree of freedom in Appendix Table 2). In the first row of Table 1.8, for example, the chi-square is calculated by comparing the observed values 5138 and 1749 with the values 5165.25 and 1721.75 expected under the 3:1 ratio

$$\begin{aligned} X^2 &= \frac{(5138-5165.25)^2}{5165.25} + \frac{(1749-1721.75)^2}{1721.75} \\ &= 0.5750 \\ X &= -0.7583 \end{aligned}$$

Edwards assigned a negative sign to X when the observed value is less than the expected value in the class with the larger expectation. For cases in which two classes have the same expectation, this rule is applied to that class with the larger number of dominant genes. When the expected ratio is true, the quantity X^2 calculated on two categories has a chi-square distribution with 1 degree of freedom (d.f.)

$$X^2 \sim \chi^2_{(1)}$$

while X has a STANDARD NORMAL distribution

$$X \sim N(0,1)$$

Large values of X^2 cause the hypothesis of the expected ratio to be rejected. In Table 1.8, no X^2 value is greater than 3.84 (which is the same as saying that no X has an absolute value exceeding 1.96). This is surprising since about four such values would be expected just by chance even if all the hypotheses being tested were true.

Some of the criticisms of Mendel's data focus on chi-square values that are unusually small as well as unusually large. The same 5% significance level can be achieved if the hypothesis is rejected for X^2 exceeding 5.02 or being less than 0.000982 (see the columns headed 0.975 and 0.025 for 1 d.f. in Appendix Table 2). The 2.5% closest and the 2.5% furthest away values from the expected ratio now cause rejection. If the 84 independent X^2 values in Table 1.8 are added, the resulting sum of 41.15 is unusually low for a chi-square with 84 d.f. Edwards said that such a chi-square may look like a "single lucky shot" but that, in fact, it is a succession of 84 shots that must be showing some kind of pattern requiring investigation.

When the hypothesized ratio is true, and the quantity X has a standard normal distribution, large positive or negative values contribute to large chi-squares, while X values close to zero contribute to small chi-squares. Among the 15 values (rows 30 – 35 and 61 – 69) that have an expected ratio of 0.6291:0.3709, there are 13 that show a positive X value and only 2 that are negative, suggesting a bias in favor of the dominant class. Fisher's criticism of Mendel for using an expected ratio of 2:1 instead of this corrected value seems borne out by the fact that the data are 57.90 times more likely to have arisen under the incorrect 2:1 ratio than under the correct one. This calculation follows from the likelihood theory taken up in Chapter 2.

Edwards looked at the 69 rows in Table 1.8 that do not have questions over expected ratios. The X values are fairly evenly distributed about zero, with 38 positive, 30 negative, and 1 zero, but there is something unusual about the range of X values. In a sample of 69, the largest three values of the absolute values of X are expected to be 2.6216, 2.2755, and 2.0828 (Edwards 1986), whereas the largest values reported by Mendel are 1.5811, 1.4237, and 1.3482. It appears that Mendel's X values are biased in the sense that there are too few extreme values, positive or negative. It is as though someone had altered figures that appeared to be too extreme.

The 69 X values are plotted in the histogram shown in Figure 1.6, along with the histogram expected for a sample of 69 values from a normal distribution. These expected values can be found from the table of values for the cumulative standard normal distribution (Appendix Table 1). That table shows that the probability of a standard normal variable falling between 0.00 and 0.20 is $0.5000 - 0.4207 = 0.0793$, so that $69 \times 0.0793 = 5.5$ values are expected in a sample of size 69.

Edwards inclined toward the explanation that some plants were misclassified (perhaps unconsciously) in the direction of conforming to the hypothesized ratios, rather than that some data were discarded. Piegorsch (1986) reviewed other published discussion of Mendel's data, concerning his apparently fortunate choice of seven unlinked characters.

SUMMARY

Discrete genetic data may take many forms, ranging from individual nucleotides in a DNA sequence to a qualitative character such as flower color, but underlying all these data is a common genetic model. The data are

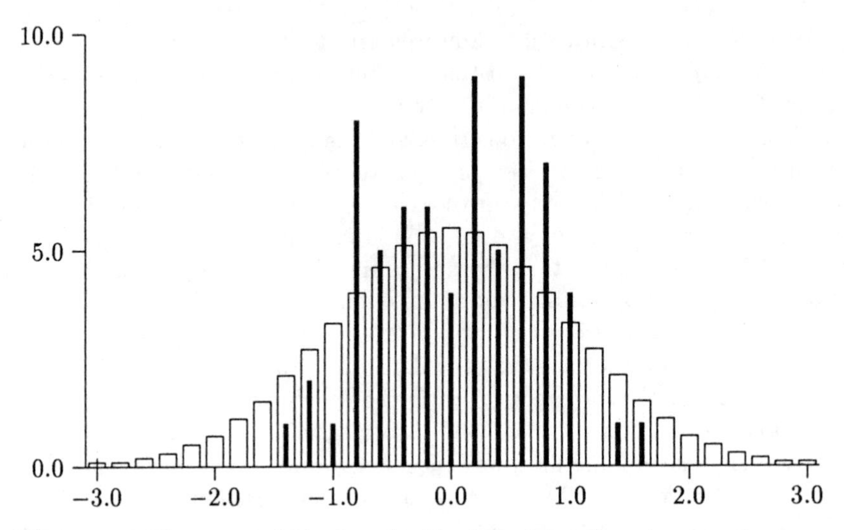

Figure 1.6 Histogram of X values for Mendel's data. Open bars are for a normal distribution, and solid bars are for Mendel's data. Source: Edwards (1986).

considered to depend on discrete heritable units that are transmitted from parent to offspring with a random choice of which parent's contribution is passed from offspring to grand-offspring. The resulting genetic variation between populations must be added to the statistical sampling within populations arising from the random sampling of individuals for observation.

Genetic parameters are unobservable and need to be estimated by statistics that are functions of the data. Estimates of parameters are indicated in notation by carets and observed values of population frequencies are indicated by tildes.

EXERCISES

Exercise 1.1

Find a published paper that presents discrete genetic data, and answer the following questions for that paper:

a. Describe the data — are they clearly genetic?
b. What genetic questions are being addressed with the data?
c. What statistical techniques are being used?
d. Has attention been paid to genetic and statistical sampling?

Exercise 1.2

Draw a sample of 69 values from a standard normal distribution, and determine the upper and lower quartiles and the median – i.e., find those three values that divide the sample into four nearly equal portions. Compare these values with those for the 69 values in Table 1.8 for which there is no doubt about the expected ratios.

Instructions on how to generate samples of random numbers are given in the Appendix.

Chapter 2

Estimating Frequencies

MULTINOMIAL GENOTYPIC COUNTS

Most population genetic data sets consist of counts of genotypes, whether at one or several loci, so that the statistical properties of such counts need to be studied. The statistical sampling model assumes that every individual in the sample has the same probability of being a particular genotype. Although this greatly simplifies the theory, it cannot be exactly true in practice because the sampling of one individual reduces the frequency of that type in the population and so reduces the probability of choosing that type for subsequent sample members. Sampling individuals without replacement requires the hypergeometric distribution, but it is generally assumed that the population being sampled is very large and sampling can be considered to be with replacement. The MULTINOMIAL DISTRIBUTION is then appropriate. It is worth noting, however, that even at the very outset of this development, an assumption is made that may not always hold. All the statistical analyses in this book are going to rest on assumptions, although they may not always be made explicit.

To set up the multinomial distribution, suppose that the genotypes define a set of k categories indexed by the subscript i. Each individual sampled has probability Q_i of being in the ith category, which is the same as saying that the population frequency of genotype i is Q_i. When each individual is sampled independently, with the category of each having no effects on the categories of the others, the probability of sampling an individual of type i and then one of type j is Q_iQ_j. Extending this argument to samples of size

n leads to the probability of the sample having n_i members in category i:

$$\Pr(n_1, n_2, ..., n_k) = \frac{n!}{\prod_{i=1}^{k} n_i!} \prod_{i=1}^{k} Q_i^{n_i} \tag{2.1}$$

where the FACTORIAL $n!$ is the product of the first n integers, and the coefficient $n!/(\prod_i n_i!)$ is the number of observably different orderings of the n individuals. Equation 2.1 defines the multinomial distribution.

When there are two categories, the multinomial distribution reduces to the BINOMIAL. If the probabilities for the two categories are written as Q and $1 - Q$, and the counts in those categories as a and $n - a$, the binomial probability is

$$\Pr(a, n-a) = \frac{n!}{a!(n-a)!} Q^a (1-Q)^{n-a} \tag{2.2}$$

To illustrate the multinomial distribution for the case of three categories, Table 2.1 shows the 45 samples of size 8 from a population in which the three genotypes AA, Aa, and aa have frequencies 0.64, 0.32, and 0.04, respectively. Samples have been ordered according to probability. The cumulative probabilities shown for each sample in Table 2.1 are the probabilities for that sample and for all less probable samples.

When multinomial categories are combined into one category versus the rest, adding the appropriate probabilities gives the binomial distribution for those two new categories. For the example in Table 2.1, the distribution of aa versus $AA + Aa$ genotypes is binomial with probabilities 0.04 and 0.96. The eight samples are shown in Table 2.2, with probabilities that can be obtained directly from the binomial formula, Equation 2.2, or from adding terms in Table 2.1.

Multinomial Moments

Although the multinomial distribution is completely characterized by Equation 2.1, most use is made of summary quantities, or MOMENTS, particularly the mean and variance. The MEAN, or expected value, of the number in any category is obtained by multiplying each possible value of that number by its probability and adding these products. Since each category can be handled with a binomial distribution it is easiest to derive the mean from the binomial. For category i the expected number in a sample of size n is

$$\mathcal{E}(n_i) = \sum_{r=0}^{n} r \Pr(n_i = r)$$

Table 2.1 Multinomial probabilities, shown to four decimal places, for trinomial with sample size of 8 and category probabilities of 0.64 (AA), 0.32 (Aa), and 0.04 (aa).

n_{AA}	n_{Aa}	n_{aa}	Prob.	Cum.	n_{AA}	n_{Aa}	n_{aa}	Prob.	Cum.
0	0	8	0.0000	0.0000	1	5	2	0.0006	0.0020
0	1	7	0.0000	0.0000	3	2	3	0.0010	0.0030
1	0	7	0.0000	0.0000	4	1	3	0.0010	0.0039
0	2	6	0.0000	0.0000	1	6	1	0.0015	0.0055
1	1	6	0.0000	0.0000	1	7	0	0.0018	0.0072
2	0	6	0.0000	0.0000	2	4	2	0.0029	0.0101
0	3	5	0.0000	0.0000	6	0	2	0.0031	0.0132
1	2	5	0.0000	0.0000	3	3	2	0.0077	0.0209
3	0	5	0.0000	0.0000	2	5	1	0.0092	0.0301
0	4	4	0.0000	0.0000	5	1	2	0.0092	0.0394
2	1	5	0.0000	0.0000	4	2	2	0.0115	0.0509
0	5	3	0.0000	0.0000	2	6	0	0.0123	0.0632
1	3	4	0.0000	0.0000	7	0	1	0.0141	0.0773
4	0	4	0.0000	0.0001	8	0	0	0.0281	0.1054
2	2	4	0.0000	0.0001	3	4	1	0.0308	0.1362
0	6	2	0.0000	0.0002	3	5	0	0.0493	0.1855
3	1	4	0.0001	0.0002	6	1	1	0.0493	0.2347
0	7	1	0.0001	0.0003	4	3	1	0.0616	0.2963
0	8	0	0.0001	0.0004	5	2	1	0.0739	0.3702
1	4	3	0.0001	0.0006	7	1	0	0.1126	0.4828
5	0	3	0.0004	0.0009	4	4	0	0.1231	0.6059
2	3	3	0.0005	0.0014	6	2	0	0.1970	0.8030
					5	3	0	0.1970	1.0000

$$= \sum_{r=0}^{n} r \frac{n!}{r!(n-r)!} Q_i^r (1-Q_i)^{n-r}$$

$$= nQ_i$$

The sample proportion for category i is $\tilde{Q}_i = n_i/n$ and, dividing the previous equations by n, the expected value of the sample proportion is found to be the population proportion

$$\mathcal{E}(\tilde{Q}_i) = Q_i \tag{2.3}$$

In other words, the sample proportion is an UNBIASED ESTIMATOR of the population proportion. Expectation here is referring to all possible samples

Table 2.2 Binomial probabilities, shown to four decimal places, for sample size of 8 and category probabilities of 0.96 ($AA + Aa$) and 0.04 (aa).

$n_{AA} + n_{Aa}$	n_{aa}	Prob.	Cum.
0	8	0.0000	0.0000
1	7	0.0000	0.0000
2	6	0.0000	0.0000
3	5	0.0000	0.0000
4	4	0.0002	0.0002
5	3	0.0029	0.0031
6	2	0.0351	0.0381
7	1	0.2405	0.2786
8	0	0.7214	1.0000

of size n from the population and so is taking account of the statistical sampling.

Some idea of the variability of the category numbers is found by taking the expected value of the squared difference of the number from its mean, which is a definition of the VARIANCE. Still using the binomial distribution for a particular category

$$\begin{aligned}
\text{Var}(n_i) &= \sum_{r=0}^{n}[r - \mathcal{E}(n_i)]^2 \Pr(n_i = r) \\
&= \sum_{r=0}^{n}(r^2 - 2rnQ_i + n^2Q_i^2)\frac{n!}{r!(n-r)!}Q_i^r(1-Q_i)^{n-r} \\
&= [n(n-1)Q_i^2 + nQ_i] - 2n^2Q_i^2 + n^2Q_i^2 \\
&= nQ_i(1-Q_i) \qquad (2.4)
\end{aligned}$$

For a sample proportion, division by n occurs inside the square brackets of the first line of Equation 2.4, and this leads to a variance of

$$\begin{aligned}
\text{Var}(\tilde{Q}_i) &= \frac{1}{n^2}\text{Var}(n_i) \\
&= \frac{1}{n}Q_i(1-Q_i)
\end{aligned}$$

In some later expressions there is a need for the expected values of the squares of counts or proportions, rather than variances, but these are easily found. For any random variable X,

$$\mathcal{E}(X^2) = [\mathcal{E}(X)]^2 + \text{Var}(X)$$

For sample proportions, manipulating the expression for the variance gives

$$\begin{aligned}\mathcal{E}(\tilde{Q}_i^2) &= [\mathcal{E}(\tilde{Q}_i)]^2 + \text{Var}(\tilde{Q}_i) \\ &= Q_i^2 + \frac{1}{n}Q_i(1 - Q_i)\end{aligned}$$

This result is exact. Approximate expressions can be found for $\tilde{Q}_i^3$ and $\tilde{Q}_i^4$ that ignore terms with squares or higher powers of sample size n in the denominator:

$$\begin{aligned}\mathcal{E}(\tilde{Q}_i^3) &= Q_i^3 + \frac{3}{n}Q_i^2(1 - Q_i) \\ \mathcal{E}(\tilde{Q}_i^4) &= Q_i^4 + \frac{6}{n}Q_i^3(1 - Q_i)\end{aligned}$$

All the results from Equation 2.2 onward were derived for the binomial distribution, and they apply to any one of the categories in a multinomial setting. The multinomial probabilities do need to be considered when the joint sampling properties for counts or proportions in two or more categories are sought. To find the expected value of the product of two counts, $n_i n_j$, it is sufficient to work with a trinomial, for categories i, j, and the amalgamation of all other categories. The expected value of the product takes account of all its values:

$$\begin{aligned}\mathcal{E}(n_i n_j) &= \sum_{r=0}^{n}\sum_{s=0}^{n-r} rs\,\Pr(n_i = r, n_j = s) \\ &= \sum_{r=0}^{n}\sum_{s=0}^{n-r} rs\frac{n!}{r!s!(n-r-s)!}Q_i^r Q_j^s(1 - Q_i - Q_j)^{n-r-s} \\ &= n(n-1)Q_iQ_j \qquad (2.5)\end{aligned}$$

and

$$\mathcal{E}(\tilde{Q}_i\tilde{Q}_j) = \frac{(n-1)}{n}Q_iQ_j \qquad (2.6)$$

The COVARIANCE of counts is defined as the expected value of the product of the deviations of counts from their means, leading to

$$\begin{aligned}\text{Cov}(n_i, n_j) &= \mathcal{E}[n_i - \mathcal{E}(n_i)][n_j - \mathcal{E}(n_j)] \\ &= \mathcal{E}(n_i n_j) - \mathcal{E}(n_i)\mathcal{E}(n_j)\end{aligned}$$

From Equations 2.5 and 2.6

$$\begin{aligned}\text{Cov}(n_i, n_j) &= -nQ_iQ_j \\ \text{Cov}(\tilde{Q}_i, \tilde{Q}_j) &= -\frac{1}{n}Q_iQ_j\end{aligned}$$

Since the sum of the counts of two categories cannot be larger than the sample size, increasing one count causes a reduction in possible values for the other count and this causes the two counts to have a negative covariance. To complete the study of a pair of counts, the CORRELATION is defined as

$$\begin{aligned}\text{Corr}(n_i, n_j) &= \frac{\text{Cov}(n_i, n_j)}{\sqrt{\text{Var}(n_i)\text{Var}(n_j)}} \\ &= -\frac{Q_iQ_j}{\sqrt{Q_i(1-Q_i)Q_j(1-Q_j)}} \\ &= \text{Corr}(\tilde{Q}_i, \tilde{Q}_j)\end{aligned}$$

Notice that in the binomial case the two probabilities sum to one, $Q_1 = 1 - Q_2$, and the correlation of the two counts or sample proportions is minus one.

Within-Population Variance of Gene Frequencies

Counts of genes are obtained by adding twice the homozygote counts to the sum of appropriate heterozygote counts. In a sample with n_{uu} homozygotes A_uA_u and n_{uv} heterozygotes A_uA_v, the number n_u of A_u genes is

$$n_u = 2n_{uu} + \sum_{v \neq u} n_{uv}$$

Throughout this book, a sum such as that in this equation is meant to include every possible heterozygote for allele A_u, but to include it only once. For three alleles, for example, the equation becomes

$$\begin{aligned}n_1 &= 2n_{11} + n_{12} + n_{13} \\ n_2 &= 2n_{22} + n_{12} + n_{23} \\ n_3 &= 2n_{33} + n_{13} + n_{23}\end{aligned}$$

As a convention, the subscripts on heterozygotes will be written in numerical (or alphabetical) order.

Writing population, or expected, frequencies of genotypes as P_{uu} for A_uA_u and P_{uv} for A_uA_v, the expected value of the number of alleles A_u is

$$\begin{aligned}\mathcal{E}(n_u) &= 2nP_{uu} + \sum_{v\neq u} nP_{uv} \\ &= 2np_u\end{aligned}$$

where p_u is the population frequency of A_u genes

$$p_u = P_{uu} + \frac{1}{2}\sum_{v\neq u} P_{uv}$$

Dividing by $2n$, the number of genes sampled, shows that the sample gene frequency $\tilde{p}_u = n_u/2n$ is also unbiased for the population value

$$\mathcal{E}(\tilde{p}_u) = p_u$$

This last relation shows that if many samples were taken from the same population the average of all the sample gene frequencies would be the gene frequency in the population. The various samples would differ from each other, and the extent of difference is indicated by the variance of the sample gene frequency. The size of this variance (of the gene count) can be found by using the result that the variance of a sum of random variables (the genotype counts) is the sum of their variances plus twice the sum of the covariances between all pairs of the variables

$$\begin{aligned}\text{Var}(n_u) &= \text{Var}(2n_{uu}) + \sum_{v\neq u}\text{Var}(n_{uv}) + 2\sum_{v\neq u}\text{Cov}(2n_{uu}, n_{uv}) \\ &\quad + 2\sum_{v\neq u}\sum_{w\neq u,v}\text{Cov}(n_{uv}, n_{uw}) \\ &= 4nP_{uu}(1-P_{uu}) + \sum_{v\neq u} nP_{uv}(1-P_{uv}) - 4\sum_{v\neq u} nP_{uu}P_{uv} \\ &\quad - 2\sum_{v\neq u}\sum_{w\neq u,v} nP_{uv}P_{uw} \\ &= 2n\left(2P_{uu} + \frac{1}{2}\sum_{v\neq u} P_{uv}\right) - n\left(2P_{uu} + \sum_{v\neq u} P_{uv}\right)^2 \\ &= 2n(p_u + P_{uu} - 2p_u^2)\end{aligned}$$

so that

$$\text{Var}(\tilde{p}_u) = \frac{1}{2n}(p_u + P_{uu} - 2p_u^2) \tag{2.7}$$

This expression involves the population frequencies p_u and P_{uu}. In practice, data are available only from a single sample. If the sample frequencies $\tilde{p}_u$ and $\tilde{P}_{uu}$ are substituted into Equation 2.7, an estimate of the variance of $\tilde{p}_u$ is obtained. This, in turn, gives an idea of the range of values within which the population frequency p_u lies. Providing the sample size is reasonably large ($n \geq 30$, say), there is about a 95% chance that the interval

$$\tilde{p}_u \pm 2\sqrt{\text{Var}(\tilde{p}_u)}$$

includes the population frequency p_u.

The process of constructing a CONFIDENCE INTERVAL for a genetic parameter ϕ, here the allelic frequency p_u, rests on an assumption that the statistic $\hat{\phi}$ used to estimate the parameter has a normal distribution. Equation 2.19 below demonstrates normality in this case. To construct an interval for which there is $(1-\alpha) \times 100\%$ confidence that it includes the parameter value, it is necessary to identify the values $z_{\alpha/2}, z_{1-\alpha/2}$ between which $(1-\alpha) \times 100\%$ of the values of the standard normal distribution lie. Such values are found in Appendix Table 1. For a 95% confidence interval, that table shows that

$$z_{0.025}, z_{0.975} = \pm 1.96 \approx 2$$

The general expression for the limits of a confidence interval for ϕ is

$$\hat{\phi} \pm z_{\alpha/2}\sqrt{\text{Var}(\hat{\phi})}$$

Equation 2.7 for the variance of a sample gene frequency is important, for it demonstrates that although each genotype number can be regarded as being binomially distributed the same is not true for gene numbers. The variance in Equation 2.7 does not have the form of a binomial variance. There is a special case of great importance, however, and that is when the population sampled is in HARDY-WEINBERG EQUILIBRIUM (see Chapter 3). This means that genotypic frequencies are equal to the products of corresponding gene frequencies:

$$\begin{aligned} P_{uu} &= p_u^2 \\ P_{uv} &= 2p_u p_v, \text{ for } u \neq v \end{aligned}$$

and Equation 2.7 reduces to

$$\text{Var}(\tilde{p}_u) = \frac{1}{2n} p_u(1 - p_u) \tag{2.8}$$

which is the variance of a binomial distribution with parameters p_u and $2n$. For Hardy-Weinberg populations, the gene numbers as well as genotype numbers are multinomially distributed. Genes as well as genotypes are sampled at random from the population since, at any locus, a genotype is just a random pair of genes.

The variance expressions derived in this section refer to repeated samples from one population, and the gene frequencies p in these expressions refer to that population.

Indicator Variables

Many expressions for the expected values of functions of gene or genotype frequencies can be found very easily by using a set of indicator variables. Such variables take the value of 1 when an event is true and 0 when it is false. To find the variance of a gene frequency, for example, an indicator variable is set up for that particular gene. Suppose the two genes at a locus within an individual are indexed with j, $j = 1, 2$, and the individuals sampled are indexed with i, $i = 1, 2, \ldots, n$. The indicator variable x_{ij} is then defined by

$$x_{ij} = \begin{cases} 1, & \text{if gene } j \text{ in individual } i \text{ is allele } A \\ 0, & \text{otherwise} \end{cases}$$

The sample frequency for allele A can then be expressed as

$$\tilde{p}_A = \frac{1}{2n} \sum_{i=1}^{n} \sum_{j=1}^{2} x_{ij}$$

Taking expectations over all possible samples from the population shows that

$$\begin{aligned} \mathcal{E}(x_{ij}) &= 1 \times \Pr(x_{ij} = 1) + 0 \times \Pr(x_{ij} = 0) \\ &= 1 \times p_A + 0 \times (1 - p_A) \\ &= p_A \end{aligned}$$

so that, once again,

$$\mathcal{E}(\tilde{p}_A) = p_A$$

To derive the variance, recognize that the product $x_{ij}x_{ij'}$ $(j \neq j')$ will be nonzero only when both genes, j and j', in individual i are allele A, whereas the square x_{ij}^2 is just the same as x_{ij} itself. In other words,

$$\begin{aligned} \mathcal{E}(x_{ij}^2) &= p_A \\ \mathcal{E}(x_{ij}x_{ij'}) &= P_{AA} \end{aligned}$$

Since individuals are sampled independently from the population, the expectation of the product of x_{ij} and $x_{i'j'}$, for genes in different individuals, is the product of expectations of these two indicator variables:

$$\begin{aligned}\mathcal{E}(x_{ij}x_{i'j'}) &= \mathcal{E}(x_{ij})\mathcal{E}(x_{i'j'}) \\ &= p_A^2\end{aligned}$$

Expanding the square of the expression for $\tilde{p}_A$, and then taking expectations leads to

$$\begin{aligned}\mathcal{E}(\tilde{p}_A^2) &= \frac{1}{4n^2}\Bigg(\mathcal{E}\sum_i\sum_j x_{ij}^2 + \mathcal{E}\sum_i\sum_j\sum_{j'\neq j} x_{ij}x_{ij'} \\ &\quad + \mathcal{E}\sum_i\sum_{i'\neq i}\sum_j\sum_{j'} x_{ij}x_{i'j'}\Bigg) \\ &= \frac{1}{4n^2}\left[2np_A + 2nP_{AA} + 4n(n-1)p_A^2\right] \\ &= p_A^2 + \frac{1}{2n}(p_A + P_{AA} - 2p_A^2) \qquad (2.9)\end{aligned}$$

and, when the square of the expected gene frequency is subtracted,

$$\text{Var}(\tilde{p}_A) = \frac{1}{2n}(p_A + P_{AA} - 2p_A^2)$$

as in Equation 2.7. The great power of this indicator variable method will become apparent in more complex cases.

It is often convenient to express genotypic frequencies in terms of gene frequencies and some measure of departure from Hardy-Weinberg equilibrium. One parameterization uses the inbreeding measure f within populations (Cockerham 1969), which is the same as the quantity F_{IS} defined by Wright (1951). For two alleles, A and a,

$$\begin{aligned}P_{AA} &= p_A^2 + p_Ap_af \\ P_{Aa} &= 2p_Ap_a(1-f) \\ P_{aa} &= p_a^2 + p_ap_Af\end{aligned}$$

and, putting these into the variance formula Equation 2.7,

$$\text{Var}(\tilde{p}_A) = \frac{1}{2n}p_A(1-p_A)(1+f) \qquad (2.10)$$

Once again the binomial variance result holds if there is Hardy-Weinberg equilibrium, $f = 0$. For any particular set of allelic frequencies, the variance increases as the amount of inbreeding increases, with an eventual doubling of the noninbred value. A completely inbred population will be homozygous, with alleles being either fixed or lost, and then the within-population variance of a sample gene frequency will be zero since $p_A(1 - p_A)$ will be zero. Remember that this variance refers to repeated samples from the same population.

Within-Population Covariance of Gene Frequencies

Indicator variables can also be used to find the covariance between frequencies of two alleles. Let x_{ij} and y_{ij} be the indicator variables for alleles A_1 and A_2, so that the two sample frequencies are

$$\begin{aligned}
\tilde{p}_1 &= \frac{1}{2n}\sum_i\sum_j x_{ij} \\
\tilde{p}_2 &= \frac{1}{2n}\sum_i\sum_j y_{ij}
\end{aligned}$$

and the expected value of their product is

$$\begin{aligned}
\mathcal{E}(\tilde{p}_1\tilde{p}_2) &= \frac{1}{4n^2}\mathcal{E}\left[(\sum_i\sum_j x_{ij})(\sum_i\sum_j y_{ij})\right] \\
&= \frac{1}{4n^2}\mathcal{E}\left(\sum_i\sum_j x_{ij}y_{ij} + \sum_i\sum_j\sum_{j'} x_{ij}y_{ij'}\right. \\
&\quad \left. + \sum_i\sum_{i'\neq i}\sum_j\sum_{j'\neq j} x_{ij}y_{i'j'}\right) \\
&= \frac{1}{4n^2}[2n \times 0 + nP_{12} + 4n(n-1)p_1p_2] \\
&= p_1p_2 + \frac{1}{4n}(P_{12} - 4p_1p_2)
\end{aligned}$$

The zero results from it not being possible for gene j in individual i to be both allele A_1 and allele A_2. Subtracting p_1p_2 provides the covariance

$$\text{Cov}(\tilde{p}_1, \tilde{p}_2) = \frac{1}{4n}(P_{12} - 4p_1p_2) \qquad (2.11)$$

For noninbred populations, with $P_{12} = 2p_1p_2$, the covariance reduces to the binomial value of

$$\text{Cov}(\tilde{p}_1, \tilde{p}_2) = -\frac{1}{2n}p_1p_2 \qquad (2.12)$$

Table 2.3 Genotypic counts for *MN* blood groups among mothers and fathers in Table 1.2.

Genotype	Father	Mother	Total
MM	26	27	53
MN	44	51	95
NN	23	15	38
Total	93	93	186

Examples

Calculation of variances and covariances of gene frequencies can be illustrated with data from Chapter 1. For the blood group data shown in Table 1.2, the genotypic counts for the *MN* locus, for mothers and fathers, are given in Table 2.3. By regarding these 93 couples as a random sample of 186 individuals from a population, sample homozygote and allele frequencies for M are calculated as

$$\begin{aligned} \tilde{P}_{MM} &= 53/186 = 0.2849 \\ \tilde{p}_M &= 201/372 = 0.5403 \end{aligned}$$

Substituting these observed values into the right-hand side of Equation 2.7 gives an estimate of the variance of $\tilde{p}_M$:

$$\begin{aligned} \text{Var}(\tilde{p}_M) &\hat{=} [0.5403 + 0.2849 - 2(0.5403)^2]/372 \\ &= 0.00064880 \end{aligned}$$

while substituting into Equation 2.8 gives a very similar result:

$$\begin{aligned} \text{Var}(\tilde{p}_M) &\hat{=} 0.5403(1 - 0.5403)/372 \\ &= 0.00066768 \end{aligned}$$

The symbol $\hat{=}$ means "is estimated by." The similarity in these two numerical results is a reflection of the genotypic frequencies for the *MN* blood group essentially being in Hardy-Weinberg proportions.

A difference in the results from Equations 2.7 and 2.8 does arise for the *Pgm* locus data in Table 1.3. The genotypic counts are shown in Table 2.4 and for allele 1 at that locus,

Table 2.4 Gene and genotype counts for *Pgm* locus in mosquito data of Table 1.3.

Genotype	Count	Gene	Count
11	9	1	19
12	1	2	26
22	5	3	17
13	0	4	18
23	7		
33	0	Total	80
14	0		
24	8		
34	10		
44	0		
Total	40		

$$\tilde{P}_{11} = 9/40 = 0.2250$$
$$\tilde{p}_1 = 19/80 = 0.2375$$

so that

$$\mathrm{Var}(\tilde{p}_1) \hat{=} \begin{cases} 0.00437109 \text{ from Equation 2.7} \\ 0.00226367 \text{ from Equation 2.8} \end{cases}$$

The two variance estimates differ because the data had more homozygotes for allele 1 than expected from the Hardy-Weinberg theorem. Turning to covariances, for alleles 1 and 2, Equation 2.11 provides

$$\mathrm{Cov}(\tilde{p}_1, \tilde{p}_2) \hat{=} -0.00177344$$

while Equation 2.12 gives

$$\mathrm{Cov}(\tilde{p}_1, \tilde{p}_2) \hat{=} -0.00096484$$

The difference is due to a deficiency of heterozygotes for alleles 1 and 2 compared to Hardy-Weinberg expectations (tests for Hardy-Weinberg are taken up in Chapter 3). The opposite situation holds for alleles 3 and 4, where there is even a difference in sign:

Table 2.5 Genotypic counts for *Esterase B* locus in barley data. Source: R. W. Allard (personal communication).

Generation	$B_{1.6}B_{1.6}$	$B_{2.7}B_{2.7}$	$B_{3.9}B_{3.9}$	$B_{1.6}B_{2.7}$	$B_{1.6}B_{3.9}$	$B_{2.7}B_{3.9}$	Total
4	58	1132	40	39	0	10	1279
5	68	1356	42	9	1	10	1486
6	91	891	24	0	0	0	1006
14	101	1709	82	30	1	5	1928
15	161	2331	227	115	0	9	2843
16	142	2132	68	27	0	0	2369
17	124	2243	92	2	0	0	2461
24	662	3518	311	68	3	25	4587
25	620	3043	262	23	0	19	3967
26	537	2274	252	10	0	10	3083

$$\text{Cov}(\tilde{p}_3, \tilde{p}_4) \hat{=} \begin{cases} 0.00036719 & \text{from Equation 2.11} \\ -0.00059766 & \text{from Equation 2.12} \end{cases}$$

A great deal of attention has been paid here to the difference between results that assume Hardy-Weinberg equilibrium and those that do not. Generally, samples from natural populations of outbreeding species can be treated as though they are indeed from Hardy-Weinberg populations, but there can be occasions when this is not the case. As a final example reference is made to some unpublished data of R. W. Allard. Genotypic counts at an esterase locus in Composite Cross V of barley, *Hordeum vulgare*, are shown in Table 2.5. The data are described in Allard et al. (1972) and consist of samples from several generations of an experimental population grown in Davis, California. The alleles are labeled by their electrophoretic migration distances. With over 99% selfing, random-mating equilibrium results cannot possibly hold. For allele $B_{1.6}$ in generation 15,

$$\tilde{P}_{1.6,1.6} = 161/2843 = 0.0566$$

$$\tilde{p}_{1.6} = 437/5686 = 0.0769$$

so that the estimated variances are

$$\text{Var}(\tilde{p}_{1.6}) \hat{=} \begin{cases} 0.00002140 & \text{from Equation 2.7} \\ 0.00001248 & \text{from Equation 2.8} \end{cases}$$

It is more appropriate to use Equation 2.7 in all situations, unless there is good reason to assume Hardy-Weinberg equilibrium.

Total Variance of Gene Frequencies

The variance of gene frequencies shown in Equations 2.7 or 2.10 referred to variation over repeated samples from the same population. To make statements about gene frequency that are not limited to one particular replicate population, the total variance of a sample gene frequency is used. This total variance must also take account of the genetic sampling that gives rise to variation between replicate populations. It is based on the variation that exists between all possible replicate samples from all possible replicate populations maintained under the same conditions. Obviously the total variance will be greater than the within-population variance, as it contains a contribution for the variation between populations.

The total variance can also be derived with indicator variables. As in the last section, x_{ij} refers to the jth gene in the ith sampled individual, with $x_{ij} = 1$ if the gene is allele A. Expectations of x_{ij}, x_{ij}^2, $x_{ij}x_{ij'}$ have the same functional form as before but the product $x_{ij}x_{i'j'}$ must be changed. Different individuals can no longer be regarded as having been sampled independently. Even if there is random mating within populations, there are differences between populations. In statistical language, the component of variance between populations is given by the covariance of individuals within populations. The genetic sampling process now plays a role, and the expectation for genes from different individuals is written as

$$\mathcal{E}(x_{ij}x_{i'j'}) = P_{A/A}$$

which is the frequency with which two individuals in one population both carry an A allele. Following the same argument in the last section, with this one change, leads to the total variance of

$$\text{Var}(\tilde{p}_A) = (P_{A/A} - p_A^2) + \frac{1}{2n}(p_A + P_{AA} - 2P_{A/A}) \tag{2.13}$$

The first term in this expression will remain even when the sample size n becomes very large, and it represents the between-population contribution to the variance. The frequencies $p_A, P_{AA}, P_{A/A}$ all refer to values expected over replicate populations — another difference from the within-population frequencies that referred to one specific population.

Equation 2.13 can be expressed more conveniently by introducing analogs of the within-population inbreeding coefficient. The quantities F and θ (Cockerham 1969) are the total inbreeding coefficient and the coancestry coefficient, respectively. They refer to pairs of genes within and between individuals, and were called F_{IT} and F_{ST} by Wright (1951). They allow the expression of the frequencies P_{AA} and $P_{A/A}$, both being frequencies expected over all samples *and* over all replicate populations (see Figure 1.4), as is the gene frequency p_A.

$$\begin{aligned} P_{AA} &= p_A^2 + p_A(1-p_A)F \\ P_{A/A} &= p_A^2 + p_A(1-p_A)\theta \end{aligned}$$

Using these expressions allows Equation 2.13 to be rewritten as

$$\text{Var}(\tilde{p}_A) = p_A(1-p_A)\left[\theta + \frac{F-\theta}{n} + \frac{1-F}{2n}\right] \qquad (2.14)$$

with three components that can be identified with the variation between populations, between individuals within populations, and between genes within individuals within populations, respectively. Notice that the between-population component depends on the relation between genes of different individuals within populations. Strictly, the component should be written as $(\theta - 0)$ where 0 is the relation, taken to be zero in this model, between genes of different populations.

In a random mating population, pairs of genes have the same relationship whether or not they are located in the same individual, so $F = \theta$ and the total variance of gene frequency becomes

$$\text{Var}(\tilde{p}_A) = p_A(1-p_A)F + p_A(1-p_A)\frac{1-F}{2n}$$

The coefficient F may simply quantify the effects of drift in random mating finite populations. As inbreeding equilibrium is reached and each population becomes fixed for one of the alleles at a locus, $F = 1$, the variance of gene frequency tends to $p_A(1-p_A)$, reflecting the possibility of the fixation of different alleles in different populations. Estimation of parameters F and θ will be treated later, but note here that it is not possible to estimate them, or the total variance of a gene frequency, from data from a single population.

The three measures of relationship that have been used for pairs of genes are related as

$$f = \frac{F-\theta}{1-\theta}$$

With random mating, $F = \theta$ and $f = 0$. Finite populations eventually become completely inbred and such limiting, or equilibrium, populations have $F = \theta = f = 1$.

Fisher's Approximate Variance Formula

Two approaches have now been used for finding variances of functions of multinomial frequencies, first by direct combination of the multinomial variances and then by use of indicator variables. Both methods give exact results, but both become unwieldy when complicated functions are considered. Furthermore, estimators of genetic parameters often involve ratios of multinomial frequencies and these do not allow exact expressions for variances to be found. There is an approximate method, often called the DELTA METHOD, based on Taylor's series expansions of the function whose variance is to be found. Suppose that the variance is wanted for some function T of variables n_i. The delta method gives the variance

$$\mathrm{Var}(T) \approx \sum_i \left(\frac{\partial T}{\partial n_i}\right)^2 \mathrm{Var}(n_i) + \sum_i \sum_j \frac{\partial T}{\partial n_i}\frac{\partial T}{\partial n_j}\mathrm{Cov}(n_i, n_j) \qquad (2.15)$$

where each of the derivatives is evaluated with the n_i's replaced by their expected values. If the variables n_i are multinomial counts, the expectations have the form nQ_i for a sample of size n. Since the variances and covariances of multinomial n_i's are of ORDER n^{-1} (meaning that they tend to zero as n tends to infinity), this expression is also of order n^{-1}. Terms that involve higher powers of n^{-1} are ignored as being of negligible size.

Because the variances and covariances of multinomial counts n_i's have a very simple form

$$\begin{aligned} \mathrm{Var}(n_i) &= nQ_i(1 - Q_i) \\ \mathrm{Cov}(n_i, n_j) &= -nQ_iQ_j \end{aligned}$$

the variance in Equation 2.15 reduces to

$$\mathrm{Var}(T) \approx n\sum_i \left(\frac{\partial T}{\partial n_i}\right)^2 Q_i - n\left(\sum_i \frac{\partial T}{\partial n_i}Q_i\right)^2$$

When T is a homogeneous function of degree zero in these counts, or in other words is either a function of (n_i/n)'s, or is a ratio of functions of the same degree in the n_i's, Fisher (1925) found the further reduction to

$$\mathrm{Var}(T) \approx n\sum_i \left(\frac{\partial T}{\partial n_i}\right)^2 Q_i - n\left(\frac{\partial T}{\partial n}\right)^2 \qquad (2.16)$$

In this expression, the derivatives of the function T are evaluated at the expected values of the n_i. The derivative with respect to the total sample size n is needed only when T explicitly involves n. It is important to note that T must be written as a function of COUNTS n_i and must be of degree zero. In other words, it must be a ratio of functions of the same order in the counts, or it must have every count divided by the total. Further discussion is given by Bailey (1961, Appendix 2).

As a simple example, consider the variance of the frequency of allele A_1 in a sample of size n for a locus with three codominant alleles. The function $T = \tilde{p}_1$, written in terms of multinomial counts, is

$$T = \frac{1}{2n}(2n_{11} + n_{12} + n_{13})$$

where n_{uv} is the number of A_uA_v genotypes in the sample. The derivatives, evaluated with the n_{uv}'s replaced by their expected values, are

$$\frac{\partial T}{\partial n_{11}} = \frac{1}{n}, \quad \frac{\partial T}{\partial n_{12}} = \frac{1}{2n}$$

$$\frac{\partial T}{\partial n_{13}} = \frac{1}{2n}, \quad \frac{\partial T}{\partial n} = -\frac{p_1}{n}$$

so that the variance becomes

$$\begin{aligned} \mathrm{Var}(\tilde{p}_1) &\approx n\left[\left(\frac{1}{n}\right)^2 P_{11} + \left(\frac{1}{2n}\right)^2 P_{12} + \left(\frac{1}{2n}\right)^2 P_{13}\right] - n\left(\frac{-p_1}{n}\right)^2 \\ &= \frac{1}{2n}(p_1 + P_{11} - 2p_1^2) \end{aligned}$$

as before (Equation 2.7). This expression made use of the relation

$$p_1 = P_{11} + \frac{1}{2}P_{12} + \frac{1}{2}P_{13}$$

Of course this result is exact and the method was not much simpler than the direct methods. For more complex functions, the method is invaluable, however, as can be seen by using it to derive an approximate variance for the quantity $\tilde{p}_A(1 - \tilde{p}_A)$:

$$\mathrm{Var}[\tilde{p}_A(1 - \tilde{p}_A)] \approx \frac{1}{2n}(1 - 2p_A)^2(p_A + P_{AA} - 2p_A^2)$$

The terms involving higher powers of n in the denominator that are ignored by Fisher's formula do not all disappear when $p_A = 0.5$. Fisher's formula gives a variance of zero when $p_A = 0.5$.

It is not possible, however, to use Fisher's formula for a function such as $n_1 = (2n_{11} + n_{12} + n_{13})$ since this is not a homogeneous function of degree zero.

LIKELIHOOD ESTIMATION

Multinomially distributed counts provide estimators of the corresponding population quantities, and functions of such counts can often be manipulated to provide estimators of other parameters. A general procedure of providing estimators that does not require such manipulation will now be considered. Provided the distribution of a random variable (i.e., the probabilities of the variable taking each of its possible values) is known, the method of MAXIMUM LIKELIHOOD can be used to estimate the parameters of that distribution. For within-population analyses, the multinomial distribution is used as a basis for likelihood estimation.

If the multinomial counts n_i depend on a set of s *known* parameters ϕ_j, so that the expected values Q_i are functions of the ϕ_j's, the probability of observing the n_i's was given in Equation 2.1. Conversely, if the parameter values are *unknown*, but the observations are at hand, then the same expression can be regarded as providing the LIKELIHOOD of any particular set of ϕ_j values. The likelihood is written as $L(\phi_1, \ldots, \phi_s)$, so that in the multinomial case

$$L(\phi_1, \ldots, \phi_s) = \frac{n!}{\prod_{i=1}^{k} n_i!} \prod_{i=1}^{k} [Q_i(\phi_1, \ldots, \phi_s)]^{n_i} \tag{2.17}$$

and the maximum likelihood estimates, MLE's, of the ϕ_j's are those values that maximize this likelihood. In this case, as in several others, it is easier and equivalent to work with $\ln L$, the natural logarithm of the likelihood, which is called the SUPPORT. The MLE's are found by setting to zero the derivatives of $\ln L$ with respect to each parameter ϕ_j, and verifying that this maximizes the likelihood. These derivatives are written as S_j and are called SCORES:

$$S_j = \frac{\partial \ln L}{\partial \phi_j}$$

To verify that the procedure leads to sensible results it will now be used to estimate allele frequencies at a two-allele locus known to be in Hardy-Weinberg equilibrium. The genotypes AA, Aa, and aa have frequencies that can be expressed in terms of the single parameter p_A

$$\begin{aligned} P_{AA} &= p_A^2 \\ P_{Aa} &= 2p_A(1-p_A) \\ P_{aa} &= (1-p_A)^2 \end{aligned}$$

and the likelihood of that parameter is

$$L(p_A) = \frac{n!}{n_{AA}!n_{Aa}!n_{aa}!}(p_A^2)^{n_{AA}}[2p_A(1-p_A)]^{n_{Aa}}[(1-p_A)^2]^{n_{aa}}$$

and the support is

$$\ln L(p_A) = \text{Constant} + (2n_{AA}+n_{Aa})\ln(p_A) + (n_{Aa}+2n_{aa})\ln(1-p_A)$$

where the constant term is the logarithm of the ratio of factorials and does not involve the parameter p_A. Differentiating with respect to p_A provides

$$S_{p_A} = \frac{2n_{AA}+n_{Aa}}{p_A} - \frac{n_{Aa}+2n_{aa}}{1-p_A}$$

and setting the score to zero leads to the MLE of

$$\hat{p}_A = \frac{1}{2n}(2n_{AA}+n_{Aa})$$

which is just the observed gene frequency $\tilde{p}_A$. Note that the score has a derivative with respect to p_A that is everywhere negative, which means that the second derivative of the likelihood is negative when $p_A = \tilde{p}_A$. This estimate does indeed maximize the likelihood. As an illustration, the log-likelihood and support curves for p_M for the data in Table 2.3 are shown in Figure 2.1. The shape of these curves makes it clear that there is no ambiguity in estimating an allelic frequency for a locus with two codominant alleles. The log likelihood has its maximum value of -5.38 when $p = 0.54$, and the score is zero at the same p value.

A less trivial case is provided by not assuming Hardy-Weinberg, and using the p_A, f parameterization

$$\begin{aligned}
P_{AA} &= p_A^2 + p_A(1-p_A)f \\
P_{Aa} &= 2p_A(1-p_A)(1-f) \\
P_{aa} &= (1-p_A)^2 + p_A(1-p_A)f
\end{aligned}$$

The likelihood and support for the two parameters are now

$$\begin{aligned}
L(p_A, f) &= \frac{n!}{n_{AA}!n_{Aa}!n_{aa}!}\{p_A[p_A+(1-p_A)f]\}^{n_{AA}} \\
&\quad \times [2p_A(1-p_A)(1-f)]^{n_{Aa}}\{(1-p_A)[(1-p_A)+p_Af]\}^{n_{aa}}
\end{aligned}$$

$$\begin{aligned}
\ln L(p_A, f) &\propto (n_{AA}+n_{Aa})\ln(p_A) + n_{AA}\ln[p_A+(1-p_A)f] \\
&\quad + (n_{Aa}+n_{aa})\ln(1-p_A) + n_{Aa}\ln(1-f) \\
&\quad + n_{aa}\ln[(1-p_A)+p_Af]
\end{aligned}$$

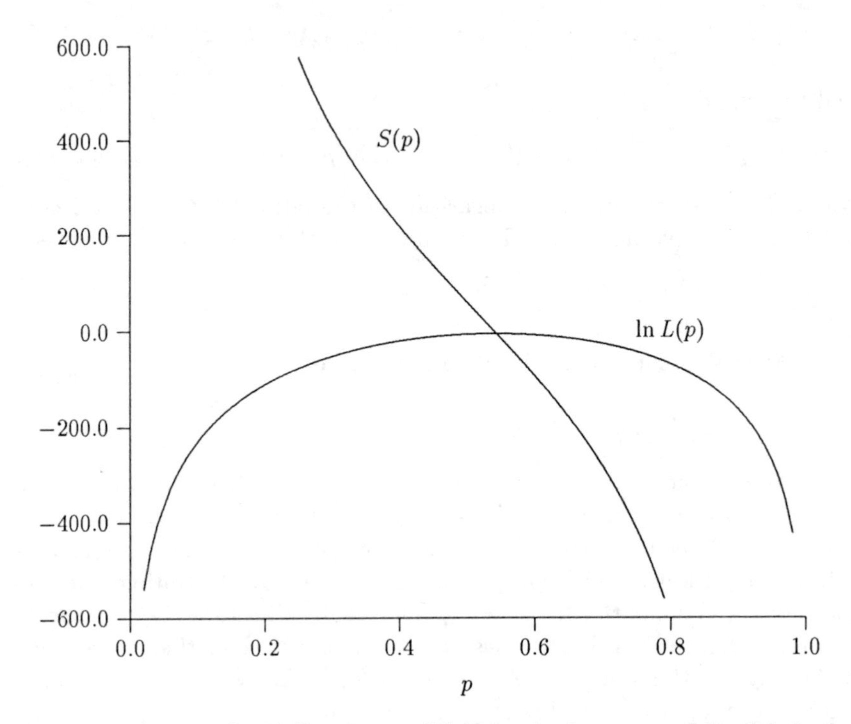

Figure 2.1 Support [$\ln L(p)$] and score [$S(p)$] for the frequency of M allele in the data of Table 2.3.

There are two scores, one for p_A and one for f:

$$\begin{aligned} S_{p_A} &= \frac{n_{AA}+n_{Aa}}{p_A} - \frac{n_{Aa}+n_{aa}}{1-p_A} + \frac{n_{AA}(1-f)}{p_A+(1-p_A)f} \\ &\quad - \frac{n_{aa}(1-f)}{(1-p_A)+p_Af} \\ S_f &= \frac{n_{AA}(1-p_A)}{p_A+(1-p_A)f} - \frac{n_{Aa}}{1-f} + \frac{n_{aa}p_A}{(1-p_A)+p_Af} \end{aligned} \qquad (2.18)$$

It is not a simple task to solve the equations $S_{p_A} = S_f = 0$, although it will be shown below that this is a case in which a procedure known as Bailey's method leads to a simpler set of equations. In general, there is often a need for numerical methods to solve likelihood equations.

Properties of Maximum Likelihood Estimates

Although the likelihood equations obtained by equating the scores to zero may be difficult to solve in some cases, the resulting estimates have a number of desirable properties, at least for large samples. These properties will be illustrated by continued reference to the multinomial distribution. More details are contained in the book on likelihood by Edwards (1972).

Estimating the multinomial parameters is relatively simple and just requires that care is taken not to attempt to estimate a set of parameters among which there is a functional relationship. Since

$$\sum_{i=1}^{k} Q_i = 1$$

it is necessary to remove the dependency by rewriting one of the Q_i's, say Q_k, as one minus the sum of the others. Then the likelihood in Equation 2.17, with the Q_i's regarded as the parameters, becomes

$$\begin{aligned} L(Q_1, \ldots, Q_{k-1}) &= \frac{n!}{n_1! \ldots n_k!}(Q_1)^{n_1} \cdots (Q_{k-1})^{n_{k-1}} \\ &\quad \times (1 - Q_1 - \cdots - Q_{k-1})^{n_k} \end{aligned}$$

and the score for Q_i is

$$S_{Q_i} = \frac{n_i}{Q_i} - \frac{n_k}{(1 - Q_1 - \cdots - Q_{k-1})}$$

Manipulating the $(k-1)$ equations $S_{Q_i} = 0$ leads to the result that the MLE's of the category probabilities Q_i are just the observed proportions $\tilde{Q}_i$:

$$\hat{Q}_i = \tilde{Q}_i$$

Now the MLE of a function of parameters is just that same function of the MLE's of the separate parameters. As an example, the MLE of a population gene frequency is the observed gene frequency since they are both the same function of population or observed genotypic frequencies. A less obvious result is that the MLE of a squared gene frequency p_i^2 for a Hardy-Weinberg population is the square of the MLE of the gene frequency p_i

$$\widehat{(p_i^2)} = (\tilde{p}_i)^2$$

From Equation 2.9 it is known that the expected value of the squared gene frequency, however, is not the square of the expected gene frequency, and this demonstrates that MLE's may be biased. In other words, the MLE of p_i^2 is $\tilde{p}_i^2$, but if this estimate was calculated in very many samples from the same population the average value of all the estimates would not be the quantity p_i^2 being estimated.

The desirable general properties of MLE's follow from them being functions of SUFFICIENT STATISTICS, which themselves are functions of data that contain all the relevant information about the parameters. In the multinomial case the counts are sufficient, whereas the complete description of the outcomes of multinomial sampling would also give the order in which each of the categories was observed. Provided there is a set of sufficient statistics for the set of parameters being estimated, the MLE's of those parameters will be unique (see Kendall and Stuart 1961 for further discussion in this area.)

Although it cannot be said that MLE's are unbiased, something can be said about their values in very large samples. It can be shown that, under very general conditions, MLE's are CONSISTENT estimators, which means that, as the sample size becomes very large, the MLE $\hat{\phi}$ for a parameter ϕ becomes arbitrarily close to ϕ. Formally,

$$\lim_{n\to\infty} \Pr(\hat{\phi} = \phi) = 1$$

Large sample sizes also allow statements to be made about the variances of MLE's, and these statements are phrased in terms of the INFORMATION. Recall that the score of a parameter is the derivative of the support, or log likelihood, with respect to that parameter. Minus the second derivative of the support is called the information. For a single parameter ϕ and likelihood $L(\phi)$, the score is

$$S_\phi = \frac{\partial \ln L(\phi)}{\partial \phi}$$

and the information is

$$I(\phi) = -\left(\frac{\partial^2 \ln L(\phi)}{\partial \phi^2}\right)$$

For large samples, the inverse of the expected value of the information provides the variance of the MLE $\hat{\phi}$. For multinomial data, the expected value is found by replacing each count n_i, wherever it occurs, by its expected value.

$$\mathrm{Var}(\hat{\phi}) = 1/\mathcal{E}[I(\phi)]$$

For any unbiased estimator (not necessarily an MLE), the inverse of the expected information provides a lower bound for the variance of an estimate for any sample size.

There is a straightforward extension to the case when several parameters are estimated, with the information becoming a form that provides a matrix of variances and covariances for the estimates. For s independent parameters ϕ_i written as a vector $\boldsymbol{\phi}$ the expected information matrix is

$$\mathcal{E}[I(\boldsymbol{\phi})] = \begin{bmatrix} -\mathcal{E}\left(\frac{\partial^2 \ln L(\boldsymbol{\phi})}{\partial \phi_1^2}\right) & -\mathcal{E}\left(\frac{\partial^2 \ln L(\boldsymbol{\phi})}{\partial \phi_1 \partial \phi_2}\right) & \cdots & -\mathcal{E}\left(\frac{\partial^2 \ln L(\boldsymbol{\phi})}{\partial \phi_1 \partial \phi_s}\right) \\ -\mathcal{E}\left(\frac{\partial^2 \ln L(\boldsymbol{\phi})}{\partial \phi_2 \partial \phi_1}\right) & -\mathcal{E}\left(\frac{\partial^2 \ln L(\boldsymbol{\phi})}{\partial \phi_2^2}\right) & \cdots & -\mathcal{E}\left(\frac{\partial^2 \ln L(\boldsymbol{\phi})}{\partial \phi_2 \partial \phi_s}\right) \\ \cdots & \cdots & \cdots & \cdots \\ -\mathcal{E}\left(\frac{\partial^2 \ln L(\boldsymbol{\phi})}{\partial \phi_s \partial \phi_1}\right) & -\mathcal{E}\left(\frac{\partial^2 \ln L(\boldsymbol{\phi})}{\partial \phi_s \partial \phi_2}\right) & \cdots & -\mathcal{E}\left(\frac{\partial^2 \ln L(\boldsymbol{\phi})}{\partial \phi_s^2}\right) \end{bmatrix}$$

Inverting this matrix gives the large-sample variances and covariances

$$\{\mathcal{E}[I(\boldsymbol{\phi})]\}^{-1} = \begin{bmatrix} \mathrm{Var}(\hat{\phi}_1) & \mathrm{Cov}(\hat{\phi}_1, \hat{\phi}_2) & \cdots & \mathrm{Cov}(\hat{\phi}_1, \hat{\phi}_s) \\ \mathrm{Cov}(\hat{\phi}_2, \hat{\phi}_1) & \mathrm{Var}(\hat{\phi}_2) & \cdots & \mathrm{Cov}(\hat{\phi}_2, \hat{\phi}_s) \\ \cdots & \cdots & \cdots & \cdots \\ \mathrm{Cov}(\hat{\phi}_s, \hat{\phi}_1) & \mathrm{Cov}(\hat{\phi}_s, \hat{\phi}_2) & \cdots & \mathrm{Var}(\hat{\phi}_s) \end{bmatrix}$$

As an illustration of these results, consider the case of estimating the gene frequencies for a locus with three alleles that is known to be in Hardy-Weinberg equilibrium. Although it is known that the estimators are the observed gene frequencies, and that these are multinomially distributed in the Hardy-Weinberg situation, it is instructive to calculate scores, information matrix, and variance-covariance matrix. Label the counts for the six possible genotypes as

A_1A_1	A_1A_2	A_2A_2	A_1A_3	A_2A_3	A_3A_3
n_1	n_2	n_3	n_4	n_5	n_6

and let n be the sums of these counts. Take p_1 and p_2 as the two independent parameters that can be estimated. The likelihood is

$$\begin{aligned} L(p_1, p_2) &\propto (p_1^2)^{n_1}(2p_1p_2)^{n_2}(p_2^2)^{n_3}[2p_1(1-p_1-p_2)]^{n_4} \\ &\quad \times [2p_2(1-p_1-p_2)]^{n_5}[(1-p_1-p_2)^2]^{n_6} \end{aligned}$$

and the two scores are

$$\begin{aligned}
S_1 &= \frac{\partial \ln L}{\partial p_1} = \frac{2n_1 + n_2 + n_4}{p_1} - \frac{2n_6 + n_4 + n_5}{1 - p_1 - p_2} \\
S_2 &= \frac{\partial \ln L}{\partial p_2} = \frac{2n_3 + n_2 + n_5}{p_2} - \frac{2n_6 + n_4 + n_5}{1 - p_1 - p_2}
\end{aligned}$$

Setting the scores to zero provides the estimates

$$\begin{aligned}
\hat{p}_1 &= \tilde{p}_1 = \frac{2n_1 + n_2 + n_4}{2n} \\
\hat{p}_2 &= \tilde{p}_2 = \frac{2n_3 + n_2 + n_5}{2n}
\end{aligned}$$

and the derivatives of the scores are

$$\frac{\partial S_1}{\partial p_1} = \frac{\partial^2 \ln L}{\partial p_1^2} = -\frac{2n_1 + n_2 + n_4}{p_1^2} - \frac{2n_6 + n_4 + n_5}{(1 - p_1 - p_2)^2}$$

$$\frac{\partial S_2}{\partial p_2} = \frac{\partial^2 \ln L}{\partial p_2^2} = -\frac{2n_3 + n_2 + n_5}{p_2^2} - \frac{2n_6 + n_4 + n_5}{(1 - p_1 - p_2)^2}$$

$$\frac{\partial S_1}{\partial p_2} = \frac{\partial S_2}{\partial p_1} = \frac{\partial^2 \ln L}{\partial p_1 \partial p_2} = -\frac{2n_6 + n_4 + n_5}{(1 - p_1 - p_2)^2}$$

Evaluating these at the expected values of the counts gives the expected information matrix

$$\mathcal{E}(I) = n \begin{bmatrix} \frac{1}{p_1} + \frac{1}{(1 - p_1 - p_2)} & \frac{1}{(1 - p_1 - p_2)} \\ \frac{1}{(1 - p_1 - p_2)} & \frac{1}{p_2} + \frac{1}{(1 - p_1 - p_2)} \end{bmatrix}$$

which has inverse

$$[\mathcal{E}(I)]^{-1} = \begin{bmatrix} \frac{p_1(1 - p_1)}{n} & -\frac{p_1 p_2}{n} \\ -\frac{p_1 p_2}{n} & \frac{p_2(1 - p_2)}{n} \end{bmatrix} = \begin{bmatrix} \mathrm{Var}(\hat{p}_1) & \mathrm{Cov}(\hat{p}_1, \hat{p}_2) \\ \mathrm{Cov}(\hat{p}_1, \hat{p}_2) & \mathrm{Var}(\hat{p}_2) \end{bmatrix}$$

The elements of this matrix are just the multinomial variances and covariances.

For large samples, MLE's are unbiased for the parameters they estimate and have variances supplied by the information, and it can further be shown that they are normally distributed. For a single parameter ϕ, the MLE $\hat{\phi}$ has the normal distribution with mean ϕ and variance $1/\mathcal{E}[I(\phi)]$

$$\hat{\phi} \sim N\left(\phi, \{\mathcal{E}[I(\phi)]\}^{-1}\right) \tag{2.19}$$

in large samples. For several parameters, the MLE's are asymptotically multivariate normal. This normality allows tests of hypotheses to be set up very easily, although these tests do require large sample sizes.

Bailey's Method for MLE's

If the number of independent parameters to be estimated equals the number of independent pieces of information, or degrees of freedom, in the data, Bailey (1951) showed that MLE's for the parameters may be found by equating observations to their expected values. Doing so often gives equations that are easier to solve than those from setting the scores equal to zero.

Suppose there are s parameters to estimate, and s degrees of freedom in the data. Bailey's rule says to write the expectation of n_i, the observed count in category i, as m_i and then solve the equations

$$m_i = n_i$$

for the s parameters ϕ_j. To see that this does indeed provide MLE's, note that, because

$$\sum_{i=1}^{k} m_i = n$$

where n is the sample size (a constant) and k is the number of categories in the data, the derivative with respect to ϕ_j is zero

$$\sum_{i=1}^{k} \frac{\partial m_i}{\partial \phi_j} = 0 \qquad j = 1, 2, \ldots, s$$

Now the log likelihood in this multinomial case is

$$\ln L = \text{Constant} + \sum_i n_i \ln m_i$$

which has derivatives

$$\begin{aligned} S_j &= \frac{\partial \ln L}{\partial \phi_j} \\ &= \sum_i n_i \frac{\partial \ln m_i}{\partial \phi_j} \\ &= \sum_i \frac{n_i}{m_i} \frac{\partial m_i}{\partial \phi_j} \end{aligned}$$

and when $m_i = n_i$ this reduces to

$$S_j = \sum_i \frac{\partial m_i}{\partial \phi_j}$$

As the sum on the right-hand side in this last equation has been shown to be zero, putting $m_i = n_i$ has made the score S_j equal to zero and the solutions must be MLE's. For the two sets of equations to have solutions, it is necessary that $s = k - 1$. The number of parameters must equal the number of independent categories.

The power of this procedure is illustrated by returning to the case of estimating gene frequency p_A and inbreeding coefficient f for a two-allele locus. The two independent categories can be taken to be AA homozygotes and Aa heterozygotes, leading to

$$\begin{aligned} n[\hat{p}_A^2 + \hat{p}_A(1-\hat{p}_A)\hat{f}] &= n_{AA} \\ \text{i.e., } m_1 &= n_1 \\ n[2\hat{p}_A(1-\hat{p}_A)(1-\hat{f})] &= n_{Aa} \\ \text{i.e., } m_2 &= n_2 \end{aligned}$$

These equations may be solved very easily:

$$\begin{aligned} \hat{p}_A &= \frac{1}{2n}(2n_{AA} + n_{Aa}) \\ \hat{f} &= 1 - \frac{n_{Aa}}{2n\hat{p}_A(1-\hat{p}_A)} \end{aligned}$$

The MLE for gene frequency is still the observed value, and that for the inbreeding coefficient is one minus the observed heterozygosity divided by the heterozygosity expected under Hardy-Weinberg equilibrium. These solutions do make the scores in Equations 2.18 equal to zero.

If genotypic frequencies were to be expressed in terms of gene frequencies and a single inbreeding coefficient for the k-allele case ($k > 2$)

$$\begin{aligned} P_{uu} &= p_u^2 + p_u(1-p_u)f \quad u = 1, 2, \ldots, k \\ P_{uv} &= 2p_u p_v(1-f) \qquad u, v = 1, 2, \ldots, k;\ u \neq v \end{aligned}$$

there are a total of k independent parameters, 1 for f and $k-1$ for the allelic frequencies. This is not the same as the number of independent categories, $k(k+1)/2 - 1$, and Bailey's method cannot be applied.

While Bailey's method provides values for MLE's very easily, it still leaves the problem of attaching variances to the estimates. Even in the relatively simple case of estimating p_A and f for a two-allele locus, it is not easy to invert the information matrix. For a large-sample variance, Fisher's variance formula (Equation 2.16) can be employed instead, after rewriting $\hat{f}$ as

$$\hat{f} = 1 - 2n\frac{n_{Aa}}{(2n_{AA} + n_{Aa})(2n_{aa} + n_{Aa})}$$

The necessary derivatives are

$$\begin{aligned}
\frac{\partial \hat{f}}{\partial n_{AA}} &= 4n\frac{n_{Aa}}{(2n_{AA} + n_{Aa})^2(2n_{aa} + n_{Aa})} \\
\frac{\partial \hat{f}}{\partial n_{Aa}} &= 2n\frac{(n_{Aa}^2 - 4n_{AA}n_{aa})}{(2n_{AA} + n_{Aa})^2(2n_{aa} + n_{Aa})^2} \\
\frac{\partial \hat{f}}{\partial n_{aa}} &= 4n\frac{n_{aa}}{(2n_{AA} + n_{Aa})(2n_{aa} + n_{Aa})^2} \\
\frac{\partial \hat{f}}{\partial n} &= -\frac{2n_{Aa}}{(2n_{AA} + n_{Aa})(2n_{aa} + n_{Aa})}
\end{aligned}$$

Evaluating these at the expected values for the three genotypic counts and simplifying the variance expression

$$\frac{1}{n}\mathrm{Var}(\hat{f}) = \left(\frac{\partial \hat{f}}{\partial n_{AA}}\right)^2 P_{AA} + \left(\frac{\partial \hat{f}}{\partial n_{Aa}}\right)^2 P_{Aa} + \left(\frac{\partial \hat{f}}{\partial n_{aa}}\right)^2 P_{aa} - \left(\frac{\partial \hat{f}}{\partial n}\right)^2$$

gives

$$\mathrm{Var}(\hat{f}) = \frac{1}{n}(1-f)^2(1-2f) + \frac{f(1-f)(2-f)}{2np_A(1-p_A)}$$

This expression, given by Fyfe and Bailey (1951), is written in terms of the parameters p_A and f, which are generally unknown, but estimated values could be substituted to estimate the variance of $\hat{f}$. The corresponding formula for the variance of the estimated allelic frequency is

$$\mathrm{Var}(\hat{p}_A) = \frac{1}{2n}(1+f)p_A(1-p_A)$$

Iterative Solutions of Likelihood Equations

There are often situations when the maximum likelihood equations obtained by setting the scores to zero do not yield analytical solutions and when Bailey's method cannot be applied. Numerical methods are needed to find the estimates for any particular set of data. The most direct method is simply to evaluate the likelihood for a range of values of the parameter(s) to be estimated, and choose the value(s) that give the largest likelihood.

This GRID SEARCH procedure can be made quite sophisticated – the range of parameter values could be divided into tenths, and then the tenth that appears to contain the maximum likelihood estimate itself could be divided into 10 parts. As many rounds of the search could be performed as significant digits are required in the solution. It is helpful to plot the likelihood to show how it is responding to changes in parameter values, and to guard against a local maximum being confused for a global maximum.

An alternative approach is given by NEWTON-RAPHSON ITERATION. Briefly, some initial value is chosen for the estimate and then this value is modified by using the score. The modified value is modified in turn and the process continues until successive iterates differ by less than some specified amount. For a parameter ϕ, write the required MLE as $\hat{\phi}$, and the initial value, or guess, as ϕ'. The score S_ϕ is to be zero at $\hat{\phi}$, and that score can be expanded by Taylor's theorem as

$$S_{\hat{\phi}} = 0 = S_{\phi'} + (\hat{\phi} - \phi')\left[\frac{\partial S_\phi}{\partial \phi}\right]_{\phi=\phi'}$$

where higher order terms in $(\hat{\phi}-\phi')$ are ignored. Rearranging this expression provides an approximate value ϕ'' for $\hat{\phi}$:

$$\begin{aligned}\phi'' &= \phi' - S_{\phi'}/\left[\frac{\partial S_\phi}{\partial \phi}\right]_{\phi=\phi'} \\ &= \phi' + S_{\phi'}/I(\phi')\end{aligned}$$

The initial value ϕ' is modified by adding to it the score divided by the information, both evaluated at the initial value. The new value then serves as an initial value for a further modification

$$\phi''' = \phi'' + S(\phi'')/I(\phi'')$$

and the iteration continues until convergence, i.e., successive values are very close to each other.

For situations with many parameters, the information $I(\phi)$ is a matrix, and the iteration procedure requires matrix inversion:

$$\phi'' = \phi' + I^{-1}(\phi')S(\phi')$$

Obviously the method breaks down if the information is zero, or the information matrix is singular. Whether or not such problems occur, it is always advisable to try several starting values and to compare the likelihoods found after convergence. This guards against problems of nonconvergence, or convergence to solutions other than maximum likelihood.

The EM Algorithm

Another iterative procedure that can lead to MLE's was discussed by Dempster et al. (1977) to handle cases in which the observations can be regarded as being incomplete data. In other words, there are more categories in the data than can be distinguished. The procedure assumes some model to estimate the frequencies of these hidden categories. Each iteration in this EM ALGORITHM consists of an Expectation step followed by a Maximization step. For multinomial data the method has also been called GENE-COUNTING (Ceppellini et al. 1955, Smith 1957).

One of the simplest situations in which the procedure may be used is the estimation of gene frequencies at a two-allele locus when one allele, A, is dominant to the other, a. Although there are three genotypic classes, AA, Aa, and aa, only two can be distinguished, AA+Aa and aa. To find the MLE of the frequency p_a of the recessive allele, the first step is to estimate the two unknown genotypic frequencies P_{AA} and P_{Aa}. These two add to $1 - p_a^2$, are in the proportion $(1 - p_a)^2$ to $2p_a(1 - p_a)$, and so are estimated to be those proportions of the estimate $(n - n_{aa})/n$ of their total. If some initial value p'_a is given to p_a and used to construct the unknown genotypic frequencies, the maximization step consists of saying that the MLE of p_a is just the "observed" frequency found using the concocted heterozygote count n^*_{Aa}:

$$\begin{aligned} \hat{p}_a &= \frac{1}{2n}(n^*_{Aa} + 2n_{aa}) \\ &= \frac{1}{2n}\left[\frac{2p'_a(1 - p'_a)}{1 - (p'_a)^2}(n - n_{aa}) + 2n_{aa}\right] \end{aligned} \quad (2.20)$$

The value found for the left-hand side of this equation is then regarded as the next iterate and is substituted into the right-hand side for the next iteration. Iteration is continued until successive values are (nearly) equal. As with Newton-Raphson iteration, several starting values should be used.

In this example of estimating the frequency of a recessive allele, an analytical solution can be found by recognizing that convergence means that the value of $\hat{p}_a$ will be unchanged by Equation 2.20:

$$\hat{p}_a = \frac{1}{n}\left[\frac{\hat{p}_a(1 - \hat{p}_a)}{1 - \hat{p}_a^2}(n - n_{aa}) + n_{aa}\right]$$

and this equation can be solved to yield

$$\hat{p}_a = \sqrt{n_{aa}/n} \quad (2.21)$$

Although this example illustrated the EM algorithm, it can be handled much more directly by noting that there are two observable classes or one degree of freedom in the data and that there is one parameter to be found. Bailey's method leads to the estimate in Equation 2.21 directly. Note that the method rests on an assumption about the missing observations: they were estimated on the assumption of Hardy-Weinberg equilibrium. Without this assumption, it is not possible to estimate gene frequencies at loci that show dominance.

Fisher's approximate variance method, Equation 2.16, applied to the estimate in Equation 2.21 shows that

$$\begin{aligned}\text{Var}(\hat{p}_a) &\approx \frac{1}{4n}(1 - P_{aa}) \\ &\hat{=} \frac{n - n_{aa}}{4n^2}\end{aligned}$$

For Hardy-Weinberg proportions, $P_{aa} = p_a^2$, this variance is larger than the value found for codominant alleles ("complete" data):

$$\text{Var}(\hat{p}_a) = \frac{1}{2n}p_a(1 - p_a)$$

reflecting the fact that less information is available. Smith (1957) related the ratio of these two variances to the limiting rate of convergence of the iterative EM process. Since explicit estimating formulas do not usually result from the EM algorithm, estimation of variances of the estimates is something of a problem. One solution is to substitute the estimates into the information matrix and invert that, as if the estimates had been obtained by solving the likelihood equations.

An interesting example is the classic one of estimating the three allelic frequencies for the ABO blood group system. With three alleles; A, B, and O, there are six genotypes but only four phenotypic classes as shown with the conventional notation in Table 2.6. Alleles A, B, O have frequencies p, q, r. Although there are only two independent parameters, EM equations can be set up for all three after estimating the missing counts for genotypes involving the recessive allele O:

$$\begin{aligned}p'' &= \frac{1}{2n}(2n_{AA}^* + n_{AO}^* + n_{AB}) \\ &= \frac{p' + r'}{p' + 2r'}\frac{n_A}{n} + \frac{n_{AB}}{2n} \\ q'' &= \frac{q' + r'}{q' + 2r'}\frac{n_B}{n} + \frac{n_{AB}}{2n}\end{aligned}$$

Table 2.6 Frequencies of genotypes and phenotypes for ABO blood groups. Alleles A,B,O have frequencies p, q, r.

Genotype	Phenotype	Count	Expected Frequency	Expected Count
AA	A	n_A	p^2	$n^*_{AA} = [p/(p+2r)]n_A$
AO			$2pr$	$n^*_{AO} = [2r/(p+2r)]n_A$
BB	B	n_B	q^2	$n^*_{BB} = [q/(q+2r)]n_B$
BO			$2qr$	$n^*_{BO} = [2r/(q+2r)]n_B$
AB	AB	n_{AB}	$2pq$	n_{AB}
OO	O	n_O	r^2	n_O

$$r'' = \frac{r'}{p'+2r'}\frac{n_A}{n} + \frac{r'}{q'+2r'}\frac{n_B}{n} + \frac{2n_O}{2n}$$

The initial values are p', q', r' and the next iterates are p'', q'', r''. A convenient set of initial values was given by Bernstein (1925):

$$p' = 1 - \sqrt{(n_O + n_B)/n},\ q' = 1 - \sqrt{(n_O + n_A)/n},\ r' = \sqrt{n_O/n}$$

Examples

Turn back to the *MNS* blood group data in Table 1.2. The system can be regarded as being determined by a codominant locus with alleles M, N and a dominant locus with alleles S, s — although other interpretations are possible, such as that of a single locus with alleles *MS, Ms, NS*, and *ns*. The data in Table 1.2 were incomplete in that family data were used to infer all three genotypes at the S locus in some families but not in others. The counts at this locus are given in Table 2.7.

Finding the MLE of the frequency of allele s under the assumption of Hardy-Weinberg equilibrium requires that the 64 dominant phenotypes be partitioned into n^*_{SS} *SS* homozygotes and n^*_{Ss} *Ss* heterozygotes, leading to an iterative equation

$$p'_s = \frac{1}{2n}(2n_{ss} + n_{Ss})$$

Table 2.7 Genotypic counts for Ss blood group in mothers and fathers of Table 1.3.

Genotype	Father	Mother	Total
SS or Ss	32	32	64
Ss	20	23	43
ss	41	38	79
Total	93	93	186

$$
\begin{aligned}
&= \frac{2 \times 79 + (43 + n^*_{Ss})}{2 \times 186} \\
&= \frac{1}{2 \times 186}\left(2 \times 79 + 43 + 64\frac{2p_s}{1+p_s}\right)
\end{aligned}
$$

that can be solved iteratively or explicitly to give $\hat{p}_s = 0.68$.

Data of Morton (1964), discussed by Yasuda and Kimura (1968), will be used to illustrate the EM algorithm for ABO frequencies. The genotypic counts they used were

$$n_A = 725, \;\; n_B = 258, \;\; n_{AB} = 72, \;\; n_O = 1073, \;\; n = 2128$$

so that the Bernstein estimators are

$$p'_A = 0.2091, \;\; p'_B = 0.0808, \;\; p'_O = 0.7101$$

The EM equations given above produce the numerical values displayed in Table 2.8 for both Bernstein's initial values and equally frequent initial values. The convergence is very rapid, and Bernstein's values are very close to the MLE's.

To complete this example, the Newton-Raphson iterative technique is applied to the likelihood equations obtained by setting the scores for frequencies p_A and p_B to zero. The likelihood is

$$
\begin{aligned}
L \;\propto\; & [p_A(2 - p_A - 2p_B)]^{n_A}[p_B(2 - 2p_A - p_B)]^{n_B}[2p_Ap_B]^{n_{AB}} \\
& \times [(1 - p_A - p_B)^2]^{n_O}
\end{aligned}
$$

and the scores are

$$S_A = \frac{n_A + n_{AB}}{p_A} - \frac{n_A}{2 - p_A - 2p_B} - \frac{2n_B}{2 - 2p_A - p_B} - \frac{2n_O}{1 - p_A - p_B}$$

Table 2.8 Numerical values of EM iterates for *ABO* data of Morton (1964).

Iterate	$\hat{p}_A$	$\hat{p}_B$	$\hat{p}_O$
	Bernstein's initial values		
0	0.20913343	0.08080208	0.71009107
1	0.20913030	0.08080095	0.71006875
2	0.20913061	0.08080100	0.71006838
3	0.20913065	0.08080101	0.71006834
4	0.20913065	0.08080101	0.71006834
5	0.20913065	0.08080101	0.71006834
	Equal initial values		
0	0.33333333	0.33333333	0.33333333
1	0.24404762	0.09774436	0.65820802
2	0.21390645	0.08172757	0.70436598
3	0.20972135	0.08086164	0.70941701
4	0.20920200	0.08080616	0.70999184
5	0.20913921	0.08080154	0.71005925
6	0.20913168	0.08080107	0.71006725
7	0.20913078	0.08080102	0.71006821
8	0.20913067	0.08080101	0.71006832
9	0.20913066	0.08080101	0.71006834
10	0.20913065	0.08080101	0.71006834

$$S_B = \frac{n_B + n_{AB}}{p_B} - \frac{2n_A}{2 - p_A - 2p_B} - \frac{n_B}{2 - 2p_A - p_B} - \frac{2n_O}{1 - p_A - p_B}$$

with derivatives

$$-\frac{\partial S_A}{\partial p_A} = \frac{n_A + n_{AB}}{p_A^2} + \frac{n_A}{(2 - p_A - 2p_B)^2} + \frac{4n_B}{(2 - 2p_A - p_B)^2} + \frac{2n_O}{(1 - p_A - p_B)^2}$$

$$-\frac{\partial S_B}{\partial p_B} = \frac{n_B + n_{AB}}{p_B^2} + \frac{4n_A}{(2 - p_A - 2p_B)^2} + \frac{n_B}{(2 - 2p_A - p_B)^2} + \frac{2n_O}{(1 - p_A - p_B)^2}$$

$$-\frac{\partial S_A}{\partial p_B} = -\frac{\partial S_B}{\partial p_A}$$

$$= \frac{2n_A}{(2-p_A-2p_B)^2} + \frac{2n_B}{(2-2p_A-p_B)^2} + \frac{2n_O}{(1-p_A-p_B)^2}$$

Writing the two parameters and two scores as vectors

$$\mathbf{p} = \begin{bmatrix} p_A \\ p_B \end{bmatrix}$$

$$\mathbf{S} = \begin{bmatrix} S_A \\ S_B \end{bmatrix}$$

and the derivatives in an information matrix

$$I = \begin{bmatrix} -\frac{\partial S_A}{\partial p_A} & -\frac{\partial S_A}{\partial p_B} \\ -\frac{\partial S_B}{\partial p_A} & -\frac{\partial S_B}{\partial p_B} \end{bmatrix}$$

allows a pair of iterative equations to be expressed in matrix form as

$$\mathbf{p}' = \mathbf{p} + I^{-1}\mathbf{S}$$

where $\mathbf{p}$ is the initial value and $\mathbf{p}'$ is the next iterate.

The information matrix is evaluated by replacing the counts with their expected values for the initial values. Numerical values for this scheme, using two sets of initial values, are shown in Table 2.9, with the variances and covariances being the elements of the inverted information matrix. The possible problems with Newton-Raphson iteration are illustrated for the case of equally frequent initial values in Table 2.9 where the iterates never do converge to the true values, and it is necessary to try other initial values.

The advantage of the Newton-Raphson procedure is that estimates of variances and covariances are obtained as part of the process. In addition to the variances shown in Table 2.9, the variance of the estimated frequency $\hat{p}_O$, that is obtained by subtraction, can be found from

$$\begin{aligned} \mathrm{Var}(\hat{p}_O) &= \mathrm{Var}(1-\hat{p}_A-\hat{p}_B) \\ &= \mathrm{Var}(\hat{p}_A) + \mathrm{Var}(\hat{p}_B) + 2\mathrm{Cov}(\hat{p}_A, \hat{p}_B) \\ &= 0.00004394 + 0.00001821 - 2(0.00000395) \\ &= 0.00005425 \end{aligned}$$

Table 2.9 Numerical values of Newton-Raphson iterates for ABO data of Morton (1964).

Iterate	$\hat{p}_A$	$\hat{p}_B$	Var($\hat{p}_A$)	Var($\hat{p}_B$)	Cov($\hat{p}_A, \hat{p}_B$)
			Bernstein's initial values		
0	0.20913343	0.08080208			
1	0.20913065	0.08080101	0.00004394	0.00001821	−0.00000395
2	0.20913065	0.08080101	0.00004394	0.00001821	−0.00000395
3	0.20913065	0.08080101	0.00004394	0.00001821	−0.00000395
4	0.20913065	0.08080101	0.00004394	0.00001821	−0.00000395
5	0.20913065	0.08080101	0.00004394	0.00001821	−0.00000395
			Equal initial values		
0	0.33333333	0.33333333			
1	0.17197290	0.00738174	0.00006364	0.00006364	−0.00002448
2	0.27589286	0.14671685	0.00008796	0.00000340	−0.00000027
3	0.14983843	0.04006071	0.00008625	0.00004542	−0.00000868
4	0.28731256	0.12231962	0.00007348	0.00001836	−0.00000125
5	0.14139559	0.05343061	0.00009247	0.00003866	−0.00000753
6	0.29056568	0.11162130	0.00006814	0.00002442	−0.00000157
7	0.13868308	0.05915438	0.00009508	0.00003569	−0.00000694
8	0.29081133	0.10679370	0.00006625	0.00002695	−0.00000171
9	0.13831752	0.06172864	0.00009611	0.00003440	−0.00000664
10	0.28992951	0.10447615	0.00006573	0.00002803	−0.00000178

Gametic Frequencies

When pairs of loci are scored, even for codominant loci, it is generally not possible to distinguish between the two types of double heterozygote for any two pairs of alleles. This makes the determination of gametic frequencies difficult. Consider loci **A** and **B** with alleles A, a and B, b, respectively. The frequency with which the pair AB is transmitted together between generations is the gametic frequency p_{AB} given by

$$p_{AB} = P_{AB}^{AB} + \frac{1}{2}(P_{Ab}^{AB} + P_{aB}^{AB} + P_{ab}^{AB}) \tag{2.22}$$

but the frequency P_{ab}^{AB} cannot be disentangled from the observed total frequency $P_{ab}^{AB} + P_{aB}^{Ab}$ of double heterozygotes. Provided the population is assumed to be mating at random, genotypic frequencies are the products of gametic frequencies and the EM algorithm may be employed. The two double heterozygote frequencies are estimated as the products of initial estimates of gametic frequencies, and then used to provide the next values of gametic frequencies. None of this affects the estimation of gene frequencies at each locus since these can be estimated directly as the observed frequencies

$$\begin{aligned}\tilde{p}_A &= \tilde{P}_{AB}^{AB} + \tilde{P}_{Ab}^{AB} + \tilde{P}_{Ab}^{Ab} + \frac{1}{2}(\tilde{P}_{aB}^{AB} + \tilde{P}_{ab}^{AB} + \tilde{P}_{aB}^{Ab} + \tilde{P}_{ab}^{Ab}) \\ \tilde{p}_B &= \tilde{P}_{AB}^{AB} + \tilde{P}_{aB}^{AB} + \tilde{P}_{aB}^{aB} + \frac{1}{2}(\tilde{P}_{Ab}^{AB} + \tilde{P}_{ab}^{AB} + \tilde{P}_{aB}^{Ab} + \tilde{P}_{ab}^{aB})\end{aligned}$$

If the initial estimate of the frequency of AB gametes is p'_{AB}, the frequencies of the three other gametes can be found from

$$\begin{aligned}p'_{Ab} &= \tilde{p}_A - p'_{AB} \\ p'_{aB} &= \tilde{p}_B - p'_{AB} \\ p'_{ab} &= 1 - \tilde{p}_A - \tilde{p}_B + p'_{AB}\end{aligned}$$

so that the frequency of AB/ab heterozygotes can be estimated as

$$P_{ab}^{AB\prime} = \frac{2p'_{AB}p'_{ab}}{2p'_{AB}p'_{ab} + 2p'_{Ab}p'_{aB}}$$

which is a function of the single unknown quantity p'_{AB}. This is the estimation step of the EM procedure. The maximization step then provides the new value p''_{AB} from Equation 2.22 of

$$p''_{AB} = \tilde{P}_{AB}^{AB} + \frac{1}{2}\left(\tilde{P}_{Ab}^{AB} + \tilde{P}_{aB}^{AB} + \frac{2p'_{AB}p'_{ab}}{2p'_{AB}p'_{ab} + 2p'_{Ab}p'_{aB}}\tilde{P}_{AaBb}\right) \quad (2.23)$$

This new value then serves as the initial value for another iteration, and the process continues until successive values are sufficiently close. The estimates obtained in this way will be maximum likelihood. There still remain the dangers inherent in all iterative schemes of the process not converging, or perhaps converging to the wrong value, and some examples of these cases are given by Weir and Cockerham (1979) (see Exercise 2.3). It appears that there is a greater chance of problems when either one of the loci does not have frequencies consistent with Hardy-Weinberg equilibrium so that the assumptions underlying the method are not valid.

One way to avoid problems in this case is to recognize that Equation 2.23 can be regarded as a cubic equation in $\hat{p}_{AB}$ if the relation $p''_{AB} = p'_{AB} = \hat{p}_{AB}$ is used and the other gametic frequencies are replaced by expressions involving $\hat{p}_{AB}$, $\hat{p}_A$, and $\hat{p}_B$. Numerical methods can be used to solve the cubic, each of the three roots checked to make sure that all gametic frequencies

Table 2.10 Two-locus genotypic counts for *Idh1* and *Mdh* loci in mosquito data of Table 1.3.

		Mdh			
		11	12	22	
Idh1	11	$n_{AABB} = 19$	$n_{AABb} = 5$	$n_{AAbb} = 0$	$n_{AA} = 24$
	12	$n_{AaBB} = 8$	$n_{AaBb} = 8$	$n_{Aabb} = 0$	$n_{Aa} = 16$
	22	$n_{aaBB} = 0$	$n_{aaBb} = 0$	$n_{aabb} = 0$	$n_{aa} = 0$
		$n_{BB} = 27$	$n_{Bb} = 13$	$n_{bb} = 0$	$n = 40$

are between zero and appropriate gene frequencies, and then the three likelihoods are calculated. The valid solution that maximizes the likelihood is the required MLE for p_{AB}. A program to perform these calculations is listed in the Appendix. The likelihood, calculated under the assumption of Hardy-Weinberg equilibrium, is

$$\begin{aligned} \ln L(p_A, p_B, p_{AB}) = & \text{ Constant} + (2n_{AABB} + n_{AABb} + n_{AaBB}) \ln(p_{AB}) \\ & + (2n_{AAbb} + n_{AABb} + n_{Aabb}) \ln(p_A - p_{AB}) \\ & + (2n_{aaBB} + n_{AaBB} + n_{aaBb}) \ln(p_B - p_{AB}) \\ & + (2n_{aabb} + n_{Aabb} + n_{aaBb}) \ln(1 - p_A - p_B + p_{AB}) \\ & + n_{AaBb} \ln[p_{AB}(1 - p_A - p_B + p_{AB}) \\ & + (p_A - p_{AB})(p_B - p_{AB})] \end{aligned} \quad (2.24)$$

As an example, consider the data shown in Table 1.3. Two-locus genotypic counts for loci A = *Idh1* and B = *Mdh* are displayed in Table 2.10 in the notation of this section. From Table 2.10 the sample allelic frequencies $\tilde{p}_A, \tilde{p}_B$ for alleles number 1 at loci *Idh1* and *Mdh* are 0.80 and 0.84. These are the MLEs of the two frequencies, and then the log-likelihood for the frequency p_{AB} of the 11 gamete is found from Equation 2.24 as

$$\begin{aligned} \ln L(p_{AB}) = & \ln(1381380) + 51 \ln(p_{AB}) + 5 \ln(0.80 - p_{AB}) \\ & + 8 \ln(0.84 - p_{AB}) + 8 \ln[p_{AB}(p_{AB} - 0.64) \\ & + (0.80 - p_{AB})(0.84 - p_{AB})] \end{aligned}$$

This support, and the score, are plotted in Figure 2.2 over the permissible range [0, 0.8] of values for p_{AB}. The score is zero at the three values 0.46,

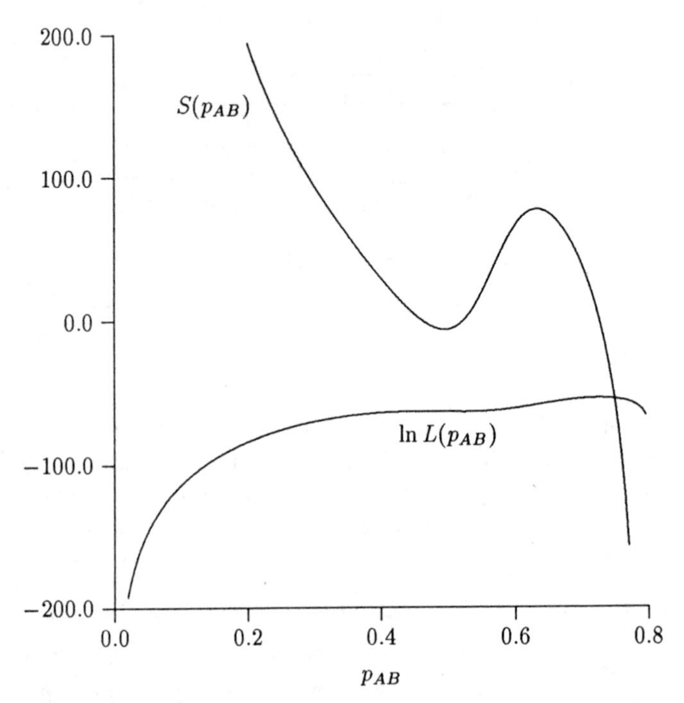

Figure 2.2 Support and score for frequency of *Idh1*-1, *Mdh*-1 gametes for mosquito data of Table 1.3.

0.52, and 0.73. The first value gives a local maximum for the support, the second value gives a local minimum, and the third maximizes the support. The maximum likelihood solution is therefore $p_{AB} = 0.73$. Care would need to be taken with iterative methods to ensure that the solution $p_{AB} = 0.46$ wasn't found.

METHOD OF MOMENTS

Throughout this chapter, estimation has been discussed in the context of maximum likelihood. Such estimates have desirable properties for large samples, although they may not be particularly good for small samples. MLE's may be biased in small samples, for example. There may be situations in which unbiasedness is of great concern, and other cases in which it is not possible to write down a likelihood because the appropriate sampling distribution is unknown. In these circumstances some other means of constructing estimates is needed, and one of the simplest is the METHOD OF MOMENTS. Briefly, statistics are sought that are unbiased for the parameters of interest, and these are used as estimates. For multinomial data, Fisher's method then provides variance estimates but for other data it may be necessary to turn to numerical resampling methods, as discussed in Chapter 3, to determine the properties of the estimates.

The method of moments is somewhat *ad-hoc*, and depends on choosing statistics whose expectations suggest a functional form for an estimator. As one example of this approach, look again at the case of estimating a single inbreeding coefficient f for a locus with more than two alleles. Although it is possible to find an MLE, it has been shown that this will require numerical solution of a nonlinear likelihood equation. From Equation 2.9, after expressing homozygote frequencies in terms of p_u and f

$$\mathcal{E}(\tilde{p}_u^2) = p_u^2 + \frac{1}{n}p_u(1-p_u)(1-f)$$

so that, adding over alleles,

$$\mathcal{E}(\sum_u \tilde{p}_u^2) = \sum_u p_u^2 + \frac{1}{n}(1-\sum_u p_u^2)(1-f)$$

while, adding over homozygotes,

$$\mathcal{E}(\sum_u \tilde{P}_{uu}) = \sum_u p_u^2 + (1-\sum_u p_u^2)f$$

These two equations lead to two expressions whose expectations differ only by a factor of f:

$$\begin{aligned}\mathcal{E}\sum_i(\tilde{P}_{uu}-\tilde{p}_u^2) + \mathcal{E}\frac{1}{n}(1-\sum_u \tilde{P}_{uu}) &= f(1-\sum_u p_u^2)\\ \mathcal{E}(1-\sum_u \tilde{p}_u^2) + \mathcal{E}\frac{1}{n}(1-\sum_u \tilde{P}_{uu}) &= (1-\sum_u p_u^2)\end{aligned}$$

Taking the ratio of these equations, the quantity $1-\sum_i p_u^2$ cancels in the right-hand side, which suggests the estimator

$$\hat{f} = \frac{\sum_u(\tilde{P}_{uu}-\tilde{p}_u^2) + \frac{1}{n}(1-\sum_u \tilde{P}_{uu})}{(1-\sum_u \tilde{p}_u^2) + \frac{1}{n}(1-\sum_u \tilde{P}_{uu})} \tag{2.25}$$

Although this ratio estimator is not itself unbiased, it is the ratio of terms that are unbiased for $f(1-\sum_u p_u^2)$ and $(1-\sum_u p_u^2)$, respectively. Accordingly, it is expected to have a low bias.

When the sample size n is large, applying Fisher's method allows a variance for this estimator to be found as described in Exercise 2.4.

SUMMARY

Genotypic counts are generally assumed to be multinomially distributed, and this forms the basis of within-population analyses. If the population is in Hardy-Weinberg equilibrium, but not otherwise, gene counts are also multinomially distributed. Variances of functions of genotypic counts can be determined exactly from multinomial theory, or approximately from Fisher's formula. Variances, including those between populations, can also be found from the use of indicator variables for each gene in the sample.

When inferences are to be extended beyond the one population sampled, it is necessary to use the total variance of a statistic. This includes the effects of both statistical and genetic sampling.

When a distribution, such as multinomial, is assumed it is possible to formulate the likelihood of a set of parameters and construct maximum likelihood estimators. Such values have desirable properties in large samples and become normally distributed. Variances of MLE's can be found from the information, which is also derived from the likelihood function. Many cases of interest require numerical determination of MLE's, although when the number of parameters equals the number of degrees of freedom, Bailey's method often gives simpler equations. Incomplete data, such as occurs with dominance, can lead to MLE's when the EM algorithm is used. The problem of distinguishing between different double heterozygotes in calculating gametic frequencies can be overcome with the EM algorithm provided random mating is assumed.

When iterative methods are used to solve likelihood equations, several starting values should be used. To choose among several valid solutions it is necessary to calculate the likelihood for each solution, and choose that solution for which the likelihood is maximized.

For cases in which a distribution cannot be assumed for genotypic counts, or when more emphasis is to be placed on unbiased estimates, the method of moments offers an alternative to maximum likelihood estimation. There is no general theory giving properties of moment estimators, and numerical methods may have to be used to find properties such as mean and variance.

EXERCISES

Exercise 2.1

For generations 5, 15, and 25 of the data in Table 2.5, find

a. genotype and gene frequencies
b. estimated variances and covariances of gene frequencies
c. a single inbreeding coefficient F

Exercise 2.2

Use both Newton-Raphson iteration and the EM algorithm to estimate the frequencies of all four alleles S, M, F, and O if the phenotypic counts are

$$n_{SS} + n_{SO} = 1149 \quad n_{MM} + n_{MO} = 36 \quad n_{FF} + n_{FP} = 17 \quad n_{OO} = 20$$
$$n_{SM} = 336 \quad n_{MF} = 25 \quad n_{SF} = 203 \quad n = 1786$$

Estimate the variances and covariances of the estimates. Assume Hardy-Weinberg equilibrium. (Data of S. Ohba, discussed by Yasuda and Kimura 1968.)

Exercise 2.3

Estimate the frequency of AB gametes for the following three sets of genotypic counts (Weir and Cockerham 1979):

a.	BB	Bb	bb	b.	BB	Bb	bb	c.	BB	Bb	bb
AA	154	81	8	AA	12	3	3	AA	12	3	3
Aa	37	14	3	Aa	3	54	3	Aa	3	51	3
aa	1	1	1	aa	12	3	3	aa	12	6	3

Exercise 2.4

Find an expression for the variance of the estimator of f given in Equation 2.25. To do this write

$$h = 1 - \sum_u p_u^2, \; H = 1 - \sum_u P_{uu}$$

so that

$$\hat{f} = \frac{(\tilde{h} - \tilde{H}) + \frac{1}{n}\tilde{H}}{\tilde{h} + \frac{1}{n}\tilde{H}}$$

To apply Fisher's formula, it is necessary to drop the $\frac{1}{n}\tilde{H}$ term in both numerator and denominator. This provides a function that is of degree zero in the counts, and will not affect the accuracy of the variance since that is of order n.

Chapter 3

Disequilibrium

HARDY-WEINBERG DISEQUILIBRIUM

Once gene and genotypic frequencies have been estimated, one of the first analyses performed on population genetic data is that of looking for associations between the two genes an individual receives at a locus. When there are no disturbing forces such as selection, mutation, or migration that would change gene frequencies over time, and when there is random mating in very large populations, these pairs of genes are known not to be associated. A consequence of this independence is that genotype frequencies are the products of gene frequencies as noted in Chapter 2:

$$\begin{aligned} P_{uu} &= p_u^2 \quad \text{for homozygotes } A_uA_u \\ P_{uv} &= 2p_up_v \quad \text{for heterozygotes } A_uA_v \end{aligned}$$

Departures from these HARDY-WEINBERG PROPORTIONS can be characterized in several ways, including the use of the within-population inbreeding coefficient f introduced in Chapter 2. With a single coefficient f, the genotypic frequencies can be written as

$$\begin{aligned} P_{uu} &= p_u^2 + p_u(1-p_u)f \\ P_{uv} &= 2p_up_v(1-f) \end{aligned}$$

and these frequencies are correctly bounded by zero and by gene frequencies

$$0 \leq P_{uu} \leq p_u$$

$$0 \leq P_{uv} \leq p_u, p_v$$

provided

$$-p_u/(1-p_u) \leq f \leq 1$$

By introducing indicator variables x_j for the jth gene of a random individual:

$$x_j = \begin{cases} 1 & \text{if gene is allele } A \\ 0 & \text{otherwise} \end{cases}$$

it emerges that f can also be regarded as the correlation of x_j and $x_{j'}, j \neq j'$. To see why this is so, note that

$$\begin{aligned} \text{Var}(x_j) &= p_A(1-p_A) \\ \text{Cov}(x_j, x_{j'}) &= P_{AA} - p_A^2 \\ &= fp_A(1-p_A) \end{aligned}$$

When there are no disturbing forces it is quite appropriate to work with just a single coefficient f, the correlation within individuals of genes regardless of their allelic form, but a more general formulation would allow as many parameters as there are degrees of freedom. For a locus with k alleles, there are k allelic frequencies and $k(k-1)/2$ heterozygotes, suggesting that the $[k(k+1)/2]$ genotypic frequencies be expressed in terms of the p_u's and a set of $k(k-1)/2$ FIXATION INDICES f_{uv}, one for each heterozygote. Coefficients f_{uu} can also be defined for the homozygotes. These fixation indices are defined by the following relations

$$\begin{aligned} P_{uu} &= p_u^2 + p_u(1-p_u)f_{uu} \\ P_{uv} &= 2p_u p_v(1-f_{uv}) \end{aligned}$$

but the relation

$$p_u = P_{uu} + \frac{1}{2}\sum_{v \neq u} P_{uv}$$

shows that the f_{uu}'s are just functions of the f_{uv}'s

$$f_{uu} = 1 - \sum_{v \neq u} \frac{p_v}{1-p_u} f_{uv}$$

Although this approach has some merit it has the disadvantage that the f parameters, containing the information about departures from Hardy-Weinberg equilibrium, are estimated as ratios of genotypic frequencies, as shown in Chapter 2. It is difficult to determine the statistical properties of ratios.

There are advantages instead in working with a composite kind of quantity that will be termed a DISEQUILIBRIUM COEFFICIENT. This is simply the difference between a frequency and its expected value when there are no associations between genes. Disequilibria will be denoted by D's and for one-locus genotypic frequencies they are defined by the relations

$$\begin{aligned} P_{uu} &= p_u^2 + D_{uu} \\ P_{uv} &= 2p_u p_v - 2D_{uv} \end{aligned}$$

There is still a dependency among the coefficients caused by the genotypic frequencies summing to gene frequencies:

$$D_{uu} = \sum_{v \neq u} D_{uv}$$

This dependency implies that there are as many independent D's as there are heterozygote types.

In the two-allele case the notation can be modified a little to use a single disequilibrium coefficient D_A

$$\begin{aligned} P_{AA} &= p_A^2 + D_A \\ P_{Aa} &= 2p_A p_a - 2D_A \\ P_{aa} &= p_a^2 + D_A \end{aligned}$$

Taking account of the range of possible genotypic frequencies, bounds on D_A are found

$$max[-p_A^2, -p_a^2] \leq D_A \leq p_A p_a$$

Estimating Disequilibrium D_A

In the two-allele case there are two parameters, which is the same as the number of degrees of freedom. Bailey's rule (Chapter 2) for maximum likelihood estimates can therefore be applied, and the MLE for the single D_A is

$$\hat{D}_A = \tilde{P}_{AA} - \tilde{p}_A^2 \tag{3.1}$$

From Equations 2.3 and 2.9, the expected value of this estimate over replicate samples from the same population is

$$\begin{aligned} \mathcal{E}(\hat{D}_A) &= D_A - \frac{1}{2n}(p_A + P_{AA} - 2p_A^2) \\ &= D_A - \frac{1}{2n}[p_A(1 - p_A) + D_A] \end{aligned} \tag{3.2}$$

showing that the estimate is biased, although the bias decreases as the sample size increases. From Fisher's variance approximation, Equation 2.16, the variance of the MLE, ignoring terms in higher powers of $(1/n)$ is

$$\mathrm{Var}(\hat{D}_A) = \frac{1}{n}[p_A^2(1-p_A)^2 + (1-2p_A)^2 D_A - D_A^2] \tag{3.3}$$

Testing for Hardy-Weinberg with D_A

When a population has Hardy-Weinberg proportions, the disequilibrium coefficient D_A is expected to be zero, suggesting that a test of the hypothesis $H_0 : D_A = 0$ is equivalent to testing for Hardy-Weinberg equilibrium (HWE). A word should be said here about the use of the word "equilibrium." Strictly, an equilibrium state is one in which properties of the population are not changing over successive generations. In the HWE case, this implies the continued absence of disturbing forces such as selection, migration, and mutation as well as the continuation of random mating. The tests presented in this chapter, on the other hand, refer to checking for consistency of sample genotypic frequencies with those expected from the Hardy-Weinberg law. Nothing is necessarily implied about future generations or about the conditions faced by the population. In spite of this distinction, it is usual to speak of testing for HWE, and this convention will be continued here.

For large samples, the MLE $\hat{D}_A$ is normally distributed

$$\hat{D}_A \sim N[\mathcal{E}(\hat{D}_A), \mathrm{Var}(\hat{D}_A)]$$

so that a standard normal variate, z, can be constructed

$$z = \frac{\hat{D}_A - \mathcal{E}(D_A)}{\sqrt{\mathrm{Var}(\hat{D}_A)}} \tag{3.4}$$

and compared to tabulated critical values (Appendix Table 1). Departures from HWE will cause $\hat{D}_A$, and hence z, to be large, and the hypothesis of HWE can be rejected if the calculated z is sufficiently large that it is unlikely to have arisen by chance. In particular, departures from HWE caused by an excess or a deficiency of homozygotes, leading to positive or negative disequilibrium values, would cause the hypothesis to be rejected at the 5% significance level when z exceeds 1.96 or is less than -1.96. Of course, if only the possibility of a deficiency *or* an excess was of interest then a one-tailed test would be used. For example, if it was suspected that there were fewer heterozygotes in a population than predicted by HWE, the null hypothesis $H_0 : D_A = 0$ would be tested against the alternative hypothesis $H_1 : D_A > 0$. The null would be rejected for a large positive D_A value, or a large positive

z value. From Appendix Table 1, a 5% significance level is attained when the hypothesis is rejected for $z > 1.64$.

An equivalent procedure depends on the result that the square of a standard normal variable is distributed as chi-square with 1 d.f., and rejects the hypothesis when

$$X^2 = z^2 \tag{3.5}$$

exceeds the 5% critical value of 3.84 (Appendix Table 2). Combining Equations 3.1 to 3.5 gives the test statistic

$$X_A^2 = \frac{n\hat{D}_A^2}{\tilde{p}_A^2(1-\tilde{p}_A)^2} \tag{3.6}$$

Large values of z, whether positive or negative, cause large values of X^2, so the chi-square test is also two-sided in that it leads to rejection of the HWE hypothesis for an excess or a deficiency of heterozygotes. Only values in the upper tail of the chi-square distribution give rejection, but one-sided alternative hypotheses cannot be accommodated as directly as they were with the z statistic.

By going through the derivation of this expression the assumptions on which it is based have been revealed. Under the hypothesis $H_0 : D_A = 0$, the expectation of the MLE $\hat{D}_A$ is zero provided the bias term $p_A(1-p_A)/2n$ is ignored. Similarly, under the hypothesis, the variance of the MLE can be approximated by the sample value $\tilde{p}_A^2(1-\tilde{p}_A^2)/n$ provided the sample is large.

Equation 3.6 can be approached from another direction, that of goodness-of-fit chi-square tests. For samples large enough that bias terms can again be ignored, the following set of expected values hold for the three genotypic frequencies when H_0 is true:

Genotype	AA	$A\bar{A}$	$\bar{A}\bar{A}$
Observed number	n_{AA}	$n_{A\bar{A}}$	$n_{\bar{A}\bar{A}}$
Expected number	$n\tilde{p}_A^2$	$2n\tilde{p}_A(1-\tilde{p}_A)$	$n(1-\tilde{p}_A)^2$
Observed−Expected	$n\hat{D}_A$	$-2n\hat{D}_A$	$n\hat{D}_A$

The notation $\bar{A}$ for "not-A" has been employed to emphasize that there is a focus on allele A but that it is not necessary for the locus to have only two alleles. The goodness-of-fit chi-square statistic is

$$\begin{aligned} X_A^2 &= \sum_{\text{genotypes}} \frac{(\text{Observed} - \text{Expected})^2}{\text{Expected}} \\ &= \frac{(n\hat{D}_A)^2}{n\tilde{p}_A^2} + \frac{(-2n\hat{D}_A)^2}{2n\tilde{p}_A(1-\tilde{p}_A)} + \frac{(n\hat{D}_A)^2}{n(1-\tilde{p}_A)^2} \end{aligned}$$

and this leads back to Equation 3.6.

Coming from this direction to Equation 3.6 indicates that the test procedure has the usual problems associated with chi-square tests, the most common one being that the tests are very sensitive to small expected values. Since the expected values occur in the denominator of X^2 they can greatly inflate the statistic when they are small, and an expected value that was less than some specified level, say five or one, would not generally be used. Other problems, such as using a continuous distribution (z or χ^2) to test hypotheses with discrete genotypic counts, may be overcome by using a continuity correction of 0.5 in the numerator of chi-square (Yates 1934),

$$X_A^2 = \sum \frac{(|\text{ Observed} - \text{Expected }| - 0.5)^2}{\text{Expected}}$$

but all problems stem from using large-sample results for small samples. A procedure that makes no such requirement is discussed in the next section.

Mention should be made here of differing philosophies behind the testing of hypotheses. The discussion just given is for a procedure that results in either rejecting, or not rejecting, an hypothesis. A SIGNIFICANCE LEVEL, α, needs to be specified, where α is the chance of rejecting a true hypothesis. The hypothesis is rejected when the test statistic exceeds the CRITICAL VALUE, or that value which is exceeded by chance with probability α when the hypothesis is true. It is more informative to give the value of the statistic and the probability, or "P-value," of obtaining such a value, or a more extreme one, when the hypothesis is true. The extent to which the data support the hypothesis is measured by P. Values of P less than α could be used to reject at the α significance level.

Exact Tests for HWE

The use of "exact" tests dates back to work of Fisher (1935), who noted that an observed sample could be used to reject an hypothesis if the total probability under the hypothesis of that sample, or a less likely one, is small. A simple way to perform such a test would be to determine the probabilities of all possible samples of the same size as the sample at hand assuming the hypothesis is true. Once the samples are ordered according to their probabilities, it is a simple matter to add the probability of the observed sample to the sum of all less probable samples and then to reject the hypothesis if that total probability is less than α. Exact tests are generally used for small sample sizes, when there is the greatest chance of having small expected numbers in the chi-square test formulation. However, if there are rare alleles at a locus expected numbers can be small even in moderately large samples, and exact tests are desirable. This was the case in a study reported by Wellso et al. (1988).

Testing for HWE is equivalent to looking at whether the observed genotypic frequencies are close enough to products of the observed gene frequencies that there can be confidence in the same relationship holding for the population frequencies. The EXACT TEST, or PROBABILITY TEST, proceeds by looking at all possible sets of genotypic frequencies for the particular observed set of gene frequencies and rejecting the hypothesis of HWE if the observed genotypic frequencies turn out to be very unusual.

The test is derived here for a locus with two alleles, A and a. Under the HWE hypothesis, the probability of the observed set of genotypic counts n_{AA}, n_{Aa}, and n_{aa} in a sample of size n is

$$\Pr(n_{AA}, n_{Aa}, n_{aa}) = \frac{n!}{n_{AA}!n_{Aa}!n_{aa}!}(p_A^2)^{n_{AA}}(2p_Ap_a)^{n_{Aa}}(p_a^2)^{n_{aa}} \quad (3.7)$$

while the gene counts n_A and n_a are binomially distributed if HWE holds:

$$\Pr(n_A, n_a) = \frac{(2n)!}{n_A!n_a!}(p_A)^{n_A}(p_a)^{n_a}$$

Putting these expressions together provides the probability of the observed genotypic frequencies, assuming HWE, conditional on the observed gene frequencies (which are a sufficient statistic under HWE):

$$\begin{aligned}
\Pr(n_{AA}, n_{Aa}, n_{aa} \mid n_A, n_a) &= \frac{\Pr(n_{AA}, n_{Aa}, n_{aa} \text{ and } n_A, n_a)}{\Pr(n_A, n_a)} \\
&= \frac{\Pr(n_{AA}, n_{Aa}, n_{aa})}{\Pr(n_A, n_a)} \\
&= \frac{n!n_A!n_a!2^{n_{Aa}}}{n_{AA}!n_{Aa}!n_{aa}!(2n)!}
\end{aligned}$$

in which the unknown gene frequencies p_A, p_a have cancelled out. Probability tests were applied to HWE testing by Haldane (1954), who pointed out that the probabilities could be expressed in terms of one of the two gene numbers and the number, $x = n_{Aa}$, of heterozygotes. The conditional probability can therefore be called $\Pr(x \mid n_A)$ and rewritten as

$$\Pr(x \mid n_A) = \frac{n!n_A!(2n-n_A)!2^x}{[(n_A-x)/2]!x![n-(n_A+x)/2]!(2n)!}$$

These probabilities are evaluated numerically at all valid values of x for a particular pair of n, n_A and then x values are ordered according to these probabilities. The least likely outcomes with a total probability of α form a rejection region of size α. An attempt is made to get α close to a conventional value such as 0.05, but this is unlikely to be met exactly.

As an example, return to the *Pgm* mosquito data displayed in Table 2.4 after collapsing to alleles 1 and not-1:

$$n_{11} = 9,\ n_{1\bar{1}} = 1,\ n_{\bar{1}\bar{1}} = 30; \quad n_1 = 19,\ n_{\bar{1}} = 61$$

The possible numbers of heterozygotes, x, when there are $n_1 = 19$ genes of allelic type 1 in a sample of size $n = 40$ are the odd numbers from 1 to 19 and the probabilities $\Pr(x|19)$ are shown in Table 3.1. It happens that the observed set of genotypic frequencies has the smallest probability of all possible samples with these gene numbers, and the HWE hypothesis would be rejected with a significance level that is extremely small. Notice that both large and small numbers of heterozygotes can lead to small probabilities, or large chi-square values, and rejection of the hypothesis. The ordering of probabilities is not the same as the ordering on the numbers of heterozygotes. The exact test is two-sided. In this particular example, though, the exact test rejection region consists only of small numbers of heterozygotes.

By adding the probabilities for 1, 3, 5, 7, and 9 heterozygotes a rejection region of size 0.0229 is found. In other words, there is a probability of 2.29% of falsely rejecting the hypothesis of HWE when it is rejected with 9 or fewer heterozygotes. This probability is the significance level or probability of a TYPE I ERROR. Adding the next largest probability, for 19 heterozygotes, would give a test of size 8.23%, which would generally be regarded as being too high. Chi-square test statistics are also shown in Table 3.1 and demonstrate that the two procedures differ even for samples as large as 40. Applying Yates' continuity correction would bring X^2 down to 2.62 for $x = 19$, below the critical value of 3.84, and the two tests would then agree.

This procedure of adding probabilities for all deviations (of the numbers of heterozygotes) from the value expected under the hypothesis regardless of sign has been criticized by Yates (1984). He advocates keeping track of sign. In the present example, the expected number of heterozygotes is $40 \times 2(19/80)(61/80) = 14.5$, and the observed data have fewer heterozygotes than expected, so that the observed disequilibrium is positive. If the

Table 3.1 Exact test for HWE at *Pgm* locus for mosquito data of Table 1.3.

Possible samples				Cumulative	Disequi-	
11	$1\bar{1}$	$\bar{1}\bar{1}$	Probability	Probability	librium	Chi-square
9	1	30*	0.0000	0.0000†	0.1686	34.67†
8	3	29	0.0000	0.0000†	0.1436	25.15†
7	5	28	0.0001	0.0001†	0.1186	17.16†
6	7	27	0.0023	0.0024†	0.0936	10.69†
5	9	26	0.0205	0.0229†	0.0686	5.74†
0	19	21	0.0594	0.0823	−0.0564	3.88†
4	11	25	0.0970	0.1793	0.0436	2.32
1	17	22	0.2308	0.4101	−0.0314	1.20
3	13	24	0.2488	0.6589	0.0186	0.42
2	15	23	0.3411	1.0000	−0.0064	0.05

*Observed sample.
†Causes rejection of HWE at 5% significance level.

rows in Table 3.1 were ordered the same as the numbers of heterozygotes, attention would be focused only on the tail with small numbers. There is a probability of 0.0229 of observing 9 or fewer heterozygotes, so that the hypothesis would be rejected with a one-tailed probability of 0.0229. In other words the hypothesis of no disequilibrium is rejected in favor of the hypothesis of positive disequilibrium at that significance level. If a two-tailed test is wanted, meaning that either positive or negative disequilibria can cause rejection, Yates says that the one-tail significance level should be doubled, to 0.0458. The case for approaching exact testing from a one-sided viewpoint does have some appeal. Which approach is taken should be clearly stated when results are presented. Note that there is no debate on normal distribution-based tests becasue of symmetry. Positive or negative values of z of the same magnitude are equally likely, wheras exact tests have different probabilities for deviations equal in size but opposite in sign.

Application of the exact test for HWE in the two-allele case is simplified by using tables published by Vithayasai (1973). He gave the rejection regions for significance levels of 0.10, 0.05, and 0.01 for samples of size $n = 20$ to 100 in steps of 5 and $n = 100$ to 200 in steps of 25. The tables show the critical numbers of heterozygotes for each possible number of the least frequent allele. For the present example, he gave as critical values those heterozygote numbers lying outside the ranges (10,18), (10,20), and (9,21) for the three significance levels 0.10, 0.05, and 0.01. With only 19 of the least frequent allele the number of heterozygotes cannot exceed 19 of course, but the tables give a quick approximation to the test.

Exact tests may also be applied to loci with more than two alleles, although there is the difficulty that the number of possible samples can increase very quickly. Instead of examining all possible samples, it may be possible to base the test just on some of the array of samples. Rather than considering all possible numbers of heterozygotes for a given set of allelic frequencies, the numbers of heterozygotes could be sampled uniformly from their ranges of values and an approximation to the distribution of probabilities of outcomes generated. Hernández and Weir (1989) found that generating 1,000 samples gave a satisfactory test. Although computationally involved, these exact tests do avoid the problems chi-square tests have when some classes have small expected numbers, as is usual for more than two alleles. Even for two alleles, there is expected to be less than one homozygote for an allele with frequency 0.1 in samples of size 50.

Likelihood Ratio Test for HWE

Just as the likelihood function provided a general means of finding estimates by the maximum likelihood method, so too does it provide a general framework for testing hypotheses. Test statistics can be found as LIKELIHOOD RATIOS. To test whether a parameter ϕ has some specific value ϕ_0, such as zero, a comparison is made of two likelihoods – that maximized with $\phi = \phi_0$, and that maximized when ϕ is not constrained. These two likelihoods are written as L_0 for the hypothesized parameter value, and L_1 for the unconstrained value. If the hypothesis is true, L_0 and L_1 should be equal in value. The less well the data support the hypothesis, the smaller L_0 will be than L_1, suggesting that the ratio be used as a test statistic. The likelihood ratio is defined as

$$\lambda = \frac{L_0}{L_1}$$

Remember that the L's refer to maximum likelihoods, and they are calculated by using the MLE's of the parameters in the two cases. If only a single parameter is being specified by the hypothesis, then a convenient approximation to the distribution of λ is available. When the hypothesis is true

$$-2\ln\lambda = -2(\ln L_0 - \ln L_1) \sim \chi^2_{(1)}$$

If L_0 and L_1 differ because of values assigned to s parameters, $-2\ln\lambda$ has a chi-square distribution with s degrees of freedom. Likelihood ratio tests for multinomial proportions have also been called G- tests (e.g., Sokal and Rohlf 1981), with G being defined as

$$G = 2\ln\left(\frac{L_1}{L_0}\right)$$

To test for HWE, the hypothesis is that $D_A = 0$. Under the unconstrained model, since there are two degrees of freedom and two parameters, Bailey's method shows that the MLE's of the genotypic frequencies are just the observed genotypic frequencies. Writing observed frequencies in terms of counts, and using the relation

$$n = n_{AA} + n_{Aa} + n_{aa}$$

the maximum likelihood L_1 is

$$L_1 = \frac{n!}{n_{AA}!n_{Aa}!n_{aa}!}\frac{(n_{AA})^{n_{AA}}(n_{Aa})^{n_{Aa}}(n_{aa})^{n_{aa}}}{n^n}$$

Under the HWE hypothesis, the MLE's of the genotypic frequencies are the appropriate products of allelic frequencies, such as

$$\begin{aligned}\hat{P}_{AA} &= (\tilde{p}_a)^2 \\ &= (n_A/2n)^2\end{aligned}$$

where

$$\begin{aligned}n_A &= 2n_{AA} + n_{Aa} \\ n_a &= 2n_{aa} + n_{Aa}\end{aligned}$$

This allows L_0 to be written as

$$L_0 = \frac{n!}{n_{AA}!n_{Aa}!n_{aa}!}\frac{(n_A)^{n_A}(n_a)^{n_a}}{(2n)^{2n}}$$

Testing for HWE can therefore be carried out with another chi-square test statistic

$$-2\ln\lambda = -2\ln\left[\frac{(2n)^{2n}(n_A)^{n_A}(n_a)^{n_a}}{(n)^n(n_{AA})^{n_{AA}}(n_{Aa})^{n_{Aa}}(n_{aa})^{n_{aa}}}\right]$$

Note that this test statistic involves only the observed counts of genes and genotypes. The unknown gene frequencies have cancelled out.

Log-Linear Models

A quite different approach to testing for HWE starts with a multiplicative instead of an additive model. Instead of representing genotypic frequencies as the sum of allelic frequency terms and coefficients of disequilibrium, they may be written as products of terms, some of which are for alleles and some of which are for disequilibrium. In other words, departures from HWE can be accommodated with a set of multiplicative coefficients. Such a representation for genotypic frequencies is

$$\begin{aligned} P_{AA} &= MM_A^2M_{AA} \\ P_{Aa} &= 2MM_AM_aM_{Aa} \\ P_{aa} &= MM_a^2M_{aa} \end{aligned}$$

M_A, M_a represent the allelic frequency contributions and M_{AA}, M_{Aa}, M_{aa} represent the associations between allelic frequencies. The term M is a mean effect. Taking logarithms of these equations to give

$$\begin{aligned} \ln P_{AA} &= \ln M + 2\ln M_A + \ln M_{AA} \\ \ln P_{Aa} &= \ln 2 + \ln M + \ln M_A + \ln M_a + \ln M_{Aa} \\ \ln P_{aa} &= \ln M + 2\ln M_a + \ln M_{aa} \end{aligned}$$

shows why such models are referred to as log-linear. It is the logarithms of frequencies that are represented as linear combinations of terms for a mean, gene, and genotypic effects. As there are four parameters for three genotypic frequencies, this formulation is overparameterized, and several ways of reducing the number of parameters can be used. One way is to set the terms involving allele a to one, leaving genotypic frequencies as

$$\begin{aligned} P_{AA} &= MM_A^2M_{AA} \\ P_{Aa} &= 2MM_A \\ P_{aa} &= M \end{aligned}$$

which shows the need for the mean term M. Since the frequencies must sum to 1,

$$M = (1 + 2M_A + M_A^2M_{AA})^{-1}$$

This results in two independent parameters, M_{AA} and M_A, for the two degrees of freedom, and Bailey's method gives estimates

$$\hat{M} = \tilde{P}_{aa}$$

$$\hat{M}_A = \frac{\tilde{P}_{Aa}}{2\tilde{P}_{aa}}$$
$$\hat{M}_{AA} = \frac{4\tilde{P}_{AA}\tilde{P}_{aa}}{\tilde{P}^2_{Aa}}$$

leading to

$$\hat{P}_{AA} = \tilde{P}_{AA}$$
$$\hat{P}_{Aa} = \tilde{P}_{Aa}$$
$$\hat{P}_{aa} = \tilde{P}_{aa}$$

for the unconstrained estimates.

HWE in this system means that the genotypic term M_{AA} is 1, and MLE's constrained by the hypothesis are

$$\hat{M} = \tilde{p}^2_a$$
$$\hat{M}_A = \frac{\tilde{p}_A}{\tilde{p}_a}$$

Therefore

$$\hat{P}_{AA} = \tilde{p}^2_A$$
$$\hat{P}_{Aa} = 2\tilde{p}_A\tilde{p}_a$$
$$\hat{P}_{aa} = \tilde{p}^2_A$$

for the estimates constrained by the hypothesis.

Testing uses the likelihood ratio procedure. A comparison is made of likelihoods maximized under models that either include M_{AA} or do not. Under the unconstrained model (or model 1), with the genotypic term M_{AA}, the maximum log-likelihood is

$$\ln L_1 = \text{Constant} + n_{AA}\ln(\tilde{P}_{AA}) + n_{Aa}\ln(\tilde{P}_{Aa}) + n_{aa}\ln(\tilde{P}_{aa})$$

Under the constrained, HWE, model (or model 0) without M_{AA}, the maximum log-likelihood is

$$\ln L_0 = \text{Constant} + n_A\ln(\tilde{p}_A) + n_a\ln(\tilde{p}_a)$$

The ratio of likelihoods measures the relative extents to which the two models fit the data. If HWE does not hold, the full model is expected to provide a higher likelihood and the hypothesis will be rejected for small values of

$$\lambda = \frac{L_0}{L_1}$$

or for large values of

$$-2\ln\lambda = -2(\ln L_0 - \ln L_1)$$

For large samples, the quantity $-2\ln\lambda$ has a chi-square distribution. In the language of log-linear models, this quantity is also called the DEVIANCE of the model 0. The chi-square distribution simplifies the determination of whether to reject the hypothesis, whereas the deviance concept allows the treatment of a series of models. The difference of two deviances also has a chi-square distribution. The deviance for the full model is zero.

Note that the likelihood ratio test statistic for the log-linear model, in this case, is the same as was previously found for the additive model, even though the two sets of parameters are different. Notice also that different restrictions could have been put on the $\hat{M}$'s in the log-linear model so that the estimates would have been different, but the final test statistic would have been the same.

Multiple Alleles

The likelihood ratio framework offers a systematic way of testing for HWE when there are more than two alleles at a locus. Each of the genotypes can differ from the Hardy-Weinberg proportions, and it may be of interest to test each of these departures separately. Specifically, each of the disequilibrium coefficients D_{uv}, for alleles A_u and A_v, can be tested. When there are k codominant alleles, the $k(k+1)/2$ genotypic frequencies provide $k(k+1)/2-1$ degrees of freedom and allow $k-1$ allelic frequencies to be estimated and $k(k-1)/2$ disequilibrium coefficients to be estimated and tested for departures from zero. The full, or completely unconstrained, model has MLE's of

$$\begin{aligned} \hat{p}_u &= \tilde{p}_u \\ \hat{D}_{uv} &= \tilde{p}_u\tilde{p}_v - \frac{1}{2}\tilde{P}_{uv} \end{aligned}$$

with a log-likelihood of

$$\ln L_1 = \text{Constant} + \sum_u n_{uu} \ln\left(\frac{n_{uu}}{n}\right) + \sum_u \sum_{v\neq u} n_{uv} \ln\left(\frac{n_{uv}}{n}\right)$$

The first sum is over all homozygotes, and the second over all heterozygotes. As is the convention in this book, the second sum involves each heterozygous class only once. For convenience, the subscripts for heterozygotes are written in numerical or alphabetical order. When the model is completely constrained by having Hardy-Weinberg proportions for all genotypes, the disequilibria are all zero while the MLE's for allelic frequencies are still the observed frequencies. The log-likelihood reduces to

$$\ln L_0 = \text{Constant} + \sum_u n_u \ln\left(\frac{n_u}{2n}\right)$$

where the allelic counts are

$$n_u = 2n_{uu} + \sum_{v \neq u} n_{uv}$$

An overall test for HWE is given by the likelihood ratio

$$G_T^2 = -2\ln\lambda$$

with

$$\lambda = \frac{\ln L_0}{\ln L_1} = \ln L_0 - \ln L_1$$

Under HWE, this quantity has a chi-square distribution with $k(k-1)/2$ degrees of freedom.

The same kind of test is found by a goodness-of-fit test on all the genotypic classes:

$$X_T^2 = \sum_u \frac{(n_{uu} - n\tilde{p}_u^2)^2}{n\tilde{p}_u^2} + \sum_u \sum_v \frac{(n_{uv} - 2n\tilde{p}_u\tilde{p}_v)^2}{2n\tilde{p}_u\tilde{p}_v}$$

This chi-square statistic also has $k(k-1)/2$ degrees of freedom. Care must be taken not to use genotypic classes with small expected numbers. Although an absolute rule cannot be given, there should be no problems if all expected numbers are above five. Cochran (1954) showed that there may even be two expectations as low as one with little effect on tests performed at the 5% significance level.

Note that X_T^2 can be expressed in terms of the estimated disequilibrium coefficients:

$$X_T^2 = \sum_u \frac{n\hat{D}_{uu}^2}{\tilde{p}_u^2} + \sum_u \sum_{v \neq u} \frac{n\hat{D}_{uv}^2}{\tilde{p}_u\tilde{p}_v}$$

where $\hat{D}_{uu}$ is the sum of the disequilibria for all heterozygotes involving the uth allele.

To test each of the D_{uv} individually, the likelihood ratio approach requires the comparison of two likelihoods – one with D_{uv} included in the model and one with D_{uv} set to zero. If D_{12} is put to zero in the three-allele case, for example, the maximum likelihood can be written as

$$L_{13,23} = max\,\{L(p_1, p_2, p_3, D_{13}, D_{23})\}$$

while for the unconstrained model

$$L_1 = max\,\{L(p_1, p_2, p_3, D_{12}, D_{13}, D_{23})\}$$

and the test statistic is G_{12}^2:

$$G_{12}^2 = -2(\ln L_{13,23} - \ln L_1)$$

This statistic has a chi-square distribution with 1 d.f. when the hypothesis $H_0 : D_{12} = 0$ is true. The difficulty with the procedure is in maximizing the likelihood under the hypothesis. As there are four parameters to estimate, two allelic frequencies and two disequilibria, from five independent genotypic classes, Bailey's method cannot be used. Numerical iterations are required to solve simultaneous equations.

An alternative procedure, which is almost as good as the likelihood ratio, was described by Hernández and Weir (1989). They used the asymptotic normality of estimated disequilibria. Fisher's approximate variance formula, Equation 2.16, leads to an expression for the variance of $\hat{D}_{uv}$:

$$\begin{aligned} 2n\mathrm{Var}(\hat{D}_{uv}) \;=\;& p_u p_v[(1-p_u)(1-p_v) + p_u p_v] \\ & - [(1-p_u-p_v)^2 - 2(p_u-p_v)^2]D_{uv} \\ & + \sum_{w \neq u,v} (p_u^2 D_{vw} + p_v^2 D_{uw}) - D_{uv}^2 \end{aligned}$$

To test that D_{uv} is zero, this condition is used in the variance formula, and the test statistic becomes

$$\begin{aligned} X_{uv}^2 &= \frac{\hat{D}_{uv}^2}{\mathrm{Var}(\hat{D}_{uv})} \\ &= \frac{2n\hat{D}_{uv}^2}{\tilde{p}_u\tilde{p}_v[(1-\tilde{p}_u)(1-\tilde{p}_v) + \tilde{p}_u\tilde{p}_v] + \sum_{w\neq u,v}(\tilde{p}_u^2\hat{D}_{vw} + \tilde{p}_v^2\hat{D}_{uw})} \end{aligned}$$

where observed and estimated values have been used for allelic frequencies and disequilibria. In the three-allele case, for example,

$$\begin{aligned} X_{12}^2 &= \frac{2n\hat{D}_{12}^2}{\tilde{p}_1\tilde{p}_2[(1-\tilde{p}_1)(1-\tilde{p}_2) + \tilde{p}_1\tilde{p}_2] + (\tilde{p}_1^2\hat{D}_{23} + \tilde{p}_2^2\hat{D}_{13})} \\ X_{13}^2 &= \frac{2n\hat{D}_{13}^2}{\tilde{p}_1\tilde{p}_3[(1-\tilde{p}_1)(1-\tilde{p}_3) + \tilde{p}_1\tilde{p}_3] + (\tilde{p}_1^2\hat{D}_{23} + \tilde{p}_3^2\hat{D}_{12})} \\ X_{23}^2 &= \frac{2n\hat{D}_{23}^2}{\tilde{p}_2\tilde{p}_3[(1-\tilde{p}_2)(1-\tilde{p}_3) + \tilde{p}_2\tilde{p}_3] + (\tilde{p}_2^2\hat{D}_{13} + \tilde{p}_3^2\hat{D}_{12})} \end{aligned}$$

These test statistics provide single degree-of-freedom chi-squares without the complicated numerical methods needed for likelihood ratio tests.

An advantage of the likelihood methods, however, is that more complicated hypotheses can be tested. The hypothesis $H_0 : D_{12} = D_{13} = 0$ can be tested by comparing the likelihoods L_{23} and L_1. The former quantity is found by maximizing the likelihood with respect to the allelic frequencies and D_{23} when the other two disequilibria are set equal to zero.

The exact test for overall departures from HWE can be established with probabilities that can be calculated from the observed genotypic counts. Following the same argument given in the two-allele case, the probability of a sample of genotype numbers (n_{uu} and n_{uv}), conditional on observed allelic numbers n_u, under the hypothesis of HWE is

$$\Pr(\{n_{uu}, n_{uv}\}|\{n_u\}) = \frac{n! 2^{(\sum_u \sum_{v>u} n_{uv})} \prod_u (n_u)!}{(2n)! \prod_u (n_{uu})! \prod_u \prod_{v>u} (2n_{uv})!}$$

A critical region of size α is identified as being the least probable fraction α of all possible samples.

When a single disequilibrium coefficient is hypothesized to be zero, however, the probability of a sample, conditional on the observed gene frequencies, depends on the true gene frequencies *and* on the other disequilibria. These parameters do not cancel out as they did for the overall test and exact tests are not available for individual disequilibria in the multiple allele case.

Power of Tests for HWE

Statistical tests of hypotheses are subject to two kinds of error: a true hypothesis may be rejected or a false hypothesis may not be rejected. The significance level measures the probability of the first kind of error, while POWER is one minus the probability of the second kind. In general, power is the probability of rejecting an hypothesis, and the particular case when the hypothesis is true gives a power value equal to the significance level of the test. For a given significance level, those tests with the greatest power (probability of rejecting a false hypothesis) are preferred. Methods for calculating power are now considered.

If a significance value is specified in advance, such as $\alpha = 0.05$, this determines the rejection region. Before any data are collected, it is known what values of the test statistic will cause the hypothesis to be rejected. Powers are calculated for each alternative value of the parameter being tested. For a given value of α in a test of $H_0 : D_{12} = 0$, the power of the test for detecting a disequilibrium value of 0.10 may be needed. In other words, what is the probability of rejecting H_0 when $D_{12} = 0.10$? The answer depends on the probabilities with which the test statistic takes its various values.

Consider first the chi-square test for HWE, with test statistic written as in Equation 3.6. Theory provides that X^2 is distributed as chi-square with 1 d.f. when the hypothesis is true, and that it is distributed as a NONCENTRAL CHI-SQUARE when the hypothesis is false (provided it is not false by too large an amount):

$$X^2 \sim \chi^2_{(1)} \text{ when } H_0 \text{ true}$$
$$X^2 \sim \chi^2_{(1,\nu)} \text{ when } H_0 \text{ false}$$

The degrees of freedom are still 1, but there is now a noncentrality parameter ν given by replacing statistics $\tilde{p}_A, \hat{D}_A^2$ with parametric values p_A, D_A in X^2:

$$\nu = \frac{nD_A^2}{p_A^2(1-p_A)^2}$$

When H_0 is true, $D_A = 0$ and so $\nu = 0$ and the distribution is central chi-square. The larger D_A is, the larger ν is and the more likely it is the hypothesis will be rejected. Recall that an X^2 value of 3.84 or greater causes rejection of H_0 for a significance level of 0.05. There is a probability of 0.05 that X^2 will exceed 3.84 when the hypothesis is true. Tables of the noncentral chi-square (Haynam et al. 1970, Appendix Table 3) show that there is a probability of 0.90 that X^2 will exceed 3.84 when the hypothesis is false if $\nu = 10.51$. This value of the noncentrality parameter depends on the critical value of 3.84 and hence on the significance level. Power and significance level are related to each other.

It is useful to turn this last argument around to see what value of the disequilibrium coefficient D_A will be detected with 90% probability when a 5% level test is used. Rearranging the expression for ν shows that

$$D_A = p_A(1-p_A)\sqrt{\frac{10.51}{n}}$$

confirming that larger samples will allow the detection of smaller levels of disequilibrium. Alternatively it can be seen that the sample size must be

$$n = 10.51\frac{p_A^2(1-p_A)^2}{D_A}$$

for a 5% test to have 90% power.

This use of the noncentral chi-square distribution should be limited to cases in which the alternative values of the parameter being tested are close to the hypothesized values – in other words, it is appropriate only for local departures from the hypothesized values. Hernández and Weir (1989) showed that, otherwise, misleadingly large power values could result from using this theory.

For exact test power calculations, an alteration is needed in Equation 3.7 for calculating the probability of each of the possible samples given the observed gene frequencies. If the true disequilibrium is D_A instead of zero,

$$\begin{aligned} \Pr(n_{AA}, n_{Aa}, n_{aa}) &= \frac{n!}{n_{AA}!n_{Aa}!n_{aa}!}(p_A^2 + D_A)^{n_{AA}} \\ &\quad \times (2p_Ap_a - 2D_A)^{n_{Aa}}(p_a^2 + D_A)^{n_{aa}} \end{aligned}$$

and $\Pr(n_A, n_a)$ needs to be calculated from appropriate sums of these probabilities since gene counts are not binomially distributed when D_A is not zero. The calculations would need to be built into the computer program used to generate the possible samples.

LINKAGE DISEQUILIBRIUM

The logical next step in a study of associations between genes is to look at the frequencies for genes at different loci. Any associations found will be referred to generally as linkage disequilibrium even though they may have nothing to do with linkage. Genes at different loci may have frequencies that show association whether or not those loci are linked.

Gametic Disequilibrium at Two Loci

In the first situation to be considered data are available on gametes and there is no need to worry about genotypic associations. This will be the situation in which single chromosomes have been sampled from natural populations (e.g., Langley et al. 1974), or when family data are available to allow gametes to be inferred from genotypes (e.g., Weir and Brooks 1986). For a pair of alleles at two loci, the procedures for defining, estimating, and testing disequilibria are entirely analogous to those for pairs of alleles at a single locus.

The disequilibrium coefficient for alleles A and B at two loci compares the gametic frequency with the product of gene frequencies

$$D_{AB} = p_{AB} - p_Ap_B$$

and inferences are based on the assumed multinomial distribution of gametes. As in the one-locus case, the MLE of D_{AB} is found directly from the observed frequencies

$$\hat{D}_{AB} = \tilde{p}_{AB} - \tilde{p}_A\tilde{p}_B$$

Taking expectations over all samples of size n from the same population shows that the MLE is biased

$$\mathcal{E}(\hat{D}_{AB}) = \frac{n-1}{n}D_{AB}$$

and that the large-sample variance is

$$\begin{aligned}\text{Var}(\hat{D}_{AB}) &= \frac{1}{n}\left[p_A(1-p_A)p_B(1-p_B)\right. \\ &\quad \left. +(1-2p_A)(1-2p_B)D_{AB} - D_{AB}^2\right] \qquad (3.8)\end{aligned}$$

with an obvious similarity to the variance of $\hat{D}_A$ shown in Equation 3.3.

A chi-square statistic for the hypothesis of no disequilibrium, $H_0 : D_{AB} = 0$, can be constructed by squaring the (asymptotically) normal variable z:

$$z = \frac{\hat{D}_{AB} - \mathcal{E}(\hat{D}_{AB})}{\sqrt{\text{Var}(\hat{D}_{AB})}}$$

after setting D_{AB} to zero in both the mean and the variance. The test statistic is written X_{AB}^2

$$\begin{aligned}X_{AB}^2 &= z^2 \\ &= \frac{n\hat{D}_{AB}^2}{\tilde{p}_A(1-\tilde{p}_A)\tilde{p}_B(1-\tilde{p}_B)} \qquad (3.9)\end{aligned}$$

The same chi-square test statistic follows from a goodness-of-fit on the four gametic classes

Gamete	AB	$A\bar{B}$	$\bar{A}B$	$\bar{A}\bar{B}$
Observed number	n_{AB}	$n_{A\bar{B}}$	$n_{\bar{A}B}$	$n_{\bar{A}\bar{B}}$
Expected number	$n\tilde{p}_A\tilde{p}_B$	$n\tilde{p}_A\tilde{p}_{\bar{B}}$	$n\tilde{p}_{\bar{A}}\tilde{p}_B$	$n\tilde{p}_{\bar{A}}\tilde{p}_{\bar{B}}$

As before, $\bar{A}, \bar{B}$ mean not-A, not-B. It is more common to arrange the four gamete frequencies in a 2×2 contingency table, as shown in Table 3.2. The test for gametic linkage disequilibrium can be regarded as a test for independence in this table – the frequency in each of the four cells is compared to the products of the row and column totals divided by the overall total.

Table 3.2 Contingency table for counts of four gametic types at two loci.

Counts		Locus B B	$\bar{B}$	Totals
Locus A	A	n_{AB}	$n_{A\bar{B}}$	n_A
	$\bar{A}$	$n_{\bar{A}B}$	$n_{\bar{A}\bar{B}}$	$n_{\bar{A}}$
Totals		n_B	$n_{\bar{B}}$	n

Exact Test for Gametic Disequilibrium

An exact test for gametic linkage disequilibrium depends on the probabilities of all possible samples of gametic numbers for the observed gene numbers. Since gametic numbers are being assumed to be multinomially distributed, the gene numbers are binomially distributed (with the same sample size). The probability of the gametic array under the hypothesis of linkage disequilibrium is

$$\Pr(n_{AB}, n_{A\bar{B}}, n_{\bar{A}B}, n_{\bar{A}\bar{B}}) = \frac{n!(p_A p_B)^{n_{AB}}(p_A p_{\bar{B}})^{n_{A\bar{B}}}(p_{\bar{A}} p_B)^{n_{\bar{A}B}}(p_{\bar{A}} p_{\bar{B}})^{n_{\bar{A}\bar{B}}}}{n_{AB}! n_{A\bar{B}}! n_{\bar{A}B}! n_{\bar{A}\bar{B}}!}$$

and the probabilities of the two gene arrays are

$$\Pr(n_A, n_{\bar{A}}) = \frac{n!}{n_A! n_{\bar{A}}!}(p_A)^{n_A}(p_{\bar{A}})^{n_{\bar{A}}}$$

$$\Pr(n_B, n_{\bar{B}}) = \frac{n!}{n_B! n_{\bar{B}}!}(p_B)^{n_B}(p_{\bar{B}})^{n_{\bar{B}}}$$

Taking the ratio of these quantities gives the probability of the gametic numbers conditional on the gene numbers

$$\Pr(n_{AB}, n_{A\bar{B}}, n_{A\bar{B}}, n_{\bar{A}\bar{B}} \mid n_A, n_B) = \frac{n_A! n_{\bar{A}}! n_B! n_{\bar{B}}!}{n_{AB}! n_{A\bar{B}}! n_{\bar{A}B}! n_{\bar{A}\bar{B}}! n!}$$

The unknown gene frequencies have cancelled out in this probability, just as they did in testing for HWE. Notice that the probabilities may just as well have been written as $\Pr(n_{AB} \mid n_A, n_B)$. Tables of significant values of n_{AB} for given values of n, n_A, and n_B have been published (Finney et al. 1963) for n up to 80.

To illustrate the exact test for gametic data, turn to the *Drosophila* restriction site data shown in Table 1.5. The presence or absence of a restriction site can be regarded as alternative alleles at a locus, and for the sites recognized by *Bam*HI and *Xho*I:

Counts		*Xho*I +	*Xho*I −	Total
*Bam*HI	+	5	6	11
	−	6	0	6
Total		11	6	17

All possible samples of size 17 with the same marginal totals as in the observed data are displayed in Table 3.3, along with the exact probabilities and chi-square statistics. The observed set of frequencies are the third least likely and lead to rejection of the hypothesis of linkage equilibrium at a significance level of 4.27%. There is a probability of 0.0427 of observing a sample as extreme, or more extreme, than 5,6,6,0. The chi-square test has a larger 5% rejection region when calculated as in Equation 3.9 since it would reject the hypothesis of linkage equilibrium for the sample 9,2,2,4 as well. However, when Yates' continuity correction is applied, the chi-square test statistic has nonsignificant values (at the 5% level) for the samples 9,2,2,4 *and* 5,6,6,0. The correction reduces the absolute value of D_{AB} by 0.5 before squaring in the numerator of X^2_{AB}. Remember that the exact test gives an accurate accounting of the chance of each possible outcome when the hypothesis is true, and so is the yardstick against which tests such as the chi-square test can be measured.

The (exact) test for HWE is two-tailed because either large positive or large negative values of the one-locus disequilibrium coefficient D_A can lead to rejection of the hypothesis. In the linkage disequilibrium case, the same situation holds. Table 3.3 illustrates that small probabilities for a sample under the hypothesis of no disequilibrium can be obtained for either positive or negative values of the coefficient D_{AB}. An alternative approach was advocated by Yates (1984). If the rows in Table 3.3 are ordered by disequilibrium values, the observed value of -0.1246 is the most extreme in the negative tail of the distribution of all possible values for these allelic frequencies. The hypothesis of no disequilibrium would be rejected in favor of the one-tailed alternative of negative disequilibrium at a significance level of 0.0373. If a two-tailed alternative is needed, Yates' rule of doubling the one-tailed prob-

Table 3.3 Linkage disequilibria for restriction sites in *Drosophila* data of Table 1.6.

Gamete*				Probability	Cumulative Probability	D_{AB}	Chi-square
+ +	+ −	− +	− −				
11	0	0	6	0.0001†	0.0001	0.2284	17.00†
10	1	1	5	0.0053†	0.0054	0.1696	9.37†
5	6	6	0‡	0.0373†	0.0427	−0.1246	5.06†
9	2	2	4	0.0667	0.1094	0.1107	4.00†
6	5	5	1	0.2240	0.3334	−0.0657	1.41
8	3	3	3	0.2666	0.6000	0.0519	0.88
7	4	4	2	0.4000	1.0000	−0.0069	0.02

* First symbol is for *Bam*HI, second is for *Xho*I.
† Causes rejection of hypothesis of no disequilibrium at 5% level.
‡ Observed data.

ability gives a significance level of 0.0746. In presenting analyses of exact tests, care is needed to state the procedure used.

Power calculations for the exact test are relatively easy since the gene numbers are binomially distributed whether or not there is disequilibrium. The probabilities will involve both the gene frequencies and the disequilibrium, however, so that power cannot be found without specifying both sets of parameters. Fu and Arnold (1990) give tables of sample sizes needed to achieve power levels of 50% or 90% when a 5% significance level is used with exact tests.

Gametic Disequilibrium with Multiple Alleles

There is no conceptual difficulty in extending the study of gametic disequilibrium at two loci to allow for several alleles at each locus. A separate coefficient D_{uv} is defined for each pair of alleles A_u and B_v

$$D_{uv} = p_{uv} - p_u p_v$$

These coefficients may be tested separately with the chi-square statistics

$$X^2_{uv} = \frac{n\hat{D}^2_{uv}}{\tilde{p}_u(1-\tilde{p}_u)\tilde{p}_v(1-\tilde{p}_v)} \quad (3.10)$$

while the overall hypothesis that none of the D_{uv}'s is different from zero can be tested with the statistic

$$X_T^2 = \sum_{u=1}^{k}\sum_{v=1}^{\ell} \frac{n\hat{D}_{uv}^2}{\tilde{p}_u\tilde{p}_v} \tag{3.11}$$

which has $(k-1)(\ell-1)$ degrees of freedom when the two loci have k and ℓ alleles, respectively.

Power calculations for the chi-squares in Equations 3.10 and 3.11 may follow from the noncentral chi-square distribution.

Variances and Covariances of Gametic Linkage Disequilibria

Primary interest here will be in associations between genes at two loci, but most data sets contain frequencies at more than two loci and many higher-order disequilibria can be defined. Even concentration on pairs of loci, however, requires a consideration of associations for three or four loci when relationships between the pairwise measures are considered. If gametic linkage disequilibria D_{AB} and D_{BC} are estimated for alleles A, B, and C at three loci, they are expected to be related in value because of the common dependence on allele B. It is possible to modify Fisher's approximate variance formula (Equation 2.16) to give approximate covariances of functions S and T of multinomial counts n_i (these counts have expected values nQ_i):

$$\frac{1}{n}\text{Cov}(S,T) = \sum_i \frac{\partial S}{\partial n_i}\frac{\partial T}{\partial n_i}Q_i - \frac{\partial S}{\partial n}\frac{\partial T}{\partial n}$$

Applying this result allows covariances of sample disequilibria to be found:

$$\begin{aligned}\text{Cov}(\hat{D}_{AB},\hat{D}_{BC}) &= \frac{1}{n}[p_B(1-p_B)D_{AC} + (1-2p_B)D_{ABC} - D_{AB}D_{BC}] \\ \text{Cov}(\hat{D}_{AB},\hat{D}_{CD}) &= \frac{1}{n}[D_{AC}D_{BD} + D_{AD}D_{BC} + D_{ABCD}]\end{aligned}$$

Note that the three- and four-locus disequilibria (defined in Equations 3.12 and 3.13 below) are involved, and that the second covariance is likely to be quite small because it involves products of two-locus disequilibria and the four-locus disequilibrium.

The relationships between disequilibria may not be particularly meaningful because of the arbitrary way in which the coefficients are generally defined. There is usually no particular reason for labeling one allele A and the others $\bar{A}$, for example, so that D_{AB} may as well have had the same value but the opposite sign

$$D_{AB} = -D_{A\bar{B}} = -D_{\bar{A}B} = D_{\bar{A}\bar{B}}$$

The arbitrary sign may lead to low covariances between disequilibria, on average, even if the quantities are quite related. There is also the feature that, in the absence of disturbing forces, disequilibrium is expected to decay to zero over time. It is common to work with squared disequilibria. This at least eliminates problems with sign, and population genetic theory shows that the ratio of $\hat{D}_{AB}^2$ to $\tilde{p}_A(1-\tilde{p}_A)\tilde{p}_B(1-\tilde{p}_B)$ remains nonzero even as populations drift to fixation for one allele at each locus. It appears then that there is a need also for variances and covariances of squared disequilibria to provide a complete picture of the relationships among two-locus linkage disequilibria (Hill and Weir 1988).

Gametic Disequilibrium at Three or Four Loci

Whether there is interest in three- and four-gene disequilibria in their own right, or merely as components of the covariances between two-locus coefficients, it is useful to define and make inferences about them. There have been alternative parameterizations suggested in the past to describe disequilibrium among alleles at more than two loci. The problem is to remove the effects of two-locus disequilibrium when describing three-locus associations, or two- and three-locus disequilibria when describing four-locus associations. Here the additive formulation of Bennett (1954), which subtracts terms for lower order disequilibria, will be used. For alleles A, B, and C at three loci:

$$D_{ABC} = p_{ABC} - p_A D_{BC} - p_B D_{AC} - p_C D_{AB} - p_A p_B p_C \tag{3.12}$$

and for alleles A, B, C, and D at four loci

$$\begin{aligned} D_{ABCD} = \; & p_{ABCD} - p_A D_{BCD} - p_B D_{ACD} - p_C D_{ABD} - p_D D_{ABC} \\ & - p_A p_B D_{CD} - p_A p_C D_{BD} - p_A p_D D_{BC} \\ & - p_B p_C D_{AD} - p_B p_D D_{AC} - p_C p_D D_{AB} \\ & - D_{AB} D_{CD} - D_{AC} D_{BD} - D_{AD} D_{BC} - p_A p_B p_C p_D \end{aligned} \tag{3.13}$$

Maximum likelihood estimates follow by replacing frequencies with their observed values, and there is a slight bias in these estimates that will be ignored. Variances of the estimates are found with Fisher's formula, although that for four loci requires a computerized algebra treatment (Weir and Golding 1990), and are expressed most succinctly with the quantities π_A and τ_A for allele A:

$$\pi_A = p_A(1-p_A), \quad \tau_A = (1-2p_A)$$

The results are

$$\begin{aligned}
\mathrm{Var}(\hat{D}_{AB}) &= \frac{1}{n}[\pi_A\pi_B + \tau_A\tau_B D_{AB} - D_{AB}^2] \\
\mathrm{Var}(\hat{D}_{ABC}) &= \frac{1}{n}[\pi_A\pi_B\pi_C + 6D_{AB}D_{BC}D_{AC} + \pi_A(\tau_B\tau_C D_{BC} - D_{BC}^2) \\
&\quad + \pi_B(\tau_A\tau_C D_{AC} - D_{AC}^2) + \pi_C(\tau_A\tau_B D_{AB} - D_{AB}^2) \\
&\quad + D_{ABC}(\tau_A\tau_B\tau_C - 2\tau_A D_{BC} - 2\tau_B D_{AC} \\
&\quad -2\tau_C D_{AB} - D_{ABC})]
\end{aligned}$$

The coefficient D_{ABC} is set to zero, and all other terms replaced by their observed values, when using the three-locus variance to test the hypothesis $H_0 : D_{ABC} = 0$ with the chi-square statistic

$$X_{ABC}^2 = \frac{\hat{D}_{ABC}^2}{\sqrt{\mathrm{Var}(\hat{D}_{ABC})}}$$

A similar procedure applies to the four-locus case, and a program is given in the Appendix that estimates and tests two-, three-, and four-locus disequilibria for gametic data.

Normalized Gametic Disequilibria

Additive disequilibrium coefficients are constrained in the values they may take since gametic frequencies cannot be negative or greater than corresponding allelic frequencies. Focusing on alleles A, B at loci **A**, **B**, the frequency of AB gametes is constrained by

$$0 \leq p_{AB} \leq p_A, p_B$$

so that the linkage disequilibrium coefficient D_{AB} is constrained by

$$-p_A p_B \leq D_{AB} \leq p_A - p_A p_B, p_B - p_A p_B$$

or

$$-p_A p_B \leq D_{AB} \leq \min(p_{\bar{A}} p_B, p_A p_{\bar{B}})$$

with $\bar{A}$ meaning not-A.

Another view of the amount of disequilibrium is provided by dividing the coefficient by its maximum numerical value, remembering that the maximum values depend on sign. This normalized value D'_{AB} is therefore defined as

$$D'_{AB} = \begin{cases} \dfrac{D_{AB}}{p_A p_B}, & D_{AB} < 0 \\ \\ \dfrac{D_{AB}}{\min(p_{\bar{A}} p_B, p_A p_{\bar{B}})}, & D_{AB} > 0 \end{cases}$$

Although D'_{AB} has a range of values from -1 to $+1$ it would be mistaken to regard it as being independent of allelic frequencies. The frequencies are very much part of the definition of the normalized coefficient, as has been pointed out recently by Lewontin (1988).

Bounds on three-locus disequilibria were given by Thomson and Baur (1984).

Genotypic Disequilibrium at Two Loci

When population genetic data are collected on genotypes, it is possible to check for associations between genes other than those at one locus or those on one gamete. Recall that HWE testing looks at two genes at the same locus, but on different gametes. Testing for linkage disequilibrium looks at two genes on the same gamete, but at different loci. There is a third alternative — two genes on different gametes and at different loci. Whatever forces are responsible for causing associations of pairs of genes in the first two cases might also cause an association in the third. Whether or not an immediate biological mechanism is suggested, it is recommended here that at least the possibility for this new type of disequilibrium be investigated. Continuing this argument, and referring back to the search for associations between frequencies of three or four genes on a single gamete, a case can be made for looking for associations between the frequencies of three or four of the four genes present at two loci in a diploid individual. Although such associations will generally be quite small, they complete the disequilibrium analysis at two loci.

When focusing on alleles A and B, there are a total of 10 possible genotypes with 9 d.f. available for estimation. The frequencies of the 10 genotypes can be used to estimate the 9 quantities listed in Table 3.4 and illustrated in Figure 3.1.

Allele frequencies and one-locus coefficients are estimated as described above and the gametic disequilibrium can also be estimated directly from observations, since it is assumed here that gametic frequencies can be recovered from the genotypic frequencies, as in

$$p_{AB} = P_{AB}^{AB} + \frac{1}{2}\left(P_{A\bar{B}}^{AB} + P_{\bar{A}B}^{AB} + P_{\bar{A}\bar{B}}^{AB}\right)$$

Table 3.4 Partitioning of the 9 d.f. available from the 10 genotypes at two loci.

allele frequencies p_A, p_B	2
one-locus disequilibria D_A, D_B	2
gametic disequilibrium D_{AB}	1
nongametic disequilibrium $D_{A/B}$	1
trigenic disequilibria D_{AAB}, D_{ABB}	2
quadrigenic disequilibrium D_{AABB}	1
Total	9

The new digenic disequilibrium, $D_{A/B}$, refers to genes at different loci and on different gametes within individuals, and is defined as

$$D_{A/B} = p_{A/B} - p_A p_B$$

with the nongametic frequency

$$p_{A/B} = P_{AB}^{AB} + \frac{1}{2}\left(P_{A\bar{B}}^{AB} + P_{\bar{A}B}^{AB} + P_{AB}^{\bar{A}B}\right)$$

So that there is no confusion over notation, an example of 10 genotypic frequencies is shown in Table 3.5, arranged to show both the one-locus geno-

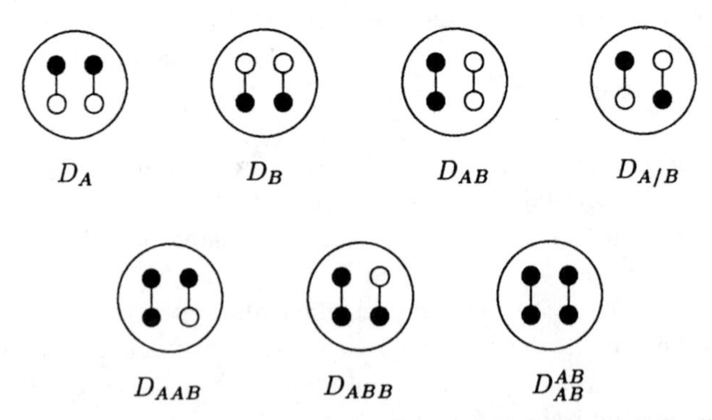

Figure 3.1 Pictorial representation of complete set of two-locus disequilibria. Vertical lines represent two gametes. Upper circles are the genes at the A locus and lower circles are the genes at the B locus. Filled circles are those for which the disequilibrium coefficient is defined.

Table 3.5 Example of frequency notation for two alleles at each of two loci.

$P^{AB}_{AB} = 0.20$	$P^{AB}_{A\bar{B}} = 0.18$	$P^{A\bar{B}}_{A\bar{B}} = 0.02$	$P^{A}_{A} = 0.40$
$P^{AB}_{\bar{A}B} = 0.26$	$P^{AB}_{\bar{A}\bar{B}} = 0.08$ $P^{A\bar{B}}_{\bar{A}B} = 0.04$	$P^{A\bar{B}}_{\bar{A}\bar{B}} = 0.02$	$P^{A}_{\bar{A}} = 0.40$
$P^{\bar{A}B}_{\bar{A}B} = 0.04$	$P^{\bar{A}B}_{\bar{A}\bar{B}} = 0.10$	$P^{\bar{A}\bar{B}}_{\bar{A}\bar{B}} = 0.06$	$P^{\bar{A}}_{\bar{A}} = 0.20$
$P^{B}_{B} = 0.50$	$P^{B}_{\bar{B}} = 0.40$	$P^{\bar{B}}_{\bar{B}} = 0.10$	1.00

$P^{AB}_{AB} = 0.20$	$\frac{1}{2}P^{AB}_{A\bar{B}} = 0.09$	$\frac{1}{2}P^{AB}_{\bar{A}B} = 0.13$	$\frac{1}{2}P^{AB}_{\bar{A}\bar{B}} = 0.04$	$P_{AB} = 0.46$
$\frac{1}{2}P^{A\bar{B}}_{AB} = 0.09$	$P^{A\bar{B}}_{A\bar{B}} = 0.02$	$\frac{1}{2}P^{A\bar{B}}_{\bar{A}B} = 0.02$	$\frac{1}{2}P^{A\bar{B}}_{\bar{A}\bar{B}} = 0.01$	$P_{A\bar{B}} = 0.14$
$\frac{1}{2}P^{\bar{A}B}_{AB} = 0.13$	$\frac{1}{2}P^{AB}_{\bar{A}\bar{B}} = 0.04$	$P^{\bar{A}B}_{\bar{A}B} = 0.04$	$\frac{1}{2}P^{\bar{A}B}_{\bar{A}\bar{B}} = 0.05$	$P_{\bar{A}/B} = 0.26$
$\frac{1}{2}P^{A\bar{B}}_{\bar{A}B} = 0.02$	$\frac{1}{2}P^{\bar{A}\bar{B}}_{A\bar{B}} = 0.01$	$\frac{1}{2}P^{\bar{A}\bar{B}}_{\bar{A}B} = 0.05$	$P^{\bar{A}\bar{B}}_{\bar{A}\bar{B}} = 0.06$	$P_{\bar{A}/\bar{B}} = 0.14$
$P_{A/B} = 0.44$	$P_{A/\bar{B}} = 0.16$	$P_{\bar{A}B} = 0.24$	$P_{\bar{A}\bar{B}} = 0.16$	1.00

typic frequencies as marginal totals, and the two-locus gametic and nongametic marginal totals.

For the trigenic coefficients, frequencies of triples of genes are compared with the products of gene frequencies, after removing any digenic disequilibria. The two trigenic frequencies are

$$\begin{aligned} p_{AAB} &= P^{AB}_{AB} + \frac{1}{2}P^{AB}_{A\bar{B}} \\ p_{ABB} &= P^{AB}_{AB} + \frac{1}{2}P^{AB}_{\bar{A}B} \end{aligned}$$

and the disequilibria are defined as

$$\begin{aligned} D_{AAB} &= p_{AAB} - p_A D_{AB} - p_A D_{A/B} - p_B D_A - p_A^2 p_B \\ D_{ABB} &= p_{ABB} - p_B D_{AB} - p_B D_{A/B} - p_A D_B - p_A p_B^2 \end{aligned}$$

Finally, the quadrigenic coefficient measures any remaining disequilibrium after the removal of all digenic and trigenic disequilibria

$$D^{AB}_{AB} = P^{AB}_{AB} - 2p_A D_{ABB} - 2p_B D_{AAB} - 2p_A p_B D_{AB} - 2p_A p_B D_{A/B}$$

$$- p_A^2 D_B - p_B^2 D_A - D_{AB}^2 - D_{A/B}^2 - D_A D_B - p_A^2 p_B^2$$

The MLE's of all seven disequilibria follow by substituting observed values into these expressions. It is now genotypic, rather than gametic, frequencies that are multinomially distributed, and this must be reflected in the sampling variances. The variances for $\hat{D}_A$ and $\hat{D}_B$ remain as in Equation 3.3, but that for $\hat{D}_{AB}$ is changed from Equation 3.8 to

$$\begin{aligned} \mathrm{Var}(\hat{D}_{AB}) &= \frac{1}{2n}\Big(\pi_A\pi_B + \tau_A\tau_B D_{AB} - D_{AB}^2 \\ &\quad + D_A D_B + D_{A/B}^2 + D_{AB}^{AB}\Big) \end{aligned}$$

for an estimate based on n genotypes (Equation 3.8 was for n gametes). This is from Fisher's approximate variance formula, as is

$$\begin{aligned} \mathrm{Var}(\hat{D}_{A/B}) &= \frac{1}{2n}\Big(\pi_A\pi_B + \tau_A\tau_B D_{A/B} - D_{A/B}^2 \\ &\quad + D_A D_B + D_{AB}^2 + D_{AB}^{AB}\Big) \end{aligned}$$

Exact results were given by Weir (1979). Further application of Fisher's formula leads to variances for the trigenic and quadrigenic coefficients, but the complexity makes it desirable that the algebraic manipulations be done on a computer. The results were derived by Lisa D. Brooks (personal communication). For the trigenic coefficient

$$\begin{aligned} 2n\mathrm{Var}(\hat{D}_{AAB}) &= (\pi_A^2 + \tau_A^2 D_A - D_A^2)(\pi_B + D_B) \\ &\quad + \pi_A\tau_A\tau_B(D_{AB} + D_{A/B}) - 2D_{AAB}^2 \\ &\quad + (1 - 5\pi_A + D_A)(D_{AB} + D_{A/B})^2 + 2\pi_A\tau_A D_{ABB} \\ &\quad + [\tau_A^2\tau_B - 2D_A\tau_B - 4\tau_A(D_{AB} + D_{A/B})]D_{AAB} \\ &\quad + (\tau_A^2 - 2D_A)(D_{AB}^{AB} - 2D_{AB}D_{A/B}) \end{aligned} \tag{3.14}$$

and for the quadrigenic coefficient

$$\begin{aligned} 2nVar(\hat{D}_{AB}^{AB}) &= (\pi_A^2 + \tau_A^2 D_A - D_A^2)(\pi_B^2 + \tau_B^2 D_B - D_B^2) \\ &\quad + \ldots - (D_{AB}^{AB})^2 \end{aligned}$$

The complete expression for this variance was given by Weir and Cockerham (1989).

These trigenic and quadrigenic variance expressions are extremely cumbersome and illustrate how quickly the additive model for disequilibrium becomes unmanageable. Extensions to three loci appear to be prohibitively complex. The results given here for two loci can be incorporated into computer programs for a complete analysis of two-locus disequilibrium, and a numerical justification of the testing procedure was given by Weir and Cockerham (1989).

The most noteworthy feature of the variance formulas just displayed is that the variances for the two digenic disequilibria calculated from genotypic data involve the higher order disequilibrium coefficients. This suggests that an appropriate testing strategy is to begin with these higher order disequilibria, and first test the hypothesis $H_0 : D_{AB}^{AB} = 0$. This is performed with the chi-square statistic

$$X_{AABB}^2 = \frac{(\hat{D}_{AB}^{AB})^2}{\sqrt{\text{Var}(\hat{D}_{AB}^{AB})}}$$

with D_{AB}^{AB} set to zero in the denominator and all other terms set to the observed or estimated values. If the hypothesis is not rejected, the quadrigenic coefficient can be set to zero in the test for trigenic and digenic disequilibria. Otherwise higher order terms must be included in test statistics for lower order disequilibria.

When quadrigenic disequilibrium can be ignored, the digenic test statistics reduce to

$$X_{AB}^2 = \frac{2n\hat{D}_{AB}^2}{\tilde{\pi}_A\tilde{\pi}_B + \tilde{D}_A\tilde{D}_B + \tilde{D}_{A/B}^2}$$

$$X_{A/B}^2 = \frac{2n\hat{D}_{A/B}^2}{\tilde{\pi}_A\tilde{\pi}_B + \tilde{D}_A\tilde{D}_B + \tilde{D}_{AB}^2}$$

showing that each test depends on the other coefficient of disequilibrium, which is somewhat unsatisfactory. A reasonable procedure would be to include each coefficient in the test for the other and then to repeat the test for a coefficient without including the other if the other is found to be not significantly different from zero. It is important to note that both tests include a term for departure from HWE at both loci. Linkage disequilibrium may be tested for whether or not there is Hardy-Weinberg equilibrium at each of the two loci.

Test statistics use estimated variances in the denominators, but these estimates incorporate the hypotheses being tested, and maybe the results of prior tests. To anticipate work in Chapter 5, note that it would not be appropriate to use variances estimated by numerical resampling since these ignore the hypotheses.

Composite Genotypic Disequilibria

When genotypes are scored, it is often not possible to distinguish between the two double heterozygotes $AB/\bar{A}\bar{B}$ and $A\bar{B}/\bar{A}B$, so that gametic frequencies cannot be inferred. Under the assumption of random mating, in which genotypic frequencies are assumed to be the products of gametic frequencies, it is possible to estimate gametic frequencies with the EM algorithm as discussed in Chapter 2, and so proceed with the full suite of genotypic disequilibria discussed in the previous section. To avoid making the random-mating assumption, however, it is possible to work with a set of composite disequilibrium coefficients.

Although the separate digenic frequencies p_{AB} and $p_{A/B}$ cannot be observed, their sum can be since

$$p_{AB} + p_{A/B} = 2P^{AB}_{AB} + P^{AB}_{A\bar{B}} + P^{AB}_{\bar{A}B} + \frac{1}{2}\left(P^{AB}_{\bar{A}\bar{B}} + P^{A\bar{B}}_{\bar{A}B}\right)$$

This can be verified in Table 3.5, where the sum is seen to depend on double heterozygotes only by their total of $P^{AB}_{\bar{A}\bar{B}} + P^{A\bar{B}}_{\bar{A}B}$. The digenic disequilibrium is measured with a composite measure Δ_{AB} defined as

$$\begin{aligned}\Delta_{AB} &= p_{AB} + p_{A/B} - 2p_A p_B \\ &= D_{AB} + D_{A/B}\end{aligned}$$

which is the sum of the gametic and nongametic coefficients.

Suppose the nine genotypic classes are numbered as

	BB	$B\bar{B}$	$\bar{B}\bar{B}$
AA	1	2	3
$A\bar{A}$	4	5	6
$\bar{A}\bar{A}$	7	8	9

and that the observed count for class i is n_i, with a total sample size of n. The digenic count for $AB + A/B$ is

$$n_{AB} = 2n_1 + n_2 + n_4 + \frac{1}{2}n_5$$

and the MLE for the composite linkage disequilibrium is

$$\hat{\Delta}_{AB} = \frac{1}{n} n_{AB} - 2\tilde{p}_A \tilde{p}_B$$

There is no problem with trigenic disequilibria when double heterozygotes cannot be distinguished, but the definitions may be simplified a little to

$$\begin{aligned} D_{AAB} &= p_{AAB} - p_A \Delta_{AB} - p_B D_A - p_A^2 p_B \\ D_{ABB} &= p_{ABB} - p_B \Delta_{AB} - p_A D_B - p_A p_B^2 \end{aligned}$$

A modification of the quadrigenic coefficient is needed to account for lack of knowledge of D_{AB} and $D_{A/B}$. A composite coefficient Δ_{AABB} is defined

$$\begin{aligned} \Delta_{AABB} &= D_{AB}^{AB} - 2D_{AB}D_{A/B} \\ &= P_{AB}^{AB} - 2p_A D_{ABB} - 2p_B D_{AAB} - 2p_A p_B \Delta_{AB} - \Delta_{AB}^2 \\ &\quad - p_A^2 D_B - p_B^2 D_A - D_A D_B - p_A^2 p_B^2 \end{aligned}$$

Substituting observed frequencies into the definition equations provides MLE's, and Fisher's formula gives approximate variances

$$\begin{aligned} n\text{Var}(\hat{\Delta}_{AB}) &= (\pi_A + D_A)(\pi_B + D_B) + \frac{1}{2}\tau_A \tau_B \Delta_{AB} \\ &\quad + \tau_A D_{ABB} + \tau_B D_{AAB} + \Delta_{AABB} \end{aligned}$$

where the divisor is now n (for n individuals) instead of $2n$. The variances for the trigenic coefficients are unchanged from Equation 3.14, although they can be simplified slightly by using the two composite coefficients Δ_{AB} and Δ_{AABB}. Finally, the quadrigenic variance (Lisa D. Brooks, personal communication) is

$$\begin{aligned} nVar(\hat{\Delta}_{AABB}) &= (\pi_A^2 + \tau_A^2 D_A - D_A^2)(\pi_B^2 + \tau_B^2 D_B - D_B^2) \\ &\quad + \ldots - \Delta_{AABB}^2 \end{aligned}$$

Details are in Weir and Cockerham (1989).

Testing employs chi-square statistics and starts by testing for the composite quadrigenic coefficient. A program is given in the Appendix. If quadrigenic and trigenic coefficients can be ignored, the test statistic for composite digenic linkage disequilibrium is

$$X_{AB}^2 = \frac{n\hat{\Delta}_{AB}^2}{(\tilde{\pi}_A + \hat{D}_A)(\tilde{\pi}_B + \hat{D}_B)}$$

Note the explicit way in which departures from Hardy-Weinberg are included in this expression. Linkage disequilibrium can be tested for, even in nonrandom mating populations.

Log-Linear Tests for Linkage Disequilibrium

To make inferences about linkage disequilibrium with a log-linear approach, a series of multiplicative models is constructed that includes different numbers of disequilibrium terms. The difference in deviances for two models that differ only by whether a particular coefficient is included provides a chi-square test statistic for that coefficient. Consider the gametic disequilibrium case first.

For alleles A and B at two loci, the multiplicative model for the four gamete frequencies, retaining only the necessary terms, can be expressed as

$$\begin{aligned} P_{AB} &= M M_A M_B M_{AB} \\ P_{A\bar{B}} &= M M_A \\ P_{\bar{A}B} &= M M_B \\ P_{\bar{A}\bar{B}} &= M \end{aligned}$$

with MLE's of the terms

$$\begin{aligned} \hat{M} &= \tilde{p}_{\bar{A}\bar{B}} \\ \hat{M}_A &= \frac{\tilde{p}_{A\bar{B}}}{\tilde{p}_{\bar{A}\bar{B}}} \\ \hat{M}_B &= \frac{\tilde{p}_{\bar{A}B}}{\tilde{p}_{\bar{A}\bar{B}}} \\ \hat{M}_{AB} &= \frac{\tilde{p}_{AB}\tilde{p}_{\bar{A}\bar{B}}}{\tilde{p}_{A\bar{B}}\tilde{p}_{\bar{A}B}} \end{aligned}$$

and maximum log-likelihood

$$\ln L_1 = n_{AB} \ln \tilde{p}_{AB} + n_{A\bar{B}} \ln \tilde{p}_{A\bar{B}} + n_{\bar{A}B} \ln \tilde{p}_{\bar{A}B} + n_{\bar{A}\bar{B}} \ln \tilde{p}_{\bar{A}\bar{B}}$$

The equation for M_{AB}, or its estimate, can be expressed in terms of the additive coefficient D_{AB} and the allelic frequencies

$$M_{AB} = \frac{(p_A p_B + D_{AB})(p_{\bar{A}} p_{\bar{B}} + D_{AB})}{(p_A p_{\bar{B}} - D_{AB})(p_{\bar{A}} p_B - D_{AB})}$$

This shows that there is a one-to-one relationship between the additive and multiplicative coefficients. As D_{AB} changes from its minimum value of $-p_A p_B$ to its maximum of $min(p_{\bar{A}} p_B, p_A p_{\bar{B}})$, the coefficient M_{AB} changes from 0 to ∞. Bounds on M_{AB} do not depend on allelic frequencies, and the coefficient is also known as the ODDS RATIO for the two by two array of gametic frequencies.

Under the hypothesis of linkage equilibrium, the model does not include the M_{AB} term and the MLE's become

$$\begin{aligned} \hat{M} &= \tilde{p}_{\bar{A}}\tilde{p}_{\bar{B}} \\ \hat{M}_A &= \frac{\tilde{p}_A}{\tilde{p}_{\bar{A}}} \\ \hat{M}_B &= \frac{\tilde{p}_B}{\tilde{p}_{\bar{B}}} \end{aligned}$$

with maximum log-likelihood

$$\ln L_0 = n_A \ln \tilde{p}_A + n_{\bar{A}} \ln \tilde{p}_{\bar{A}} + n_B \ln \tilde{p}_B + n_{\bar{B}} \ln \tilde{p}_{\bar{B}}$$

Minus twice the difference of the log-likelihoods provides a test statistic for linkage disequilibrium.

$$X^2_{AB} = -2(\ln L_0 - \ln L_1)$$

With more loci, explicit expressions cannot be given for the MLE's for all terms under the various models, and there is a need to use numerical methods for estimates and maximum likelihoods.

For genotypic disequilibria, details for the two-allele, two-locus case were given by Weir and Wilson (1986). Changing their conditions on the terms, and setting all terms for alleles $\bar{A}$ or $\bar{B}$ equal to 1, the 10 genotypic frequencies can be expressed as

$$\begin{aligned} P^{AB}_{AB} &= MM_A^2M_B^2M_{AA}M_{BB}M_{AB}^2M_{A/B}^2M_{AAB}^2M_{ABB}^2M^{AB}_{AB} \\ P^{AB}_{A\bar{B}} &= 2MM_A^2M_BM_{AA}M_{AB}M_{A/B}M_{AAB} \\ P^{A\bar{B}}_{A\bar{B}} &= MM_A^2M_{AA} \\ P^{AB}_{\bar{A}B} &= 2MM_AM_B^2M_{BB}M_{AB}M_{A/B}M_{ABB} \\ P^{AB}_{\bar{A}\bar{B}} &= 2MM_AM_BM_{AB} \\ P^{A\bar{B}}_{\bar{A}B} &= 2MM_AM_BM_{A/B} \\ P^{A\bar{B}}_{\bar{A}\bar{B}} &= 2MM_A \\ P^{\bar{A}B}_{\bar{A}B} &= MM_B^2M_{BB} \\ P^{\bar{A}B}_{\bar{A}\bar{B}} &= 2MM_B \\ P^{\bar{A}\bar{B}}_{\bar{A}\bar{B}} &= M \end{aligned}$$

When only nine genotypic classes can be distinguished, it is sufficient to use

$$M, M_A, M_B, M_{AA}, M_{BB}, S_{AB}, Q_{AB}, M_{AAB}, M_{ABB}, M^{AB}_{AB}$$

where

$$S_{AB} = \frac{1}{2}(M_{AB} + M_{A/B}),\ Q_{AB} = M_{AB}M_{A/B}$$

are the sum and product of digenic disequilibria. To reduce the number of parameters to nine, the absence of quadrigenic disequilibrium could be assumed, $M_{AB}^{AB} = 1$, which is similar to assuming $\Delta_{AABB} = 0$ in the additive model to allow a test of $\Delta_{AB} = 0$. A computer program is needed (Weir and Wilson 1986) that considers a series of models of increasing complexity and calculates the deviance for each model. The simplest model, model 0, includes only the mean term, M, and this has MLE of $\hat{M} = 1/16$ and maximum log-likelihood

$$\begin{aligned} \ln L_0 &= \text{Constant} \\ &\quad + n_1 \ln(\hat{M}) + n_2 \ln(2\hat{M}) + n_3 \ln(\hat{M}) \\ &\quad + n_4 \ln(2\hat{M}) + n_5 \ln(4\hat{M}) + n_6 \ln(2\hat{M}) \\ &\quad + n_7 \ln(\hat{M}) + n_8 \ln(2\hat{M}) + n_9 \ln(\hat{M}) \end{aligned}$$

where the constant term is the logarithm of the multinomial coefficient and the nine genotypic classes are numbered as they were in the previous section.

The full model, with all nine parameters, equates expected and observed genotypic frequencies and has a maximum log-likelihood of

$$\begin{aligned} \ln L_8 &= \text{Constant} \\ &\quad + n_1 \ln(\tilde{P}_1) + n_2 \ln(\tilde{P}_2) + n_3 \ln(\tilde{P}_3) \\ &\quad + n_4 \ln(\tilde{P}_4) + n_5 \ln(\tilde{P}_5) + n_6 \ln(\tilde{P}_6) \\ &\quad + n_7 \ln(\tilde{P}_7) + n_8 \ln(\tilde{P}_8) + n_9 \ln(\tilde{P}_9) \end{aligned}$$

The deviance of the model with mean only is

$$Dev_0 = -2(\ln L_0 - \ln L_8)$$

The quantity Dev_0 is distributed as chi-square with 8 d.f. when model 0 is true. It will almost certainly be very large, indicating that the genotypic frequencies cannot be fitted just by a constant average frequency. The deviance for the full model, Dev_8, is zero since this model fits the data perfectly. The next pair of models, 1a and 1b, includes the mean and a term for either allele A or B. With M and M_A in the model, since the frequencies sum to 1,

$$M = \frac{1}{[2(M_A + 1)]^2}$$

and the MLE of M_A can be written as

$$\hat{M}_A = \frac{\tilde{p}_A}{\tilde{p}_{\bar{A}}}$$

so that the maximum log-likelihood is

$$\begin{aligned}\ln L_{1a} &= \text{Constant} \\ &\quad + n_1 \ln(\hat{M}\hat{M}_A^2) + n_2 \ln(2\hat{M}\hat{M}_A^2) + n_3 \ln(\hat{M}\hat{M}_A^2) \\ &\quad + n_4 \ln(2\hat{M}\hat{M}_A) + n_5 \ln(4\hat{M}\hat{M}_A) + n_6 \ln(2\hat{M}\hat{M}_A) \\ &\quad + n_7 \ln(\hat{M}) + n_8 \ln(2\hat{M}) + n_9 \ln(\hat{M})\end{aligned}$$

and the deviance

$$Dev_{1a} = -2(\ln L_{1a} - \ln L_8)$$

provides a seven d.f. chi-square statistic for testing how well a model with only A gene effects fits the data.

The first model likely to be of interest, Model 2, has terms for allele frequencies at each locus, but does not allow for any disequilibria. This test has 6 d.f. and should be performed prior to the single-degree-of-freedom tests for individual disequilibria. The first model not to allow explicit MLE's for the terms is Model 5, with terms M, M_A, M_{AA}, M_B, M_{BB}, and S_{AB} as the two loci cannot now be treated separately. Numerical methods are needed to solve the likelihood equations.

Example

To illustrate the log-linear models approach, Table 3.6 shows the results of analyzing some *MNS* blood group data of Mourant et al. (1976). The data, in the present notation, are

		Locus "B"		
		SS	Ss	ss
Locus "A"	MM	$n_1 = 91$	$n_2 = 147$	$n_3 = 85$
	MN	$n_4 = 32$	$n_5 = 78$	$n_6 = 75$
	NN	$n_7 = 5$	$n_8 = 17$	$n_9 = 7$

Table 3.7 shows an analysis of the same data with the additive set of disequilibrium coefficients.

Table 3.6 Two-locus log-linear analysis of *MNS* data of Mourant et al. (1976).

Model	Effects in Model	Deviance	d.f.
0	M	366.1	8
1*a*	M, M_A	26.49	7
1*b*	M, M_B	361.14	7
2	M, M_A, M_B	20.82	6
3*a*	M, M_A, M_B, M_{AA}	20.69	5
3*b*	M, M_A, M_B, M_{BB}	16.08	5
4	$M, M_A, M_B, M_{AA}, M_{BB}$	15.95	4
5	$M, M_A, M_B, M_{AA}, M_{BB}, S_{AB}$	14.98	3
6	$M, M_A, M_B, M_{AA}, M_{BB}, S_{AB}, Q_{AB}$	6.88	2
7*a*	$M, M_A, M_B, M_{AA}, M_{BB}, S_{AB}, Q_{AB}, M_{ABB}$	3.62	1
7*b*	$M, M_A, M_B, M_{AA}, M_{BB}, S_{AB}, Q_{AB}, M_{AAB}$	1.91	1
8	$M, M_A, M_B, M_{AA}, M_{BB}, S_{AB}, Q_{AB}, M_{ABB}, M_{AAB}$	0	0

Successive deviances in Table 3.6 can be subtracted to provide chi-square statistics for testing the significance of the term by which the two models differ. Subtracting Dev_{3a} from Dev_2 gives a 1 d.f. chi-square for testing for HWE at locus A (i.e., locus M here). This statistic is 0.13, as listed in

Table 3.7 Two-locus disequilibrium coefficient analysis of *MNS* data of Mourant et al. (1976).

Estimate	SD(Est.)	Test Statistic
$\tilde{p}_A = 0.7737$		
$\tilde{p}_B = 0.4637$		
$\hat{D}_A = 0.0028$	0.0077	$X^2_A = 0.14$
$\hat{D}_B = 0.0234$	0.0107	$X^2_B = 4.74^*$
$\hat{\Delta}_{AB} = 0.0273$	0.0090	$X^2_{AB} = 8.21^*$
$\hat{D}_{ABB} = 0.0063$	0.0026	$X^2_{ABB} = 5.00^*$
$\hat{D}_{AAB} = 0.0022$	0.0032	$X^2_{AAB} = 0.46$
$\hat{\Delta}_{AABB} = -0.0034$	0.0020	$X^2_{AABB} = 3.00$

*Estimate significantly different from zero at the 5% level.

Table 3.7, and it could also have been found from Equation 3.6. For the other locus, the chi-square is 4.74 (Dev_{3a} subtracted from Dev_2), indicating that blood group S is not in HWE for this population. Similarly, evidence is seen for linkage disequilibrium, and trigenic disequilibrium for two M genes and one S gene. All these log-linear tests rest on the assumed absence of a quadrigenic effect. Table 3.7 shows general agreement with the log-linear analysis, although the disequilibria between loci are handled differently. It is not known in Table 3.7 whether the nonsignificance of the composite quadrigenic component implies an absence of a four-gene association or whether the product $2D_{AB}D_{A/B}$ is nearly equal to but of opposite sign to the four-gene term D_{AB}^{AB}.

One major difference in the two approaches is that the estimates of a term such as M_A differ according to which model is being fitted, while the additive model estimates the gene frequency p_A by the observed gene frequency, regardless of how many disequilibrium coefficients are also being fitted.

MULTIPLE TESTS

When testing for the presence of associations within or between loci, a problem arises when the data set contains several loci. If there is interest in whether HWE holds at each of these particular loci, then the tests as presented are appropriate. If, however, there is interest in whether HWE holds in the population, the loci at hand serve to give multiple tests of the same hypothesis. Among a set of L tests the largest chi-square will exceed 3.84 more than 5% of the time simply because it is the largest. The significance level of the set of tests is the probability that one or more of them causes rejection of the hypothesis when it is true. This EXPERIMENTWISE error rate α' is

$$\begin{aligned} \alpha' &= \Pr(\text{at least one test causes rejection } |H_0 \text{ true}) \\ &= 1 - \Pr(\text{no test causes rejection } |H_0 \text{ true}) \\ &= 1 - [\Pr(\text{one test does not cause rejection } |H_0 \text{ true})]^L \\ &= 1 - (1 - \alpha)^L \qquad (3.15) \end{aligned}$$

where α is the significance level for an individual test. This argument, called the BONFERRONI PROCEDURE, assumes that all the tests are independent, which is not strictly true, but the error in making this assumption is not major. With a 5% level used for 10 tests, the actual significance level for the set of tests is

$$\begin{aligned}\alpha' &= 1 - 0.95^{10} \\ &= 0.40\end{aligned}$$

which is substantially larger. To avoid these spurious rejections of the hypothesis, each individual test needs to be more stringent. For an overall $\alpha' = 0.05$, it is necessary that individual values are $\alpha = 0.005$. This individual α value is obtained from

$$\alpha = 1 - (1 - \alpha')^{1/L}$$

For single degree-of-freedom chi-squares this translates into a rejection value of 7.84 instead of 3.84. Tables can be constructed to provide these values (Rohlf and Sokal 1981). Equation 3.15 is known as Sidak's multiplicative inequality.

TESTS FOR HOMOGENEITY

Use has been made in this chapter of chi-square goodness-of-fit tests. Often data will be available from several samples, and it is generally desirable to combine such data to perform a goodness-of-fit test on all the information available. Before doing so, it is necessary to verify that the samples are homogeneous and can indeed be combined. A test for homogeneity is essentially a test for independence of rows (samples) in a contingency table.

For Mendel's data on seed shape from 10 F_2 plants, Table 1.8, a 2×10 contingency table can be constructed. The 10 rows are for the samples, and the two columns for seed shape. If the 10 samples are homogeneous, each has the same proportion of round seeds, and the common proportion is estimated by the overall proportion of 336/437. Under the hypothesis of homogeneity, the expected count of round seeds for plant 1 is $57 \times 336/437 = 43.83$. These calculations are given in Table 3.8, and the chi-square for homogeneity found by adding the [(observed − expected)^2/expected] values over all 20 cells in the contingency table. This quantity has the value 4.71, and is from a chi-square distribution with 9 d.f. under the homogeneity hypothesis. The hypothesis is not rejected, and information from all 10 plants can be pooled. In the pooled data the ratio of round to wrinkled is observed to be 336:101 and expected to be 327.75:109.25. The single degree-of-freedom chi-square test statistic is

$$\begin{aligned}X^2 &= \frac{(336 - 327.75)^2}{327.75} + \frac{(101 - 109.25)^2}{109.25} \\ &= 0.83\end{aligned}$$

Table 3.8 Testing for homogeneity of Mendel's data on seed shape in 10 F_2 plants from Table 1.8.

	Observed		Expected*			Expected†		
Plant	Round	Wrinkled	Round	Wrinkled	X^2	Round	Wrinkled	X^2
1	45	12	42.75	14.25	0.47	43.83	13.17	0.14
2	27	8	26.25	8.75	0.09	26.92	8.08	0.00
3	24	7	23.25	7.75	0.10	23.84	7.16	0.00
4	19	10	21.75	7.25	1.39	22.30	6.70	2.11
5	32	11	32.25	10.75	0.01	33.06	9.94	0.15
6	26	6	24.00	8.00	0.67	24.60	7.40	0.34
7	88	24	84.00	28.00	0.76	86.11	25.89	0.18
8	22	10	24.00	8.00	0.67	24.60	7.40	1.19
9	28	6	25.50	8.50	0.98	26.14	7.86	0.57
10	25	7	24.00	8.00	0.17	24.60	7.40	0.03
Totals	336	101			5.31	336	101	4.71

* Under the hypothesis of a 3:1 ratio in each sample.

† Under the hypothesis of unspecified equal proportions in each sample.

which indicates that the data support the hypothesis of a 3:1 ratio.

There are also occasions where there are genetic reasons to hypothesize a specified ratio for all samples. This is the case for Mendel's data on seed shape. Under Mendel's theory, each sample should exhibit a 3:1 ratio of round versus wrinkled seed shape. In Table 3.8 the observed counts are compared with those expected under a 3:1 ratio. If each sample is tested separately for the 3:1 hypothesis, none of the chi-square values, with 1 d.f., is significant. Their sum of 5.31, which is a chi-square value with 10 d.f., is also nonsignificant. Note that the various chi-squares are not additive ($5.31 \neq 4.71 + 0.83$), wheras the G-test statistics of Sokal and Rohlf (1981) are additive.

Another approach to testing for homogeneity can be used for the case of testing linkage disequilibrium. It requires that the disequilibrium coefficients be transformed to correlation coefficients by Fisher's z transformation (Fisher 1925). If D_{AB} is the gametic linkage disequilibrium, then the ratio

$$r_{AB} = \frac{\tilde{D}_{AB}}{\sqrt{\tilde{p}_A(1-\tilde{p}_A)\tilde{p}_B(1-\tilde{p}_B)}}$$

behaves like a correlation coefficient. As such, it can be transformed to a normal variable z by

$$z = \frac{1}{2}\ln\left(\frac{1+r}{1-r}\right)$$

Under the hypothesis of no disequilibrium

$$z \sim N\left(0, \frac{1}{n-3}\right)$$

for a sample of size n gametes. In other words, $(n-3)z^2$ has a chi-square distribution with 1 d.f.

When linkage disequilibrium estimates are available from several samples, i, and $\bar{z}$ is the average of the z_i values, the identity

$$\sum_{i=1}^{r}(n_i-3)z_i^2 = \sum_{i=1}^{r}(n_i-3)(z_i-\bar{z})^2 + \bar{z}^2\sum_{i=1}^{r}(n_i-3)$$

can be written as

$$X_m^2 = X_{m-1}^2 + X_1^2$$

The total chi-square, with m d.f. is partitioned into a component with $(m-1)$ d.f. for testing for homogeneity, and a component with 1 d.f. for testing for zero disequilibrium on the pooled data if the samples are homogeneous.

The same approach applies to the composite disequilibrium coefficient Δ_{AB} estimated from genotypic data. Provided there are no higher order disequilibria, the correlation becomes

$$r_{AB} = \frac{\tilde{\Delta}_{AB}}{\sqrt{[\tilde{p}_A(1-\tilde{p}_A)+\tilde{D}_A][\tilde{p}_B(1-\tilde{p}_B)+\tilde{D}_B]}}$$

and the appropriate sample size is now the number of individuals. These statistics were used to test for homogeneity of samples by Laurie-Ahlberg and Weir (1979). They worked with isozymes in *Drosophila melanogaster*, and calculations for loci *EST-C* and *EST-6* are shown in Table 3.9. The seven samples represented in that table are clearly not homogeneous.

Table 3.9 Testing for homogeneity of linkage disequilibrium estimates for data of Laurie-Ahlberg and Weir (1979).

Sample(i)	r_i	z_i	$n_i - 3$	$(n_i - 3)z_i$	$(n_i - 3)z_i^{2*}$	$(n_i - 3)(z_i - \bar{z})^2$
1	0.066	0.066	96	6.336	0.418	0.431
2	−0.053	−0.053	107	−5.671	0.301	0.289
3	0.429	0.459	95	43.605	20.015	20.102
4	−0.167	−0.169	92	−15.548	2.628	2.597
5	0.079	0.079	97	7.663	0.605	0.621
6	−0.116	−0.117	112	−13.104	1.533	1.507
7	−0.192	−0.194	125	−24.250	4.705	4.656
Totals			724†	−0.969†	30.205	30.203

* These are single d.f. chi-squares, and indicate when disequilibria are significantly different from zero.

† Weighted average $\bar{z} = -0.969/724 = -0.001$.

SUMMARY

In studying associations between sets of genes, whether at one or several loci, a convenient procedure is to define measures of association as differences between the joint frequency of the set and the frequency that would result if there was no association. For large samples, the maximum likelihood estimates of these measures of association are normally distributed, and this allows the construction of test statistics based on the expected variance of the estimates under the hypothesis of no association. For smaller samples, exact tests can be constructed that reject the hypothesis when the observed or less likely outcomes have a very low probability.

An alternative to the additive models of association is provided by multiplicative, or log-linear models. These give hypothesis tests with the likelihood ratio procedure, and have a straightforward extension to multiple loci, although the genetic interpretation of the terms in the models is not as clear as in the additive models.

EXERCISES

Exercise 3.1

Suppose that 10 individuals are scored for a locus with two alleles A, a and 6 AA, 3 Aa, 1 aa individuals are found. Test the sample for Hardy-Weinberg equilibrium with both a chi-square and an exact test.

Exercise 3.2

Suppose 20 individuals are scored for two loci, each with two alleles, and the following genotypic counts are found:

	BB	Bb	bb
AA	9	2	1
Aa	3	2	1
aa	1	0	1

Test the sample for linkage disequilibrium, using the gametic measure D_{AB} (assuming HWE) and using the composite measure Δ_{AB} (not assuming HWE).

Exercise 3.3

A locus has three alleles, A_1, A_2, A_3 in frequencies p_1, p_2, p_3. Three disequilibrium coefficients can be defined in terms of the homozygote frequencies:

$$D_{11} = P_{11} - p_1^2,\ D_{22} = P_{22} - p_2^2,\ D_{33} = P_{33} - p_3^2$$

Find expressions for the maximum possible values of these homozygote disequilibria. How large must the sample be for the overall test of HWE to have 90% power and 5% significance when all three alleles are equally frequent and the disequilibria are at half their maximum values?

Chapter 4

Diversity

INTRODUCTION

The study of evolution is essentially that of characterizing the extent and causes of genetic variation. For the present discussion, different ways of measuring variation will be considered. The simplest descriptors are just the frequencies of alleles or genotypes, but emphasis in this chapter will be given to heterozygosity and gene diversity. The frequency of heterozygotes is important, since each heterozygote carries different alleles and represents the existence of variation. There are situations, however, such as in selfing species, where variation results from the continued presence of different homozygotes and gene diversity is then a more appropriate measure. Sampling properties for each of these two measures are covered in this chapter.

HETEROZYGOSITY

A simple measure of genetic variation in a population is the amount of heterozygosity observed, and this is often reported for a single locus or as an average over several loci. Properties of the statistic will be investigated in both situations.

If $n_{\ell uv}$ is the observed count of $A_u A_v$ heterozygotes, $u \neq v$, at locus ℓ in a sample of size n, then the sample heterozygote frequency at this locus is

$$\tilde{H}_\ell = \sum_u \sum_{v \neq u} \frac{n_{\ell uv}}{n}$$

If m loci are scored, the average heterozygosity is

$$\tilde{H} = \frac{1}{m} \sum_{\ell=1}^{m} \tilde{H}_\ell$$

Within-Population Variance of Heterozygosity

As $\tilde{H}_\ell$ is the sum of counts that are multinomially distributed over samples from the same population, it is binomially distributed and

$$\begin{aligned} \mathcal{E}(\tilde{H}_\ell) &= H_\ell \\ \text{Var}(\tilde{H}_\ell) &= \frac{1}{n} H_\ell(1 - H_\ell) \end{aligned} \qquad (4.1)$$

where H_ℓ is the population frequency of heterozygotes at locus ℓ. Note that H_ℓ refers to the frequency of heterozygotes in the one population being sampled.

It is helpful to introduce indicator variables, as was done for gene frequencies in Chapter 2. Define $x_{j\ell}$ for locus ℓ in individual j as

$$x_{j\ell} = \begin{cases} 1 & \text{if individual is heterozygous at locus } \ell \\ 0 & \text{otherwise} \end{cases}$$

without regard to what the (heterozygous) genotype is at that locus. The sample heterozygosity is the average of the indicator variables

$$\tilde{H}_\ell = \frac{1}{n} \sum_{j=1}^{n} x_{j\ell}$$

with properties determined by the expected values of functions of the x's. Taking expectations over all samples for one population

$$\begin{aligned} \mathcal{E}(x_{j\ell}) &= H_\ell \\ \mathcal{E}(x_{j\ell}^2) &= H_\ell \\ \mathcal{E}(x_{j\ell} x_{j'\ell}) &= H_\ell^2 \end{aligned}$$

with the last expression reflecting the independence of individuals within a sample. Using the indicator variables leads to the same mean and variance of $\tilde{H}_\ell$ as given in Equations 4.1.

For average heterozygosity, the observed value remains unbiased

$$\mathcal{E}(\tilde{H}) = H = \frac{1}{m} \sum_{\ell} H_\ell$$

but the variance requires account to be taken of the covariance between heterozygosities at different loci, ℓ and ℓ', caused by them both depending on the frequency of double heterozygotes $\tilde{H}_{\ell\ell'}$ at those loci. This covariance is

$$\mathrm{Cov}(\tilde{H}_\ell, \tilde{H}_{\ell'}) = \frac{1}{n}(H_{\ell\ell'} - H_\ell H_{\ell'})$$

as may be found from the expectation

$$\mathcal{E}(x_{j\ell}x_{j\ell'}) = H_{\ell\ell'}$$

i.e.,

$$\begin{aligned}
\mathrm{Cov}(\tilde{H}_\ell, \tilde{H}_{\ell'}) &= \mathcal{E}\left(\frac{1}{n}\sum_j x_{j\ell}\frac{1}{n}\sum_{j'} x_{j'\ell'}\right) - H_\ell H_{\ell'} \\
&= \mathcal{E}\left[\frac{1}{n^2}\left(\sum_j x_{j\ell}x_{j\ell'} + \sum_j\sum_{j'\neq j} x_{j\ell}x_{j'\ell'}\right)\right] - H_\ell H_{\ell'} \\
&= \frac{1}{n^2}\left[nH_{\ell\ell'} + n(n-1)H_\ell H_{\ell'}\right] - H_\ell H_{\ell'} \\
&= \frac{1}{n}(H_{\ell\ell'} - H_\ell H_{\ell'})
\end{aligned}$$

The variance of average heterozygosity within populations is, therefore,

$$\begin{aligned}
\mathrm{Var}(\tilde{H}) &= \frac{1}{m^2}\left[\sum_\ell \mathrm{Var}(\tilde{H}_\ell) + \sum_\ell\sum_{\ell'\neq\ell} \mathrm{Cov}(\tilde{H}_\ell, \tilde{H}_{\ell'})\right] \\
&= \frac{1}{nm^2}\sum_\ell H_\ell(1 - H_\ell) \\
&\quad + \frac{1}{nm^2}\sum_\ell\sum_{\ell'\neq\ell}(H_{\ell\ell'} - H_\ell H_{\ell'}) \qquad (4.2)
\end{aligned}$$

This expression illustrates the general result that the variance of an average of single-locus statistics must involve two-locus parameters. In itself, this provides a justification for determining the properties of two-locus measures.

It may be thought possible to use the variance among the sample heterozygosities at individual loci to estimate the variance of average heterozygosity, but this ignores the covariance between heterozygosities and it also ignores the differences among mean values of the individual heterozygosities. The sample variance of single-locus heterozygosities is

$$\begin{aligned}
s_H^2 &= \frac{1}{m-1}\sum_\ell(\tilde{H}_\ell - \tilde{H})^2 \\
&= \frac{1}{m}\sum_\ell \tilde{H}_\ell^2 - \frac{1}{m(m-1)}\sum_\ell\sum_{\ell'\neq\ell}\tilde{H}_\ell\tilde{H}_{\ell'}
\end{aligned}$$

which has expectation over all samples from that population of

$$\mathcal{E}(s_H^2) = \frac{1}{m}\sum_\ell \left[H_\ell^2 + \frac{1}{n}H_\ell(1-H_\ell)\right] - \frac{1}{m(m-1)}\sum_\ell\sum_{\ell'\neq\ell}\left[H_\ell H_{\ell'} + \frac{1}{n}(H_{\ell\ell'} - H_\ell H_{\ell'})\right] \quad (4.3)$$

This equation used the results

$$\mathcal{E}(\tilde{H}_\ell^2) = H_\ell^2 + \text{Var}(\tilde{H}_\ell)$$
$$\mathcal{E}(\tilde{H}_\ell\tilde{H}_{\ell'}) = H_\ell H_{\ell'} + \text{Cov}(\tilde{H}_\ell, \tilde{H}_{\ell'})$$

If the sample heterozygosities at each locus had the same distribution, the variance of the average heterozygosity would be $1/m$ times the variance of one of them. In other words s_H^2/m would provide an estimate of $\text{Var}(\tilde{H})$. However, for Equations 4.3, divided by m, and 4.2 to be equal, it is necessary that all loci have the same heterozygosity, $H_\ell = H$ for all ℓ, and that the heterozygosities at different loci are independent, $H_{\ell\ell'} = H_\ell H_{\ell'}$ for all $\ell \neq \ell'$. In that case, the variance of average heterozygosity reduces to a binomial form

$$\text{Var}(\tilde{H}) = \frac{H(1-H)}{nm}$$

Otherwise, writing $m_1 = m(m-1)$,

$$\mathcal{E}(s_H^2) - \text{Var}(\tilde{H}) = \frac{1}{m_1}\left[\sum_\ell (H_\ell - H)^2 - \frac{1}{n}\sum_\ell\sum_{\ell'\neq\ell}(H_{\ell\ell'} - H_\ell H_{\ell'})\right]$$
$$= \frac{1}{m_1}\left[\sum_\ell (H_\ell - H)^2 - \frac{1}{n}\sum_\ell\sum_{\ell'\neq\ell}\text{Cov}(\tilde{H}_\ell, \tilde{H}_{\ell'})\right] \quad (4.4)$$

Total Variance of Heterozygosity

With variation between replicate populations, expectations of the products of indicator variables in different individuals are no longer the products of the expectations. To accommodate the dependence between different sample members, caused by genetic sampling, write

$$\mathcal{E}(x_{j\ell}x_{j'\ell}) = M_\ell,\ j \neq j'$$

where M_ℓ is the probability that two individuals in a population are heterozygous. This quantity is playing a role analogous to that of $P_{A/A}$ for the frequency with which two individuals both carry allele A, when the total variance of the frequency of A was being discussed in Chapter 2. The frequencies H and M refer to values expected over all populations. Using these expectations gives the mean and variance of $\tilde{H}_\ell$:

$$\begin{aligned}\mathcal{E}(\tilde{H}_\ell) &= H_\ell \\ \mathrm{Var}(\tilde{H}_\ell) &= \frac{1}{n^2}\mathcal{E}\left(\sum_j x_{j\ell}^2 + \sum_j \sum_{j\neq j'} x_{j\ell}x_{j'\ell}\right) - [\mathcal{E}(\tilde{H}_\ell^2)] \\ &= (M_\ell - H_\ell^2) + \frac{1}{n}(H_\ell - M_\ell)\end{aligned}$$

There is a between-population component to this variance that remains even when the sample size is very large, unless all individuals in the population are independent and $M_\ell = H_\ell^2$. This will not be the case in finite populations or when there is any other association between individuals.

Averaging over all m loci scored

$$\tilde{H} = \frac{1}{m}\sum_\ell \tilde{H}_\ell = \frac{1}{nm}\sum_j \sum_\ell x_{j\ell}$$

In addition to the two-locus heterozygosity $H_{\ell\ell'}$

$$\mathcal{E}(x_{j\ell}x_{j\ell'}) = H_{\ell\ell'}$$

which is the probability that a random individual is heterozygous at loci ℓ and ℓ', there is a need for a two-locus version of between-individual heterozygosity, M_ℓ

$$\mathcal{E}(x_{j\ell}x_{j'\ell'}) = M_{\ell\ell'}$$

The quantity $M_{\ell\ell'}$ is the probability that two random individuals are heterozygous, one at locus ℓ and one at locus ℓ'. These new frequencies lead to

$$\mathcal{E}(\tilde{H}) = H = \frac{1}{m}\sum_\ell H_\ell$$

$$\begin{aligned}\mathrm{Var}(\tilde{H}) = \frac{1}{m^2n^2}\mathcal{E}\Bigg(&\sum_j\sum_\ell x_{j\ell}^2 + \sum_j\sum_{j'\neq j}\sum_\ell x_{j\ell}x_{j'\ell} \\ &+ \sum_j\sum_\ell\sum_{\ell'\neq\ell} x_{j\ell}x_{j\ell'} + \sum_j\sum_{j'\neq j}\sum_\ell\sum_{\ell'\neq\ell} x_{j\ell}x_{j'\ell'}\Bigg) - H^2\end{aligned}$$

$$= \frac{1}{m^2}\left[\sum_{\ell}(M_{\ell} - H_{\ell}^2) + \sum_{\ell}\sum_{\ell' \neq \ell}(M_{\ell\ell'} - H_{\ell}H_{\ell'})\right]$$
$$+ \frac{1}{m^2 n}\left[\sum_{\ell}(H_{\ell} - M_{\ell}) + \sum_{\ell}\sum_{\ell' \neq \ell}(H_{\ell\ell'} - M_{\ell\ell'})\right] \quad (4.5)$$

This last expression contains four terms that can be rearranged to show just how populations, loci, and individuals contribute to the variance of average heterozygosity. The rearrangement can be found by setting out the calculations slightly differently, in a framework similar to that used for analysis of variance.

Although the distributional properties of the various sums of squares of the indicator variables are not needed, the analysis of variance format simplifies calculations and the method of moments is used to estimate components of variance. An index i is added to the heterozygosity indicator variable to denote the population being sampled, so that $x_{ij\ell}$ is 1 when individual j from population i is heterozygous at locus ℓ and is 0 otherwise. The observations fall in a nested scheme of individuals within populations, but as the same set of loci is scored in all individuals, loci are crossed with populations. The whole scheme has an analysis similar to that of a split-plot design (Steel and Torrie 1980, Chapter 16), with populations playing the role of whole plot treatments and loci that of split plot treatments. The analysis of variance format is shown in Table 4.1 for the case of m loci scored in n individuals taken from each of r populations.

In Table 4.1 sums of squares of indicator variables are constructed to reflect the sampling structure, and corresponding expected mean squares are written down as though the variables could be represented by the linear model

$$x_{ij\ell} = \alpha_i + \beta_{ij} + \gamma_{\ell} + (\alpha\gamma)_{i\ell} + (\beta\gamma)_{ij\ell}$$

This equation has terms α_i for populations, β_{ij} for individuals within populations, γ_{ℓ} for loci, $(\alpha\gamma)_{i\ell}$ for populations by loci, and $(\beta\gamma)_{ij\ell}$ for loci by individuals within populations. The term for loci is regarded as a fixed effect since the same loci are repeatedly scored. All other effects must be regarded as being random if inferences are to be extended to all individuals and to all populations. Loci should also be regarded as random effects if inferences are to be extended to all loci in the genome, and then a mean would be added to the model.

The expected values of all the random terms are 0, and γ_{ℓ} equals H_{ℓ}. The variances of α_i, β_{ij}, $(\alpha\gamma)_{ij\ell}$, and $(\beta\gamma)_{ij\ell}$ are σ_p^2, $\sigma_{i/p}^2$, $\sigma_{p\ell}^2$, and $\sigma_{\ell i/p}^2$,

Table 4.1 Analysis of variance format for variable that indicates heterozygosity.*

Source	d.f.	Sum of Squares	Expected Mean Square
Populations	$r-1$	$SS_1 - C$	$\sigma^2_{\ell i/p} + m\sigma^2_{i/p} + n\sigma^2_{p\ell} + mn\sigma^2_p$
Individuals within populations	$r(n-1)$	$SS_2 - SS_1$	$\sigma^2_{\ell i/p} + m\sigma^2_{i/p}$
Loci	$m-1$	$SS_3 - C$	$\sigma^2_{\ell i/p} + n\sigma^2_{p\ell} + L$
Loci by populations	$(r-1)(m-1)$	$SS_4 - SS_1$ $-SS_3 + C$	$\sigma^2_{\ell i/p} + n\sigma^2_{\ell p}$
Loci by individuals within populations	$r(n-1)(m-1)$	$SS_5 - SS_2$ $-SS_4 + SS_1$	$\sigma^2_{\ell i/p}$

$^*SS_1 = \frac{1}{mn}\sum_i x^2_{i\cdot\cdot}$ $\quad SS_2 = \frac{1}{m}\sum_i\sum_j x^2_{ij\cdot}$ $\quad SS_3 = \frac{1}{rn}\sum_\ell x^2_{\cdot\cdot\ell}$

$SS_4 = \frac{1}{n}\sum_i\sum_\ell x^2_{i\cdot\ell}$ $\quad SS_5 = \sum_i\sum_j\sum_\ell x^2_{ij\ell}$ $\quad C = \frac{1}{mnr}x^2_{\cdot\cdot\cdot}$

$x_{ij\cdot} = \sum_\ell x_{ij\ell}$ $\quad x_{i\cdot\ell} = \sum_j x_{ij\ell}$ $\quad x_{i\cdot\cdot} = \sum_j\sum_\ell x_{ij\ell}$

$x_{\cdot\cdot\ell} = \sum_i\sum_j x_{ij\ell}$ $\quad x_{\cdot\cdot\cdot} = \sum_i\sum_j\sum_\ell x_{ij\ell}$

respectively. Expectations of the terms in the sums of squares, defined in the footnote to Table 4.1, are

$$\begin{aligned}
\mathcal{E}(x^2_{ij\ell}) &= \sigma^2_p + \sigma^2_{i/p} + \gamma^2_\ell + \sigma^2_{p\ell} + \sigma^2_{\ell i/p} \\
\mathcal{E}(x^2_{ij\cdot}) &= m^2\sigma^2_p + m^2\sigma^2_{i/p} + (\sum_\ell H_\ell)^2 + m\sigma^2_{p\ell} + m\sigma^2_{\ell i/p} \\
\mathcal{E}(x^2_{i\cdot\ell}) &= n^2\sigma^2_p + n\sigma^2_{i/p} + n^2H^2_\ell + n^2\sigma^2_{p\ell} + n\sigma^2_{\ell i/p} \\
\mathcal{E}(x^2_{i\cdot\cdot}) &= n^2m^2\sigma^2_p + nm^2\sigma^2_{i/p} + n^2(\sum_\ell H_\ell)^2 + n^2m\sigma^2_{p\ell} + nm\sigma^2_{\ell i/p} \\
\mathcal{E}(x^2_{\cdot\cdot\ell}) &= rn^2\sigma^2_p + rn\sigma^2_{i/p} + r^2n^2H^2_\ell + rn^2\sigma^2_{p\ell} + rn\sigma^2_{\ell i/p}
\end{aligned}$$

$$\mathcal{E}(x_{...}^2) = rn^2m^2\sigma_p^2 + rnm^2\sigma_{i/p}^2 + r^2n^2(\sum_\ell H_\ell)^2 + rn^2m\sigma_{p\ell}^2 + rnm\sigma_{\ell i/p}^2$$

so that

$$\begin{aligned}
\mathcal{E}(SS_1) &= rnm\sigma_p^2 + rm\sigma_{i/p}^2 + rn(\sum_\ell H_\ell)^2/m + rn\sigma_{p\ell}^2 + r\sigma_{\ell i/p}^2 \\
\mathcal{E}(SS_2) &= rnm\sigma_p^2 + rnm\sigma_{i/p}^2 + \frac{rn}{m}(\sum_\ell H_\ell)^2 + rn\sigma_{p\ell}^2 + rn\sigma_{\ell i/p}^2 \\
\mathcal{E}(SS_3) &= nm\sigma_p^2 + m\sigma_{i/p}^2 + rn\sum_\ell H_\ell + nm\sigma_{p\ell}^2 + m\sigma_{i/p}^2 \\
\mathcal{E}(SS_4) &= rnm\sigma_p^2 + rm\sigma_{i/p}^2 + rn\sum_\ell H_\ell + rnm\sigma_{p\ell}^2 + rm\sigma_{\ell i/p}^2 \\
\mathcal{E}(SS_5) &= rnm\sigma_p^2 + rnm\sigma_{i/p}^2 + rn\sum_\ell H_\ell + rnm\sigma_{p\ell}^2 + rnm\sigma_{\ell i/p}^2 \\
\mathcal{E}(C) &= nm\sigma_p^2 + m\sigma_{i/p}^2 + \frac{rn}{m}(\sum_\ell H_\ell)^2 + n\sigma_{p\ell}^2 + \sigma_{\ell i/p}^2
\end{aligned}$$

and these expressions lead to the expected mean squares in Table 4.1. The quantity L

$$L = \frac{1}{m-1}\sum_\ell (H_\ell - H)^2, \; m > 1$$

is analogous to a variance among heterozygosities.

For a single locus, the split-plot component of the analysis disappears and the only variance components are those for populations and individuals within populations. In that case the components are

populations

$$\sigma_p^2 = M_\ell - H_\ell^2$$

individuals within populations

$$\sigma_{i/p}^2 = H_\ell - M_\ell$$

For more than one locus, however, the four components are

populations

$$\sigma_p^2 = \frac{1}{m(m-1)}\sum_\ell \sum_{\ell' \neq \ell}(M_{\ell\ell'} - H_\ell H_{\ell'})$$

individuals within populations

$$\sigma^2_{i/p} = \frac{1}{m(m-1)} \sum_{\ell} \sum_{\ell'} (H_{\ell\ell'} - M_{\ell\ell'})$$

populations by loci

$$\sigma^2_{p\ell} = \frac{1}{m} \sum_{\ell} (M_{\ell} - H^2_{\ell}) - \frac{1}{m(m-1)} \sum_{\ell} \sum_{\ell' \neq \ell} (M_{\ell\ell'} - H_{\ell} H_{\ell'})$$

loci by individuals within populations

$$\sigma^2_{\ell i/p} = \frac{1}{m} \sum_{\ell} (H_{\ell} - M_{\ell}) - \frac{1}{m(m-1)} \sum_{\ell} \sum_{\ell'} (H_{\ell\ell'} - M_{\ell\ell'})$$

Variances of means can be expressed very simply in terms of the four variance components. For the average heterozygosity at a single locus ℓ in one population

$$\begin{aligned} \text{Var}(\tilde{H}_{\ell}) &= \text{Var}\left(\frac{1}{n} \sum_{j} x_{ij\ell}\right) \\ &= \sigma^2_p + \frac{1}{n} \sigma^2_{i/p} \end{aligned} \tag{4.6}$$

while for the average heterozygosity over all loci in one population

$$\begin{aligned} \text{Var}(\tilde{H}) &= \text{Var}\left(\frac{1}{nm} \sum_{j} \sum_{\ell} x_{j\ell}\right) \\ &= \sigma^2_p + \frac{1}{n} \sigma^2_{i/p} + \frac{1}{m} \sigma^2_{p\ell} + \frac{1}{mn} \sigma^2_{\ell i/p} \end{aligned} \tag{4.7}$$

Equation 4.7 allows the planning of surveys of populations in such a way as to achieve a desirable level of variance of sample heterozygosity. If the variance components are known, then adjustments can be made to the number, m, of loci scored and the number, n, of individuals sampled for the variance to be the size required. No juggling of sample sizes, however, can remove the effects of variation between replicate populations. Note that the variances in Equations 4.5 and 4.7 are the same.

A complete discussion of the effects of mating system, allelic frequency distributions, and unbalanced data is given by Weir et al. (1990), but note that for finite random mating populations the variance components σ_p^2 and $\sigma_{i/p}^2$ are quite small and the variance of average heterozygosity is reduced most quickly by increasing the number of loci scored per individual. For infinite random mating populations in which there is a constant amount of selfing, the variance components σ_p^2 and $\sigma_{p\ell}^2$ are zero, and then the best strategy for reducing variance is to increase the number of individuals sampled.

As in the within-population situation, it may be desired to use the variation of heterozygosity among loci to estimate the total variance of average heterozygosity. Once again this procedure ignores the differences between heterozygosities at different loci, and it assumes that single-locus heterozygosities are independent. To see this, suppose that the sample variance of single-locus heterozygosities, as before, is written as

$$s_H^2 = \frac{1}{m-1}\sum_{\ell}(\tilde{H}_\ell - \tilde{H})^2$$

and if each locus had the same expected heterozygosity, the variance of the mean heterozygosity $\tilde{H}$ could be estimated by s_H^2/m. The same arguments as in the within-population case lead to

$$\mathcal{E}(s_H^2) - \mathrm{Var}(\tilde{H}) = \frac{1}{m_1}\left[\sum_{\ell}(H_\ell - H)^2 - \frac{1}{n}\sum_{\ell}\sum_{\ell'\neq\ell}\mathrm{Cov}(\tilde{H}_\ell, \tilde{H}_{\ell'})\right]$$

This expression, in which $m_1 = m(m-1)$, has the same form as Equation 4.4, although the frequencies and covariances now apply to all replicate populations, and not just to samples from a single population. They therefore involve the M frequencies.

Although the variance among loci does not provide an unbiased estimate of the total variance of average heterozygosity, it is all that is available when data are taken from only a single population. Such data do not allow the proper estimation of the between-population component of variance.

For large data sets, involving more than one population, analysis of heterozygosity is most conveniently handled by a statistical computer package. The data are recoded to consist of ones for every heterozygous locus, and zeros for every homozygous locus. The resulting set of ones and zeros is analyzed as being from a split-plot experiment. Commercial packages generally allow estimation of variance components from analyses of such designs.

GENE DIVERSITY

An alternative measure of variation, often referred to loosely as average heterozygosity, but more properly known as GENE DIVERSITY, is formed from the sum of squares of gene frequencies. It is a more appropriate measure of variability for inbred populations where there are very few heterozygotes, but there may be several different homozygous types. For random mating populations, it will be close in value to heterozygosity.

Since gene diversity is calculated from gene frequencies, it can be found from tables that give only these frequencies and do not give genotypic frequencies. Although this may be an advantage in some situations, it is well to remember that the data are generally collected at the genotypic level. Sampling properties of gene diversity will depend on the genotype as well as the gene frequencies.

If $p_{\ell u}$ is the frequency of the uth allele at the ℓth locus, the gene diversity at this locus is

$$D_\ell = 1 - \sum_u p_{\ell u}^2$$

and, as an average over m loci,

$$D = 1 - \frac{1}{m} \sum_\ell \sum_u p_{\ell u}^2$$

Sample gene frequencies in these equations provide MLE's of gene diversity:

$$\hat{D} = 1 - \frac{1}{m} \sum_\ell \sum_u \tilde{p}_{\ell u}^2$$

Within-Population Variance of Gene Diversity

Among samples from within a single population, the expected values of squared gene frequencies have been calculated in Chapter 2, and these allow the expected value of gene diversity at a locus to be found as

$$\begin{aligned} \mathcal{E}(\tilde{D}_\ell) &= 1 - \sum_\ell p_{\ell u}^2 - \frac{1}{2n} \sum_u p_{\ell u}(1 - p_{\ell u})(1 + f) \\ &= \left(1 - \frac{1+f}{2n}\right) D_\ell \end{aligned}$$

There is a small bias factor of $(2n-1)/2n$ for noninbred populations, but a more serious bias otherwise. The presence of the f term indicates that the expected value of D depends on genotype as well as gene frequencies.

The variance requires the sum of variances and covariances of squared gene frequencies of alleles at the same locus. From Fisher's variance formula

(Chapter 2 and Weir 1989a),

$$\begin{aligned}\mathrm{Var}(\tilde{p}_{\ell u}^2) &= \frac{1}{n}2p_{\ell u}^3(1-p_{\ell u})(1+f)\\ \mathrm{Cov}(\tilde{p}_{\ell u}^2, \tilde{p}_{\ell v}^2) &= \frac{1}{n}2p_{\ell u}^2 p_{\ell v}^2(1+f),\ v \neq u\end{aligned}$$

so that

$$\begin{aligned}\mathrm{Var}(\hat{D}_\ell) &= \sum_\ell \mathrm{Var}(\tilde{p}_{\ell u}^2) + \sum_\ell \sum_u \sum_{v\neq u} \mathrm{Cov}(\tilde{p}_{\ell u}^2, \tilde{p}_{\ell v}^2)\\ &= \frac{2(1+f)}{n}\left[\sum_u p_{\ell u}^3 - (\sum_u p_{\ell u}^2)^2\right]\end{aligned}$$

For multiple loci, there is no problem in calculating allelic frequencies, and the genotypic frequencies at each separate locus are multinomially distributed. There is a covariance between sample allele frequencies at different loci, caused by the frequencies being estimated from the same individuals, and hence there is also a covariance between squared sample frequencies. Multinomial moments show that the covariance between frequencies $\tilde{p}_{\ell u}$ and $\tilde{p}_{\ell' v}$ of alleles u and v at loci ℓ and ℓ', respectively, is given by the composite coefficient of linkage disequilibrium for these two alleles

$$\mathrm{Cov}(\tilde{p}_{\ell u}, \tilde{p}_{\ell' v}) = \frac{1}{n}\Delta_{\ell u, \ell' v}$$

and Fisher's formula gives the covariance between squared gene frequencies at different loci to be approximately

$$\mathrm{Cov}(\tilde{p}_{\ell u}^2, \tilde{p}_{\ell' v}^2) = \frac{2}{n}p_{\ell u}p_{\ell' v}\Delta_{\ell u, \ell' v}$$

This last expression provides the covariance of diversities at different loci:

$$\mathrm{Cov}(\tilde{D}_\ell, \tilde{D}_{\ell'}) = \frac{2}{n}\sum_u \sum_v p_{\ell u}p_{\ell' v}\Delta_{\ell u, \ell' v}$$

The variance of average gene diversity over samples from the same population then follows as

$$\begin{aligned}\mathrm{Var}(\tilde{D}) &= \sum_\ell \sum_u \mathrm{Var}(\tilde{p}_{\ell u}^2) + \sum_\ell \sum_u \sum_{v\neq u} \mathrm{Cov}(\tilde{p}_{\ell u}^2, \tilde{p}_{\ell v}^2)\\ &\quad + \sum_\ell \sum_{\ell \neq \ell'} \sum_u \sum_v \mathrm{Cov}(\tilde{p}_{\ell u}^2, \tilde{p}_{\ell' v}^2)\\ &= \frac{2}{m^2 n}\sum_\ell (1+f_\ell)\left[\sum_u p_{\ell u}^3 - (\sum_u p_{\ell u}^2)^2\right]\\ &\quad + \frac{2}{m^2 n}\sum_\ell \sum_{\ell' \neq \ell} \sum_u \sum_v p_{\ell u}p_{\ell' v}\Delta_{\ell u, \ell' v} \qquad (4.8)\end{aligned}$$

This variance depends on the association f_ℓ between genes at locus ℓ as well the associations $\Delta_{\ell u,\ell' v}$ between alleles at different loci. Simply taking the variance of single-locus diversities would ignore the between-locus associations, as well as ignoring the differences in expected diversity between loci.

For loci that all have two alleles, Equation 4.8 simplifies. If p_ℓ is the frequency of one of the two alleles at locus ℓ, and $\Delta_{\ell\ell'}$ is the composite linkage disequilibrium between these identified alleles at loci ℓ and ℓ'

$$\begin{aligned}
\text{Var}(\tilde{D}) &= \frac{2}{m^2 n}\sum_{\ell}\left[(1+f_\ell)p_\ell(1-p_\ell)(1-2p_\ell)^2\right] \\
&\quad + \frac{2}{m^2 n}\sum_{\ell}\sum_{\ell'\neq\ell}(1-2p_\ell)(1-2p_{\ell'})\Delta_{\ell\ell'} \\
&= \frac{1}{m^2 n}\sum_{\ell}(D_\ell - \frac{1}{2}H_\ell)(1-2p_\ell)^2 \\
&\quad + \frac{2}{m^2 n}\sum_{\ell}\sum_{\ell'\neq\ell}(1-2p_\ell)(1-2p_{\ell'})\Delta_{\ell\ell'}
\end{aligned}$$

Total Variance of Gene Diversity

When expectations are taken over replicate populations as well as replicate samples from the same population, the argument that led to Equation 2.13 shows that

$$\mathcal{E}(\tilde{p}_{\ell u}^2) = p_{\ell u}^2 + p_{\ell u}(1-p_{\ell u})\theta + \frac{1}{2n}p_{\ell u}(1-p_{\ell u})(1+F-2\theta)$$

and the total expectation of gene diversity is

$$\mathcal{E}(\tilde{D}_\ell) = D_\ell\left[(1-\theta) - \frac{1}{2n}(1+F-2\theta)\right]$$

Cockerham (1967) defined a group coancestry coefficient θ_L that refers to the identity of a random pair of genes among the $2n$ genes of n individuals. This coefficient allows the expected gene diversity to be written simply as

$$\mathcal{E}(\tilde{D}_\ell) = (1-\theta_L)D_\ell$$

The total variance requires the expectations of squares and products of squared gene frequencies. In Chapter 2, it was necessary to introduce the joint frequency $P_{A|A}$ for the total expectation of $\tilde{p}_A^2$. Here, joint frequencies for three or four genes are needed. Dropping the locus subscript ℓ, the expectations for alleles u and v are (Weir 1989a)

$$\begin{aligned}
\mathcal{E}(\tilde{p}_u^4) &= P_{u|u|u|u} + \frac{3}{n}(P_{u|u|u} + P_{uu|u|u} - 2P_{u|u|u|u}) \\
\mathcal{E}(\tilde{p}_u^2\tilde{p}_v^2) &= P_{u|u|v|v} + \frac{1}{2n}\Big(P_{u|v|v} + P_{u|u|v} + 4P_{uv|u|v} + \\
&\quad + P_{uu|v|v} + P_{vv|u|u} - 12P_{u|u|v|v}\Big), \quad v \neq u
\end{aligned}$$

A quantity such as $P_{uu|v|v}$ is the frequency with which one individual carries two copies of allele u and two other individuals each has a copy of allele v. The total variance of sample gene diversity at one locus then becomes

$$\begin{aligned}
\mathrm{Var}(\tilde{D}_\ell) &= \left[\sum_u\sum_v P_{u|u|v|v} - \left(\sum_u P_{u|u}\right)^2\right] \\
&\quad + \frac{1}{n}\left[2\sum_u P_{u|u|u} - \sum_u P_{uu}\sum_u P_{u|u} + 2\left(\sum_u P_{u|u}\right)^2\right. \\
&\quad \left. + \sum_u\sum_v(2P_{uv|u|v} + P_{uu|v|v} - 6P_{u|u|v|v})\right]
\end{aligned}$$

For diversity averaged over loci, the total expectations necessary are quite complicated. The notation for joint frequencies of genes at different loci separates genes in different individuals with a vertical rule and genes on different gametes within individuals by a slash. With this convention, the expectations needed are

$$\begin{aligned}
\mathcal{E}(\tilde{p}_{\ell u}\tilde{p}_{\ell' v}) &= P_{\ell u|\ell' v} + \frac{1}{2n}(P_{\ell u,\ell v} + P_{\ell u/\ell' v} - 2P_{\ell u|\ell' v}) \\
\mathcal{E}(\tilde{p}_{\ell u}^2\tilde{p}_{\ell' v}^2) &= P_{\ell u|\ell u|\ell' v|\ell' v} + \frac{1}{2n}\Big[P_{\ell u|\ell u|\ell' v} + P_{\ell u|\ell' v|\ell' v} \\
&\quad + 4(P_{\ell u,\ell' v|\ell u|\ell' v} + P_{\ell u/\ell' v|\ell u|\ell' v}) \\
&\quad - 12P_{\ell u|\ell u|\ell' v|\ell' v}\Big]
\end{aligned}$$

The total variance for gene diversity is a very cumbersome expression.

Estimating the Variance of Diversity

For data from a single population, the variance of gene diversity is estimated by substituting observed frequencies into Equation 4.8. A more convenient computing formula for the within-population variance is

$$\begin{aligned}\mathrm{Var}(\tilde{D}) &= \frac{2}{m^2n}\left\{\sum_\ell\left[\sum_u(p_{\ell u}^3 + p_{\ell u}^2 P_{\ell u,\ell u} - 2p_{\ell u}^4)\right.\right. \\ &\quad \left. + \frac{1}{2}\sum_u\sum_{v\neq u} p_{\ell u}p_{\ell v}(P_{\ell u,\ell v} - 4p_{\ell u}p_{\ell v})\right] \\ &\quad \left. + \sum_\ell\sum_{\ell'}\sum_u\sum_v p_{\ell u}p_{\ell' v}\Delta_{\ell u,\ell' v}\right\}\end{aligned}$$

For two alleles this becomes

$$\begin{aligned}\mathrm{Var}(\tilde{D}) &= \frac{2}{m^2n}\left\{\sum_\ell\left[p_{\ell 1}^3 + p_{\ell 2}^3 - 2(p_{\ell 1}^2 + p_{\ell 2}^2)^2\right.\right. \\ &\quad \left. + p_{\ell 1}^2 P_{\ell 1,\ell 1} + p_{\ell 2}^2 P_{\ell 2,\ell 2} + p_{\ell 1}p_{\ell 2}P_{\ell 1,\ell 2}\right] \\ &\quad \left. + \sum_\ell\sum_{\ell'}(1 - 2p_{\ell 1})(1 - 2p_{\ell' 1})\Delta_{\ell 1,\ell' 1}\right\} \qquad (4.9)\end{aligned}$$

For populations that have Hardy-Weinberg proportions at each locus, the genotypic frequencies can be replaced by products of allelic frequencies and then Equation 4.9 becomes

$$\begin{aligned}\mathrm{Var}(\tilde{D}) &= \frac{2}{m^2n}\left\{\sum_\ell\left[p_{\ell 1}p_{\ell 2}(p_{\ell 1} - p_{\ell 2})^2\right]\right. \\ &\quad \left. + \sum_\ell\sum_{\ell'}(1 - 2p_{\ell 1})(1 - 2p_{\ell' 1})\Delta_{\ell 1,\ell' 1}\right\}\end{aligned}$$

Estimation of the total variance requires data from more than one population. It is often suggested that different loci can play the role of replicate populations, however, so that the variance among diversities at different loci could serve as an estimate of the total variance. The same strategy was discussed for heterozygosity in the previous section, and it has merit in some situations.

With this approach, the total variance of $\tilde{D}$ is taken to be $1/m$ times the variance among the $\tilde{D}_\ell$'s. If this estimate is written as s_D^2, then

$$s_D^2 = \frac{1}{m(m-1)}\sum_\ell(\tilde{D}_\ell - \tilde{D})^2$$

but taking total expectations shows that, writing $m_1 = m(m-1)$,

$$\mathcal{E}(s_D^2) = \mathrm{Var}(\tilde{D}) + \frac{1}{m_1}\left[\sum_\ell(D_\ell - D)^2 + \sum_\ell\sum_{\ell'\neq\ell}\mathrm{Cov}(\tilde{D}_\ell, \tilde{D}_{\ell'})\right]$$

Table 4.2 Two-locus genotypic counts for three esterase loci in barley data of Weir et al. (1972).

	BB	*Bb*	*bb*	Total
AA	520	3	20	543
Aa	72	6	1	79
aa	502	38	72	612
Total	1094	47	93	1234

	CC	*Cc*	*cc*	Total
AA	375	25	143	543
Aa	44	27	8	79
aa	391	61	160	612
Total	810	113	311	1234

	CC	*Cc*	*cc*	Total
BB	770	83	241	1094
Bb	4	18	25	47
bb	36	12	45	93
Total	810	113	311	1234

Just as for heterozygosity, s_D^2 can serve as an estimator of the total variance of average diversity only when each locus has the same expected diversity and when the diversities at different loci have zero covariance. These conditions are met for random-mating populations and independent loci at equilibrium for drift and mutation (Weir 1989a). A good indication of when the approach is not valid would be provided by evidence of linkage disequilibrium between loci. In the case of a plant population with mixed selfing and random outcrossing, the covariance between diversities at two loci is directly proportional to the (composite) linkage disequilibrium between the loci, and s_D^2 should not be used to estimate the total variance of average diversity if the disequilibria are found to be significantly different from zero.

Example

Gene diversity is a better indication of variability than heterozygosity for selfing species, as will now be demonstrated for some data on esterase loci **A**, **B**, and **C** (*Est*1, *Est*2, and *Est*4) in generation 4 of Barley Composite Cross V (Weir et al. 1972). Table 4.2 shows two-locus counts for each of the three pairs of loci, obtained by collapsing the alleles at each locus into the most common one versus the rest.

From Table 4.2, the heterozygosities at one and two loci are:

ℓ	$\tilde{H}_\ell$	ℓ, ℓ'	$\tilde{H}_{\ell,\ell'}$
A	0.0640	**A, B**	0.0049
B	0.0381	**A, C**	0.0219
C	0.0916	**B, C**	0.0146

and the average over loci is $\tilde{H} = 0.0646$. The estimate of the variance of this average, over replicate samples from the same population, is found from Equation 4.2.

$$\begin{aligned}
\text{Var}(\tilde{H}) &= \frac{1}{m^2}\left[\sum_\ell \text{Var}(\tilde{H}_\ell) + \sum_\ell \sum_{\ell' \neq \ell} \text{Cov}(\tilde{H}_\ell, \tilde{H}_{\ell'})\right] \\
&\hat{=} \frac{1}{9}\left[\frac{0.0640 \times 0.9360}{1234} + \frac{0.0381 \times 0.9619}{1234} + \frac{0.0916 \times 0.9084}{1234}\right. \\
&\quad + \frac{0.0049 - 0.0640 \times 0.0381}{1234} + \frac{0.0219 - 0.0640 \times 0.0916}{1234} \\
&\quad \left. + \frac{0.0146 - 0.0381 \times 0.0916}{1234}\right] \\
&= \frac{0.2093}{9 \times 1234} \\
&= 0.000018846
\end{aligned}$$

The estimated standard deviation of the average heterozygosity is, therefore, 0.0043. The standard deviation found as the square root of the variance among the three single-locus heterozygosities is 0.0154. For the within-population variance, the difference among single-locus heterozygosities and the covariances between them make s_H^2 an inappropriate estimator.

The total variance, applicable to inferences for more than one population, cannot be estimated from data on one population. If counts from several populations were available, the analysis of variance framework of Table 4.1 would allow estimation of the components of variance, and hence of the average heterozygosity. The variance among single-locus heterozygosities is also inappropriate for the total variance.

For gene diversity, the single-locus estimates and the pairwise composite linkage disequilibria are

ℓ	$\tilde{D}_\ell$	ℓ, ℓ'	$\hat{\Delta}_{\ell,\ell'}$
A	0.4984	**A, B**	0.0478
B	0.1710	**A, C**	0.0117
C	0.4182	**B, C**	0.0540

The diversities are substantially larger than the heterozygosities, reflecting the fact that this population has a great deal of polymorphism, but that it is almost entirely homozygous. The significant linkage disequilibria between loci **A, B** and **B, C** precludes the between-locus variance of diversity being of any help in estimating the variance of average diversity, within or between populations.

The average gene diversity is 0.3625, with a within-population variance estimated by using the following frequencies

ℓ	u	$\tilde{p}_{\ell u}$	$\tilde{P}_{\ell u,\ell u}$	u, v	$\tilde{P}_{\ell u,\ell v}$
A	1	0.4720	0.4400	1,2	0.0640
	2	0.5280	0.4959		
B	1	0.9056	0.8784	1,3	0.0381
	2	0.0944	0.0754		
C	1	0.7022	0.6564	2,3	0.0916
	2	0.2978	0.2520		

The variance is estimated from Equation 4.9 as

$$\mathrm{Var}(\tilde{D}) \doteq \frac{2}{9 \times 1234}\Big[(0.4720)^3 + (0.5280)^3 - 2(0.5016)^2$$

$$+ (0.4720)^2(0.4400) + (0.5280)^2(0.4959) + (0.4720)(0.5280)0.0640$$
$$+ (0.9056)^3 + (0.0944)^3 - 2(0.8290)^2$$
$$+ (0.9056)^2(0.8784) + (0.0944)^2(0.0754) + (0.9056)(0.0944)0.0381$$
$$+ (0.7022)^3 + (0.2978)^3 - 2(0.5818)^2$$
$$+ (0.7022)^2(0.6564) + (0.2978)^2(0.2520) + (0.7022)(0.2978)0.0916$$
$$+ (0.0560)(-0.8122)(0.0478)$$
$$+ (0.0560)(-0.4044)(0.0117)$$
$$+(-0.8112)(-0.4044)(0.0540)]$$
$$= 0.000030767$$

and the standard deviation estimated as 0.0055. There is no estimate of the total variance.

SUMMARY

Variation in populations can be characterized with heterozygosity or gene diversity, the latter being more appropriate for inbred populations. The variance of these measures when averaged over loci needs to take account of different levels of variation at different loci, and the associations of variation levels at different loci.

EXERCISES

Exercise 4.1

Perform an "analysis of variance" of heterozygosity for the following data:

	Individual	Locus 1	Locus 2	Locus 3
Population 1	1	*BB*	*AA*	*AA*
	2	*BD*	*AC*	*AA*
	3	*BB*	*AB*	*AA*
	4	*DD*	*AA*	*AA*
	5	*BB*	*AB*	*AA*
Population 2	1	*BD*	*BB*	*AA*
	2	*BB*	*AD*	*AA*
	3	*BB*	*AB*	*AA*
	4	*DE*	*AA*	*AA*
	5	*BE*	*AC*	*AA*

Set out the calculations as in Table 4.1 and replace any negative estimates by zero. Estimate the variance of average heterozygosity within each population, and the total variance of average heterozygosity. Compare these values to those obtained by taking the variances among the heterozygosities at each of the three loci.

Exercise 4.2

Using the same data as in Exercise 4.1, estimate the gene diversities at each locus and in each population. Estimate the total variance of average gene diversity, and compare that to the variance among gene diversities at individual loci.

Chapter 5

Population Structure

INTRODUCTION

This chapter considers ways of characterizing genetic variation within and between populations. Analyses are established for the case in which data are available in samples from different populations, or different subdivisions of the same population. The discussion of variation in such data, quantified with F-statistics, leads naturally into a treatment of genetic distance.

As discussed in Chapter 1, unless a population is invariant for the loci being studied, different samples from the population will show different levels of genetic variation. This is simply a consequence of the "statistical sampling" that results in each sample having a different set of individuals. Analyses need to be set up that allow statements to be made about the population, based on the sample at hand, with this sampling variation accommodated.

In population genetics there is another level of sampling to be considered. Each generation of a population is formed by the union of gametes chosen from among those produced by the previous generation. There is a "genetic" sampling process here that would cause the population to look different if the formation of a new generation was replicated. Population genetic theory depends on the concept of replicate populations that are maintained under the same conditions, but that will differ because of genetic sampling. It is possible to derive variances for statistics of interest that include both types of variation.

One use for such total variances is in predicting future values. From a specified population, it is possible to predict the expected value of a statistic, such as gene frequency or heterozygosity or linkage disequilibrium, in a future population, but not to specify the exact value of the statistic. For a neutral gene in a finite population, for example, it is known that the allelic frequencies are *expected* to remain constant, although in any particular population the frequency may have drifted to any value between zero and one. Statements about the statistic in some future sample must therefore take account of the variation between replicate populations, as well as between replicate samples from any one population.

Difficulty arises in that the magnitude of between-population variation cannot be estimated with a sample from a single population. One way around this problem is sometimes afforded by the availability of several loci in the data set. To the extent that different loci may be regarded as being independent, they may also be regarded as playing the role of separate populations. The loci have each been exposed to the genetic sampling forces between generations, and each locus can have a different pedigree.

The distinction between statistical and genetic sampling can also be phrased in terms of fixed and random effects, to show how the intended scope of inference affects the sampling properties of genetic statistics. If there is interest only in the one particular population sampled, then the genetic sampling between populations is not of consequence. It is necessary to take account of the statistical sampling for repeated samples only from this one "fixed" population. Future samples would be taken from the population as it is presently constituted. Comparisons between different fixed populations can be phrased in terms of means, and it will be shown that procedures of numerical resampling are of use in estimating variances for a single population.

A different situation arises when the sample is to be used to make inferences about the species as a whole. There is then less interest in the particular population sampled, which can now be regarded as being "random." Future samples may very well be drawn from a different population, so that both statistical and genetic sampling variation needs to be considered. The distinction between fixed and random effects arises in statistics. In the analysis of variance context it is easier to detect differences between means in a fixed effects situation because a smaller mean square is used in the denominator for the F test statistic. It is only one specific set of means that is being compared, and not some population of means for which the means at hand are just a sample.

FIXED POPULATIONS

Gene Frequencies

With data collected for genotypes, the first set of descriptors of a population is simply the genotypic frequencies. Under the fixed-population approach, different populations for the same species can be compared simply by comparing mean frequencies. When HWE (Hardy-Weinberg Equilibrium — see Chapter 3) cannot be assumed, this requires the comparison of genotypic frequencies. The most straightforward procedure is to use contingency table chi-square tests. With v alleles at a locus, the genotypic counts in each of r samples are arranged in an $v(v+1)/2 \times r$ contingency table and a chi-square statistic with $[v(v+1)/2 - 1] \times (r - 1)$ degrees of freedom is calculated. In practice, the method has problems when some cells have small expected counts, since this can give test statistics that are spuriously large, and it may be necessary to collapse the least frequent classes.

When HWE is assumed, it is sufficient to compare the allelic arrays in each of the populations. The contingency table is then only $v \times r$, and problems of small expected counts are less likely.

An alternative to the contingency table approach is provided by NUMERICAL RESAMPLING (Efron 1982). This is a means of making inferences about the population gene frequencies from the sample frequencies. Variances can be estimated and confidence intervals can be constructed, both referring to repeated sampling from the same populations. Briefly, numerical resampling mimics the drawing of new samples by resampling the one sample at hand from each population. Two methods are commonly used: JACKKNIFING and BOOTSTRAPPING.

The Jackknife

The simplest of the numerical resampling techniques is the jackknife, first discussed by Quenouille (1956). For a parameter ϕ and a set of observations $X_1, X_2, \ldots, X_n$ there is some procedure that leads to an estimate $\hat{\phi}$. The estimate may result from an explicit formula, such as those given below for the F-statistic estimates, or it may result from an iterative technique as shown in likelihood estimation in Chapter 2. The jackknife procedure requires that n new estimates $\phi_{(i)}$ be formed – the ith new estimate after removal of the ith observation, $i = 1, 2, \ldots, n$. Each of the new estimates is therefore based on $(n - 1)$ observations instead of n, and the average of the new estimates is

$$\hat{\phi}_{(\cdot)} = \frac{1}{n} \sum_i \hat{\phi}_{(i)}$$

Two uses are made of this set of new estimates. First, they provide a new estimator $\hat{\phi}_{\mathrm{J}}$ that should have less bias than did the original $\hat{\phi}$

$$\hat{\phi}_{\mathrm{J}} = n\hat{\phi} - (n - 1)\hat{\phi}_{(\cdot)}$$

and, second, they provide an estimate $\text{Var}(\hat{\phi})_\text{J}$ of the variance of $\hat{\phi}$

$$\text{Var}(\hat{\phi})_\text{J} = \frac{n-1}{n}\sum_i \left[\hat{\phi}_{(i)} - \hat{\phi}_{(\cdot)}\right]^2$$

The justification for the bias correction rests on assuming that the expected value of the original estimator is

$$\mathcal{E}(\hat{\phi}) = \phi + \frac{1}{n}a_1 + \frac{1}{n^2}a_2 + \cdots$$

so that the bias is a series of terms in increasing powers of $(1/n)$. The statistic $\hat{\phi}_\text{J}$ will also be biased, but it does not have the first-order term in the bias. Jackknifing removes the first-order bias. To see this, notice that

$$\mathcal{E}(\hat{\phi}_{(i)}) = \phi + \frac{1}{n-1}a_1 + \frac{1}{(n-1)^2}a_2 + \cdots$$

and that, therefore

$$\mathcal{E}(\hat{\phi}_{(\cdot)}) = \phi + \frac{1}{n-1}a_1 + \frac{1}{(n-1)^2}a_2 + \cdots$$

Substituting these terms into the definition of the jackknife estimator shows that the a_1 term cancels out

$$\mathcal{E}(\hat{\phi}_\text{J}) = \phi - \frac{1}{n(n-1)}a_2 + \cdots$$

Tukey (1958) motivated the variance estimate by introducing a set of n "pseudovariables"

$$\hat{\phi} + (n-1)(\hat{\phi} - \hat{\phi}_{(i)})$$

It is these quantities that have mean $\hat{\phi}_\text{J}$ and variance $\text{Var}(\hat{\phi})_\text{J}$.

As a simple example, consider the estimation of the mean μ of a variable **X** when a sample of size n is available. The initial estimator is the sample mean

$$\hat{\mu} = \frac{1}{n}\sum_i X_i = \bar{X}$$

and omitting the ith observation gives

$$\hat{\mu}_{(i)} = \frac{1}{n-1}\sum_{j\neq i} X_j$$

with a mean of

$$\hat{\mu}_{(\cdot)} = \frac{1}{n}\sum_i \left(\frac{1}{n-1}\sum_{j\neq i} X_j\right) = \bar{X}$$

Evidently the jackknife estimator $\hat{\mu}_J$ is just the sample mean $\bar{X}$ again. Using the jackknife procedure to estimate the sampling variance of the estimate

$$\begin{aligned}
\text{Var}(\hat{\mu})_J &= \frac{n-1}{n}\sum_i \left(\hat{\mu}_{(i)} - \hat{\mu}_{(\cdot)}\right)^2 \\
&= \frac{n-1}{n}\sum_i \left[\left(\frac{1}{n-1}\sum_{j\neq i} X_j\right) - \bar{X}\right]^2 \\
&= \frac{1}{n(n-1)}\sum_i (X_i - \bar{X})^2 \\
&= \frac{1}{n}s^2
\end{aligned}$$

where s^2 is the sample variance of the n observations. The jackknife gives the usual variance estimate for the sample mean — the sample variance divided by the sample size. This quantity has an expected value of σ^2/n where σ^2 is the variance of $\mathbf{X}$.

As a genetic example, consider the estimation of the gametic linkage disequilibrium coefficient D_{AB} from a sample of n gametes. In Chapter 3, the MLE was found to be

$$\hat{D}_{AB} = \tilde{p}_{AB} - \tilde{p}_A\tilde{p}_B$$

This can be written in terms of indicator variables x_j and y_j that take the value 1 if the jth gamete carries alleles A or B, respectively, and are 0 otherwise:

$$\hat{D}_{AB} = \frac{1}{n}\sum_{j=1}^{n} x_j y_j - \left(\frac{1}{n}\sum_{j=1}^{n} x_j\right)\left(\frac{1}{n}\sum_{j=1}^{n} y_j\right)$$

Omitting gamete j from the sample gives a new estimate

$$\hat{D}_{AB(j)} = \frac{1}{n-1}\sum_{k\neq j} x_k y_k - \left(\frac{1}{n-1}\sum_{k\neq j} x_k\right)\left(\frac{1}{n-1}\sum_{k\neq j} y_k\right)$$

and the average of these n new estimates is

$$\hat{D}_{AB(\cdot)} = \tilde{p}_{AB} - \frac{1}{n^2}\left[n(n-1)\tilde{p}_A\tilde{p}_B + \tilde{p}_{AB}\right]$$

so that the jackknife estimate becomes

$$\hat{D}_{AB_J} = \frac{n}{n-1}\hat{D}_{AB}$$

From Chapter 3 it can be shown that this jackknife estimate removes the bias in the likelihood estimate $\hat{D}_{AB}$. The analytical expression for $\text{Var}(\hat{D}_{AB})_J$ is quite cumbersome but the numerical value of this quantity can be used as an estimate for the sampling variance of $\hat{D}_{AB}$.

For F-statistics, jackknife estimation of variances has been found by simulation studies (Reynolds et al. 1983, Dodds 1986) to perform well, but careful attention must be paid to the unit of resampling. If inferences are to be drawn for the particular set of populations sampled, then the variation among samples from just those populations is required. The appropriate resampling unit is the individual, and jackknifing requires the formation of $\sum_i n_i$ new estimates when the data set contains n_i individuals from the ith population. Each of the new estimates is found by omitting one individual from the data set. Note that this omits two genes at each locus. When inferences are to be extended to all sets of populations that may have arisen from the same founder population, account must be taken of the genetic sampling processes. This is accomplished, to a good approximation, by resampling loci since events at a set of independent loci are affected by genetic forces. Provided data are available for several loci, and all these loci are being used to estimate the F-statistics, new estimates are formed by ignoring each locus in turn as will be discussed below in the section on random populations.

An alternative situation arises when several loci are scored. The F-statistics estimated for each locus may be compared in the hopes of detecting the effects of forces such as selection that may be affecting loci differentially. It is then desirable to assign a sampling variance to the estimates for each locus. Resampling individuals will once again provide within-population variances, while resampling populations may provide a satisfactory estimate of the total variance at each locus.

The Bootstrap

The jackknife provides estimates of bias and variance for genetic parameter estimates, but little information for the distribution of the estimates. The bootstrap, on the other hand, is not limited by the number of resampling units, such as individuals or loci, and can provide as many new estimates as needed to give a good approximation to the distribution of the original estimator. Instead of forming new samples by omitting one observation at a time, bootstrapping operates by drawing random samples of the same size as the original sample from that sample. This MONTE CARLO sampling process requires the use of random numbers and does not allow explicit expressions for the new estimates.

If a sample consists of n observations $X_i, i = 1, 2, \ldots, n$, a BOOTSTRAP sample is a set of n numbers drawn at random, with replacement, from this set in such a way that every one of the original observations has an equal chance of being chosen at any stage. Some of the original sample elements will not, therefore, appear in any particular bootstrap sample while some may appear many times. The generation of "random" numbers to allow such choices is discussed in the Appendix. Many of these bootstrap samples are drawn, and the parameter of interest estimated for each one. Many estimates become available for determining the sampling properties of the estimator.

To show some of the theoretical consequences of this procedure, the population mean μ is estimated again from a sample of size n. The original estimator is the sample mean, and the mean is also calculated for each bootstrap sample. Properties of the original estimator are determined by calculating the mean and variance of this collection of bootstrap means. Suppose that a bootstrap sample has original observation X_i represented r_i times, which means that $\sum_i r_i = n$. Because each observation is equally likely to be chosen at each draw, r_i is a binomial variable and the set of r_i values is multinomially distributed

$$r_i \sim B(n, 1/n)$$

with moments

$$\begin{aligned}
\mathcal{E}(r_i) &= 1 \\
\text{Var}(r_i) &= \frac{n-1}{n} \\
\text{Cov}(r_i, r_{i'}) &= -\frac{1}{n}, \ i' \neq i
\end{aligned}$$

The sample mean of the bootstrap sample is

$$\bar{X}_{\text{B}} = \frac{1}{n}\sum_i r_i X_i$$

and a double set of expectations is needed to determine its properties. First regard the X_i's as fixed quantities and the r_i's as variables and take expectations $\mathcal{E}_{\text{B}}$ over all possible r_i values

$$\mathcal{E}_{\text{B}}(\bar{X}_{\text{B}}) = \frac{1}{n}\sum_i X_i = \bar{X}$$

which is just the original sample mean again. Still taking into account only the variation of variables r_i, the variance of a bootstrap mean is

$$\begin{aligned}
\mathrm{Var}_\mathrm{B}(\bar{X}_\mathrm{B}) &= \frac{1}{n^2}\left[\sum_i \frac{n-1}{n}X_i^2 + \sum_i \sum_{i' \neq i}\left(-\frac{1}{n}X_i X_{i'}\right)\right] \\
&= \frac{n-1}{n^2}s^2
\end{aligned}$$

To complete the argument, expectations are taken over all sets of X_i values to find total mean and variance:

$$\begin{aligned}
\mathcal{E}(\bar{X}_\mathrm{B}) &= \mathcal{E}(\bar{X}) \\
&= \mu \\
\mathrm{Var}(\bar{X}_\mathrm{B}) &= \mathcal{E}\left(\frac{n-1}{n^2}s^2\right) \\
&= \frac{n-1}{n^2}\sigma^2
\end{aligned}$$

Whereas jackknifing gave an estimate of the mean with correct mean and variance, bootstrapping gives an estimate with correct mean but slightly biased variance. There is also some judgment called for in applying the bootstrap technique. It is necessary to take a sufficient number of samples so that the mean and variance over samples give good approximations to the theoretical values, but not such a large number that computing time becomes prohibitive. For variances, about 100 bootstrap samples are generally regarded as being sufficient, but for the distribution of the estimator it may be better to use 1000 samples.

When a sample distribution is obtained, confidence intervals for the estimator can be constructed by ordering the estimates. A 95% confidence interval for the parameter being estimated, for example, can be constructed as the interval between the 26th and the 975th of 1000 ordered bootstrap estimates.

For the gametic linkage disequilibrium coefficient the estimate from a bootstrap sample is

$$\tilde{D}_{AB_\mathrm{B}} = \frac{1}{n}\sum_i r_i X_i Y_i - \frac{1}{n^2}\sum_i r_i X_i \sum_i r_i Y_i$$

and the following expectations are needed

$$\begin{aligned}
\mathcal{E}(r_i) &= 1 \\
\mathcal{E}(r_i^2) &= \frac{2n-1}{n} \\
\mathcal{E}(r_i r_{i'}) &= \frac{n-1}{n}, \; i' \neq i
\end{aligned}$$

so that the bootstrap expectation of the disequilibrium is

$$\begin{aligned}\mathcal{E}_{\mathrm{B}}(\tilde{D}_{AB_{\mathrm{B}}}) &= \frac{1}{n}\sum_i \mathcal{E}(r_i)X_iY_i \\ &\quad - \frac{1}{n^2}\left[\sum_i \mathcal{E}(r_i^2)X_iY_i + \sum_i\sum_{i'\neq i}\mathcal{E}(r_ir_{i'})X_iY_{i'}\right] \\ &= \frac{n-1}{n}\tilde{D}_{AB}\end{aligned}$$

Taking the total expectation shows that the bootstrap estimator is more biased than the original estimator

$$\mathcal{E}(\tilde{D}_{AB_{\mathrm{B}}}) = \frac{(n-1)^2}{n^2}D_{AB}$$

The expected value of the bootstrap variance is not presented here. Note that the advantage of bootstrapping is that it will lead to the entire distribution of the linkage disequilibrium estimator.

Bootstrapping within each of a set of samples can provide confidence intervals for gene frequencies. These intervals do not require assumptions such as the gene frequencies are binomially distributed, or that there is Hardy-Weinberg equilibrium. Two populations can be judged to have different allelic frequencies if the estimated (observed) frequencies have non-overlapping confidence intervals. In other words, numerical resampling provides a convenient way of making inferences when there is not the basis for expectations over populations provided by random models.

Analysis of Variance

Comparison of means in a statistical setting is generally performed by an analysis of variance. Different variances are computed, which, when the means are the same, are expected to be estimating the same quantity. For normally distributed variables, the ratio of between- to within-population variances has an F distribution under the hypothesis of equal population means. The same approach is not strictly applicable here, but this is a convenient place to introduce the concepts and notation that are needed later for random populations. For gene j in the sample from population i, a variable x_{ij} can be defined by

$$x_{ij} = \begin{cases} 1 \text{ if gene is } A \\ 0 \text{ if gene is not } A \end{cases}$$

Table 5.1 Analysis of variance layout for variable indicating allele A in fixed populations.*

Source	d.f.	Sum of Squares	Expected Mean Square
Between populations	$r-1$	$\sum_i \frac{x_{i\cdot}^2}{n_i} - \frac{x_{\cdot\cdot}^2}{n_\cdot}$	$\frac{1}{r-1}\sum_i k_1 p_{Ai}(1-p_{Ai})$
		$= \sum_i n_i(\tilde{p}_{Ai} - \tilde{p}_{A\cdot})^2$	$+\frac{1}{r-1}\sum_i n_i(p_{Ai} - \bar{p}_{A\cdot})^2$
Within populations	$\sum_i(n_i - 1)$	$\sum_{i=1}^r \sum_{j=1}^{n_i} x_{ij}^2 - \sum_i \frac{x_{i\cdot}^2}{n_i}$	$\frac{1}{\sum_i(n_i-1)}\sum_i k_2 p_{Ai}(1-p_{Ai})$
	$= n_\cdot - r$	$= \sum_i n_i\tilde{p}_{Ai}(1-\tilde{p}_{Ai})$	

$^*k_1 = \left(1 - \frac{n_i}{\sum_i n_i}\right)$, $k_2 = (n_i - 1)$

The mean of the x_{ij} values for population i is just the gene frequency p_{Ai} in that population, so that the hypothesis of equal gene frequencies among populations would appear to be addressed by an analysis of variance of the x's. The analysis of variance table has the structure shown in Table 5.1. In this table, the average of the sample allelic frequencies over the r samples is

$$\tilde{p}_{A\cdot} = \frac{\sum_i n_i\tilde{p}_{Ai}}{\sum_i n_i}$$

while the average of the population frequencies is

$$\bar{p}_{A\cdot} = \frac{\sum_i n_i p_{Ai}}{\sum_i n_i}$$

The expected values in Table 5.1 are calculated on the basis of the n_i genes from population i being independent. When the r populations all have the same allelic frequencies, $p_{Ai} = \bar{p}_{A\cdot} = p_A$, the expected mean squares for between- and within-populations are the same. Although an analysis of variance layout is used here, there is no justification for regarding the ratio of the two mean squares as having the F distribution since the underlying variables x_{ij} are not normally distributed. There is no direct use for this approach in the fixed model framework, but it serves as the basis for analyses in the random model. It also provides further illustration of the difference in the two models.

RANDOM POPULATIONS

Under the random model, the populations sampled are considered to represent the species and therefore they have a common evolutionary history. Even though the populations may have been distinct for some time, the analysis is built on the assumption that there is a single ancestral population. Expectations now need to be taken over repeated samples from the populations *and* over replicate populations. In the absence of disturbing forces, all populations are expected to have the same allelic frequencies.

Underlying the analysis of differentiation in the random model is the notion that genetic sampling causes different genes in a population to be dependent, or related. Even though individuals, or genes, may be sampled randomly, the process of taking expectations must recognize that they are dependent through their shared ancestry.

Interest is generally centered on the extent to which different populations within the species have differentiated over the time since the ancestral population. The process of genetic sampling, or drift, between successive generations will result in intraspecific differentiation, and this differentiation is conveniently quantified with the F-STATISTICS of Wright (1951), or the analogous measures of Cockerham (1969, 1973).

Haploid Data

If data are available on genes directly, then the analysis is in terms of gene frequencies. Shared ancestry means that the expected value of a squared sample frequency from a sample of size n_i needs to be written as

$$\mathcal{E}(\tilde{p}_{Ai}^2) = p_A^2 + p_A(1-p_A)\theta + \frac{1}{n}p_A(1-p_A)(1-\theta)$$

where p_A is the frequency in the ancestral population and θ, or F_{ST}, is the coancestry coefficient. The derivation of this expression was given in Chapter 2, following Equation 2.14 with $F = \theta$.

The basic model assumes the absence of disturbing forces, so that gene frequencies are expected to remain constant over all generations and all populations

$$\mathcal{E}(\tilde{p}_{Ai}) = p_A$$

The coancestry will increase over time, and so it indicates the extent of differentiation between populations. It is worth stressing that this measure of between-population differentiation is a consequence of the relatedness of genes within populations.

Table 5.2 Analysis of variance layout for variable indicating allele A in random populations.

Source	d.f.	Sum of Squares	Expected Mean Square*
Between populations	$r-1$	$\sum_{i=1}^{r}\frac{x_{i\cdot}^2}{n_i}-\frac{x_{\cdot\cdot}^2}{n_\cdot}$ $=\sum_{i=1}^{r}n_i(\tilde{p}_{Ai}-\tilde{p}_{A\cdot})^2$	$p_A(1-p_A)[(1-\theta)+n_c\theta]$
Within populations	$\sum_{i=1}^{r}(n_i-1)$ $=n_\cdot-r$	$\sum_{i=1}^{r}\sum_{j=1}^{n_i}x_{ij}^2-\sum_i\frac{x_{i\cdot}^2}{n_i}$ $=\sum_i n_i\tilde{p}_{Ai}(1-\tilde{p}_{Ai})$	$p_A(1-p_A)(1-\theta)$

$^*n_c=\frac{1}{r-1}\left(\sum_{i=1}^{r}n_i-\frac{\sum_i n_i^2}{\sum_i n_i}\right)$

Estimation of θ can follow from the analysis of variance layout mentioned above, which removes the need to know the expected population frequencies p_A. The different expectation framework results in expected mean squares different from those given in Table 5.1. Instead, Table 5.2 is appropriate. The two tables differ in that Table 5.2 has a common expectation for each $\tilde{p}_{Ai}$. The more substantial difference between the two tables, however, is that Table 5.2 explicitly allows different genes in a sample to be related to an extent θ. In Table 5.1, the genes are all independent.

From the method of moments, the parameters are estimated by equating the observed and the expected mean squares. Writing the observed mean square for between populations as *MSP* and for genes within populations as *MSG*:

$$\begin{aligned} p_A(1-p_A)(1-\theta) &\hat{=} \frac{1}{\sum_i(n_i-1)}\sum_i n_i\tilde{p}_{Ai}(1-\tilde{p}_{Ai}) \\ &= MSG \end{aligned}$$

$$\begin{aligned} p_A(1-p_A)(1-\theta)+n_cp_A(1-p_A)\theta &\hat{=} \frac{1}{r-1}\sum_i n_i(\tilde{p}_{Ai}-\tilde{p}_{A\cdot})^2 \\ &= MSP \end{aligned}$$

(Recall that $\hat{=}$ means "is estimated by.")

It may be convenient to identify two COMPONENTS OF VARIANCE. For genes within populations:

$$\sigma_{\mathrm{G}}^2 = p_A(1-p_A)(1-\theta)$$

This component can be estimated, by the method of moments, as

$$\hat{\sigma}_{\mathrm{G}}^2 = MSG$$

A component for populations is defined as

$$\sigma_{\mathrm{P}}^2 = p_A(1-p_A)\theta$$

and is estimated as

$$\hat{\sigma}_{\mathrm{P}}^2 = \frac{1}{n_c}(MSP - MSG)$$

Note carefully that the two components of variance have been defined in terms of the expected mean squares in Table 5.2. They were not defined as variances, in spite of the conventional σ^2 notation. Instead, the component σ_{P}^2 is actually a covariance (for genes in the same population) and there is no requirement that this quantity be positive.

The sum of the components of variance for populations and genes within populations involves only the gene frequency p_A, which allows the quantity $p_A(1-p_A)$ to be eliminated, and an estimate of θ can be found as

$$\begin{aligned}\hat{\theta} &= \frac{\hat{\sigma}_{\mathrm{P}}^2}{\hat{\sigma}_{\mathrm{P}}^2 + \hat{\sigma}_{\mathrm{G}}^2} \\ &= \frac{MSP - MSG}{MSP + (n_c - 1)MSG}\end{aligned}$$

This can also be expressed in terms of the variance of gene frequencies over populations

$$s_A^2 = \frac{1}{(r-1)\bar{n}}\sum_i n_i(\tilde{p}_{Ai} - \tilde{p}_{A\cdot})^2$$

where

$$\bar{n} = \frac{1}{r}\sum_{i=1}^{r} n_i$$

The general formula is

$$\hat{\theta} = \frac{T_1}{T_2}$$

with

$$\begin{aligned} T_1 &= s_A^2 - \frac{1}{\bar{n}-1}\left[\tilde{p}_{A\cdot}(1-\tilde{p}_{A\cdot}) - \frac{r-1}{r}s_A^2\right] \\ T_2 &= \frac{n_c-1}{\bar{n}-1}\tilde{p}_{A\cdot}(1-\tilde{p}_{A\cdot}) + \left[1+\frac{(r-1)(\bar{n}-n_c)}{\bar{n}-1}\right]\frac{s_A^2}{r} \end{aligned}$$

For a large number of large samples, when both $1/\bar{n}$ and $1/r$ can be ignored, this estimate reduces to

$$\hat{\theta} = \frac{s_A^2}{\tilde{p}_{A\cdot}(1-\tilde{p}_{A\cdot})}$$

but there seems to be no need to use this instead of the more general result.

The estimation of θ has been presented in terms of one of the alleles, A, at a locus. If there are only two alleles at the locus, then the same estimate would be obtained if the other allele is used. For more than two alleles, however, a different *estimate* will result with every allele. Since the *parameter* θ is the same for every allele under the basic model of no disturbing forces, all these estimates are estimating the same quantity and an appropriate average is needed to give the best single estimate. For the uth allele, the estimate could be written as

$$\hat{\theta}_u = \frac{T_{1u}}{T_{2u}}$$

and then an overall estimate with good properties is given by combining the information from all v alleles

$$\hat{\theta} = \frac{\sum_{u=1}^{v} T_{1u}}{\sum_{u=1}^{v} T_{2u}}$$

There is an additional extension to cover the case where several loci are scored. Once again, under the neutral model, every allele at every locus provides an estimate of the same quantity. Indexing the loci by ℓ and alleles within loci by u, the individual estimates are

$$\hat{\theta}_{\ell u} = \frac{T_{1\ell u}}{T_{2\ell u}}$$

and the overall estimate from all m loci is

$$\hat{\theta} = \frac{\sum_{\ell=1}^{m}\sum_{u=1}^{v} T_{1\ell u}}{\sum_{\ell=1}^{m}\sum_{u=1}^{v} T_{2\ell u}}$$

With θ estimated, there is a quantification of the degree of divergence among a set of r populations. For a pair of populations, θ may serve as a measure of distance. Under a strict neutral model, $\ln(1-\theta)$ will increase linearly with the time since the two populations diverged from the ancestral population.

If other forces are involved, such as mutation, then the allelic frequencies no longer remain constant over time, and it is not appropriate to regard p_A as a constant in the expectation of $\tilde{p}^2_{Ai}$. Instead, pairs of genes can be compared for allelic state within and between populations. In the notation of Cockerham (1984), Q is the probability that two different genes within a population are the same allele, and q is the probability that genes in different populations are the same allele. These probabilities can be expressed in terms of indicator variables, letting x_{uij} take the value 1 if the jth gene from the ith population is allele A_u, and the value 0 otherwise. The similarity probabilities are therefore defined from the sum over all alleles

$$Q = \sum_u \mathcal{E}(x_{uij}x_{uij'}),\ j' \neq j \tag{5.1}$$

$$q = \sum_u \mathcal{E}(x_{uij}x_{ui'j'}),\ i' \neq i \tag{5.2}$$

The quantities Q, q can be related to the variance components from the analysis of variance layout. Subscripting the variance components by u for allele A_u,

$$\begin{aligned} 1 - Q &= \sum_u \sigma^2_{\mathrm{G}_u} \\ Q - q &= \sum_u \sigma^2_{\mathrm{P}_u} \end{aligned}$$

The mean squares in Table 5.2, after summing over alleles, therefore can be written as

$$\begin{aligned} \mathcal{E}(\sum_u MSP_u) &= (1-Q) + n_c(Q-q) \\ \mathcal{E}(\sum_u MSG_u) &= (1-Q) \end{aligned}$$

By analogy to the discussion on θ, a measure of differentiation between populations is provided by

$$\begin{aligned} \beta &= \frac{\sum_u MSP_u}{\sum_u (MSP_u + MSG_u)} \\ &= \frac{Q-q}{(Q-q)+(1-Q)} \end{aligned}$$

$$= \frac{Q-q}{1-q}$$

The symbol β was used by Cockerham and Weir (1987) in a discussion that included the effects of mutation and migration. Without such forces, β reduces to θ, which had been introduced to accommodate drift. Regardless of the forces acting, β serves as a measure of differentiation between populations. However, Weir and Basten (1990) showed that, in a model with drift and mutation between a finite number of alleles, β could discriminate among populations only for a relatively short time.

With the measures Q, q defined in terms of similarity over all alleles at a locus, it may be more convenient to estimate them in the following way. Two statistics, X, Y, are defined in terms of sample allelic frequencies $\tilde{p}_{ui}$ for allele A_u in the sample of size n_i genes from population i

$$X = \sum_u \sum_{i=1}^{r} n_i \tilde{p}_{ui}^2 \qquad (5.3)$$

$$Y = \sum_u \Big(\sum_{i=1}^{r} n_i \tilde{p}_{ui}\Big)^2 \qquad (5.4)$$

Manipulating the mean squares in Table 5.2 provides the following unbiased estimates of the similarities

$$\hat{Q} = \frac{X-r}{r(\bar{n}-1)} \qquad (5.5)$$

$$\hat{q} = \frac{1}{r(r-1)\bar{n}n_c}\left(Y - \frac{\bar{n}(n_c-1)}{\bar{n}-1}X\right) + \frac{\bar{n}-n_c}{n_c(\bar{n}-1)}\left(1 - \frac{1}{r-1}X\right) \qquad (5.6)$$

with great simplification when the samples are all of equal size, $n_i = n = \bar{n} = n_c$

$$\hat{Q} = \frac{X-r}{r(n-1)} \qquad (5.7)$$

$$\hat{q} = \frac{1}{r(r-1)n^2}(Y - nX) \qquad (5.8)$$

It may be more convenient, in the multiple alleles case, to use the statistics from Equations 5.3 and 5.4 in Equations 5.5 and 5.6 and then estimate θ from

$$\hat{\theta} = \frac{\hat{Q}-\hat{q}}{1-\hat{q}}$$

than to use the analysis of variance format in Table 5.2.

If the indicator variables x_{ij} were normally distributed, then the hypothesis $H_0 : \theta = 0$ could be tested with the ratio *MSP/MSG* in the analysis of variance layout. Since the x_{ij}'s are either 0 or 1, however, this ratio will not have an F distribution, and will not provide a test statistic. Making inferences about θ beyond simply estimating it may instead be accomplished by numerical resampling.

Because θ is a parameter appropriate for the random model, numerical resampling cannot be performed by resampling genes at a locus within populations, as was done in the fixed model. Resampling must mimic both the genetic sampling that causes replicate populations to differ, and the choice of genes for observation from each population. Two possibilities are suggested.

In the first place resampling may be done over loci. For a study in which m loci are scored, jackknifing can be performed by omitting each locus, L, in turn to provide m new estimates (in the analysis of variance terminology)

$$\hat{\theta}_{(L)} = \frac{\sum_{\ell \neq L} \sum_u T_{1\ell u}}{\sum_{\ell \neq L} \sum_u T_{2\ell u}}$$

and the variance of $\hat{\theta}$ estimated as

$$\text{Var}(\hat{\theta}) \hat{=} \frac{m-1}{m} \sum_{L=1}^{m} \left(\hat{\theta}_{(L)} - \frac{1}{m} \sum_{L=1}^{m} \hat{\theta}_{(L)} \right)^2$$

This procedure makes use of the fact that loci are expected to provide nearly independent replicates of the genetic sampling process. Jackknifing therefore provides an estimate of the variance of $\hat{\theta}$. It may be preferable to bootstrap over loci to construct a confidence interval for θ. Each bootstrap sample consists of a set of m pairs of values $\sum_u T_{1\ell u}, \sum_u T_{2\ell u}$ drawn with replacement from the m calculated values, and the combined estimate formed from this new collection of T's. As before, the middle 95% of these new estimates will provide a 95% confidence interval.

It is also possible to resample populations, and this may be done for each locus separately. In that way the estimates of θ for each locus could be compared. Loci that did not give overlapping confidence intervals for θ may be supposed to be showing the presence of disturbing forces. Numerically resampling over populations requires that there are several populations, just as resampling over loci supposes that several loci have been scored. A program is given in the Appendix.

Finally, it is necessary to comment on the possibility that the estimate of θ may be negative. There are two situations that are likely to give this outcome. It may be that the true value of θ is positive but small. Since the estimate has fairly low bias, the estimate is about as likely to be below as above the true value. Estimates less than the true value will often be less than zero when the true value is close to zero. The second situation is that the parameter may be negative. In statistical language, this corresponds to a negative INTRACLASS CORRELATION. In genetic language, it means that genes are more related between than within populations. A more complete discussion of biological causes for a negative component of variance between populations, $\sigma_P^2 < 0$ was given by Cockerham (1973) for the analysis of diploid data. Any mating system, such as the avoidance of self-mating (e.g., any system with separate sexes), that allows genes to be more alike between individuals than within individuals can cause this phenomenon, as will become evident in the next section.

Diploid Data

When observations are made on diploid individuals, the analysis should be performed at the diploid level. The same general approach is followed, but it is now possible to estimate the degree of relationship F between genes within individuals, as well as the degree θ between genes of different individuals. The haploid analysis essentially dropped the distinction between F and θ. Differentiation between populations is still measured in terms of θ, reflecting the relatedness of individuals within populations, but the analysis also provides an estimate of the overall inbreeding coefficient F. In Wright's notation, $F = F_{IT}, \theta = F_{ST}$. The degree of inbreeding within populations, f, or F_{IS}, can be expressed as

$$f = \frac{F - \theta}{1 - \theta}$$

The expected value of the squared sample allelic frequencies now must reflect the two levels of relatedness of different genes within populations — both a consequence of prior genetic sampling:

$$\mathcal{E}(\tilde{p}_{Ai}^2) = p_A^2 + p_A(1 - p_A)\theta + \frac{1}{2n_i}p_A(1 - p_A)(1 + F - 2\theta)$$

Once again, the estimation procedure follows naturally from an analysis of variance layout. There are now three sources of variation: populations, individuals within populations, and genes within populations. The sums of squares are constructed with gene *and* genotypic frequencies, as shown in Table 5.3. The first form of each sum of squares is the most convenient for computations with data.

The expected mean squares could have been written in terms of variance components for populations

Table 5.3 Analysis of variance layout for genotypic data in random populations.

Source	d.f.	Sum of Squares	Expected Mean Square*
Between populations	$r-1$	$2\sum_{i=1}^{r} n_i(\tilde{p}_{Ai}-\tilde{p}_{A\cdot})^2$	$p_A(1-p_A)\,[(1-F)$
		$=2(r-1)\bar{n}s_A^2$	$+2(F-\theta)+2n_c\theta]$
Individuals in populations	$\sum_{i=1}^{r}(n_i-1)$	$\sum_{i=1}^{r} n_i(\tilde{p}_{Ai}+\tilde{P}_{AAi}-2\tilde{p}_{Ai}^2)$	$p_A(1-p_A)\,[(1-F)$
	$=n_{\cdot}-r$	$=2r\bar{n}\tilde{p}_{A\cdot}(1-\tilde{p}_{A\cdot})-\frac{1}{2}r\bar{n}\tilde{H}_{A\cdot}$	$+2(F-\theta)]$
		$-2(r-1)\bar{n}s_A^2$	
Genes in individuals	$\sum_{i=1}^{r} n_i$	$\sum_{i=1}^{r} n_i(\tilde{p}_{Ai}-\tilde{P}_{AAi})$	$p_A(1-p_A)(1-F)$
	$=n_{\cdot}$	$=\frac{1}{2}r\bar{n}\tilde{H}_{A\cdot}$	

$$^*n_c=\frac{1}{r-1}\left(\sum_{i=1}^{r} n_i-\frac{\sum_i n_i^2}{\sum_i n_i}\right)$$

$$\sigma_{\mathrm{P}}^2=p_A(1-p_A)\theta$$

for individuals within populations

$$\sigma_{\mathrm{I}}^2=p_A(1-p_A)(F-\theta)$$

and for genes within individuals

$$\sigma_{\mathrm{G}}^2=p_A(1-p_A)(1-F)$$

From the method of moments, unbiased estimates of the components are functions of the observed mean squares *MSP*, *MSI*, and *MSG* for populations, individuals, and genes, respectively:

$$\begin{aligned}\hat{\sigma}_{\mathrm{G}}^2 &= MSG\\ \hat{\sigma}_{\mathrm{I}}^2 &= \frac{1}{2}(MSI-MSG)\\ \hat{\sigma}_{\mathrm{P}}^2 &= \frac{1}{2n_c}(MSP-MSI)\end{aligned}$$

In Table 5.3, the sample sizes n_i are for the numbers of individuals in each sample — in Tables 5.1 and 5.2 they were the numbers of genes. Also in

Table 5.3, $\tilde{P}_{AAi}$ is the frequency of AA homozygotes in the ith sample, while $\tilde{H}_{A\cdot}$ is the average frequency of individuals, over all samples, of heterozygous individuals that have allele A. In particular

$$\begin{aligned}\tilde{H}_{A\cdot} &= \frac{\sum_i n_i \tilde{H}_{Ai}}{\sum_i n_i} \\ &= \frac{1}{\sum_i n_i}\sum_i 2n_i(\tilde{p}_{Ai} - \tilde{P}_{AAi})\end{aligned}$$

Measures of differentiation can be estimated as

$$\begin{aligned}\hat{F} &= 1 - \frac{\hat{\sigma}^2_{\mathrm{G}}}{\hat{\sigma}^2_{\mathrm{P}} + \hat{\sigma}^2_{\mathrm{I}} + \hat{\sigma}^2_{\mathrm{G}}} \\ &= 1 - \frac{2n_c MSG}{MSP + (n_c - 1)MSI + n_c MSG} \\ &= 1 - \frac{S_3}{S_2}\end{aligned}$$

$$\begin{aligned}\hat{\theta} &= \frac{\hat{\sigma}^2_{\mathrm{P}}}{\hat{\sigma}^2_{\mathrm{P}} + \hat{\sigma}^2_{\mathrm{I}} + \hat{\sigma}^2_{\mathrm{G}}} \\ &= \frac{MSP - MSI}{MSP + (n_c - 1)MSI + n_c MSG} \\ &= \frac{S_1}{S_2}\end{aligned}$$

These are probably the most convenient computing formulas, but it is possible to give explicit expressions for the terms S_1, S_2, and S_3:

$$\begin{aligned}S_1 &= s^2_A - \frac{1}{\bar{n} - 1}\left[\tilde{p}_{A\cdot}(1 - \tilde{p}_{A\cdot}) - \frac{r-1}{r}s^2_A - \frac{1}{4}\tilde{H}_{A\cdot}\right] \\ S_2 &= \tilde{p}_{A\cdot}(1 - \tilde{p}_{A\cdot}) - \frac{\bar{n}}{r(\bar{n} - 1)}\left\{\frac{r(\bar{n} - n_c)}{\bar{n}}\tilde{p}_{A\cdot}(1 - \tilde{p}_{A\cdot})\right. \\ &\quad \left. - \frac{1}{\bar{n}}[(\bar{n} - 1) + (r - 1)(\bar{n} - n_c)]s^2_A - \frac{\bar{n} - n_c}{4n^2_c}\tilde{H}_{A\cdot}\right\} \\ S_3 &= \frac{n_c}{\bar{n}}\tilde{H}_{A\cdot}\end{aligned}$$

It is worth stressing that, with data collected for genotypes, estimates of the F-statistics F, θ are functions of genotypic frequencies.

These estimates have all been presented in terms of one particular allele A. In practice, several alleles at several loci will be available, and each will provide an estimate of the same parameters under a strictly neutral model. To combine estimates over all alleles, numerators and denominators are combined separately, as in the haploid case. For the uth allele at the ℓth locus, the estimates may be expressed as

$$\hat{F}_{\ell u} = 1 - \frac{S_{3\ell u}}{S_{2\ell u}}$$

$$\hat{\theta}_{\ell u} = \frac{S_{1\ell u}}{S_{2\ell u}}$$

and then the combined estimates follow from

$$\hat{F} = 1 - \frac{\sum_\ell \sum_u S_{3\ell u}}{\sum_\ell \sum_u S_{2\ell u}}$$

$$\hat{\theta} = \frac{\sum_\ell \sum_u S_{1\ell u}}{\sum_\ell \sum_u S_{2\ell u}}$$

Also as in the haploid case, numerical resampling over loci provides the means for making inferences about the parameters F and θ, while resampling over populations allows comparisons between loci.

Note that large numbers of large samples allow approximate expressions to be found for the estimates from each allele:

$$\hat{F} = 1 - \frac{\tilde{H}_{A\cdot}}{\tilde{p}_{A\cdot}(1 - \tilde{p}_{A\cdot})}$$

$$\hat{\theta} = \frac{s_A^2}{\tilde{p}_{A\cdot}(1 - \tilde{p}_{A\cdot})}$$

but the usual use of computers for data analysis makes this level of approximation unnecessary.

POPULATION SUBDIVISION

It is often the case that individuals are sampled in a nested classification such as transects and sites within transects (Barker et al. 1986), or even more complex structures such as drainages, rivers within drainages, and sites within rivers (P. Kukuk, personal communication). Each recognizable level of such hierarchies adds another level to the degrees of relatedness of genes. Appropriate nested analysis of variance layouts suggest the procedures for estimating the various measures of differentiation. Details will be given now for three- and four-level hierarchies.

Table 5.4 Analysis of variance layout for a three-level sampling hierarchy in random populations.

Source	d.f.	Mean Square	Expected Mean Square*
Between populations	$r-1$	MSP	$(1-F)+2(F-\theta_S)$ $+2n_{c1}(\theta_S-\theta_P)+2n_{c2}\theta_P$
Subpopulations in populations	$\sum_{i=1}^r(s_i-1)$ $=s.-r$	MSS	$(1-F)+2(F-\theta_S)$ $+2n_{c3}(\theta_S-\theta_P)$
Individuals in subpopulations	$\sum_{i=1}^r\sum_{j=1}^{s_i}(n_{ij}-1)$ $=n..-s.$	MSI	$(1-F)+2(F-\theta_S)$
Genes in individuals	$\sum_{i=1}^r\sum_{j=1}^{s_i}n_{ij}$ $=n..$	MSG	$(1-F)$

*to be multiplied by $p_A(1-p_A)$

Three-Level Hierarchy

Suppose that n_{ij} individuals are sampled from the jth subpopulation of the ith population, and that there are s_i subpopulations sampled from the ith population. There are still r populations. Pairs of genes may be related by virtue of being within individuals, between individuals within subpopulations, or between subpopulations within populations. The three degrees of relatedness are quantified by F, θ_S, and θ_P, respectively, and these parameters give the measures of differentiation among individuals, subpopulations, or populations. Genes in different populations are assumed to be unrelated and all genes sampled have the same expectation, p_A, of being of allelic type A. The analysis of variance layout is given in Table 5.4.

Although no explicit use of them will be made, components of variance can be defined for populations

$$\sigma_P^2 = p_A(1-p_A)\theta_P$$

subpopulations within populations

$$\sigma_S^2 = p_A(1-p_A)(\theta_S-\theta_P)$$

individuals within subpopulations

$$\sigma_{\rm I}^2 = p_A(1 - p_A)(F - \theta_{\rm S})$$

and genes within subpopulations

$$\sigma_{\rm G}^2 = p_A(1 - p_A)(1 - F)$$

Calculations of the mean squares in Table 5.4 make use of the following sums:

$$\begin{aligned}
n_{i\cdot} &= \sum_{j=1}^{s_i} n_{ij} \\
n_{\cdot\cdot} &= \sum_{i=1}^{r} n_{i\cdot} \\
s_{\cdot} &= \sum_{i=1}^{r} s_i \\
n_{c1} &= \frac{1}{r-1} \sum_{i=1}^{r} \sum_{j=1}^{s_i} \left[\frac{(n_{\cdot\cdot} - n_{i\cdot}) n_{ij}^2}{n_{i\cdot} n_{\cdot\cdot}} \right] \\
n_{c2} &= \frac{1}{r-1} \left(n_{\cdot\cdot} - \frac{1}{n_{\cdot\cdot}} \sum_{i=1}^{r} n_{i\cdot}^2 \right) \\
n_{c3} &= \frac{1}{s_{\cdot} - r} \left[n_{\cdot\cdot} - \sum_{i=1}^{r} \sum_{j=1}^{s_i} \left(\frac{n_{ij}^2}{n_{i\cdot}} \right) \right]
\end{aligned}$$

and mean allelic frequencies

$$\begin{aligned}
\tilde{p}_{Ai\cdot} &= \frac{1}{n_{i\cdot}} \sum_{j=1}^{s_i} n_{ij} \tilde{p}_{Aij} \\
\tilde{p}_{A\cdot\cdot} &= \frac{1}{n_{\cdot\cdot}} \sum_{i=1}^{r} \sum_{j=1}^{s_i} n_{ij} \tilde{p}_{Aij} \\
&= \frac{1}{n_{\cdot\cdot}} \sum_{i=1}^{r} n_{i\cdot} \tilde{p}_{Ai\cdot}
\end{aligned}$$

where $\tilde{p}_{Aij}$ is the allelic frequency of A in subpopulation j of population i.

The four mean squares use the allelic frequencies and the heterozygote frequencies — $\tilde{H}_{Aij}$ is the frequency in the sample from subpopulation j of population i of heterozygous individuals that contain allele A.

$$MSP = \frac{2}{r-1} \sum_{i=1}^{r} n_{i\cdot} (\tilde{p}_{Ai\cdot} - \tilde{p}_{A\cdot\cdot})^2$$

$$
\begin{aligned}
MSS &= \frac{2}{s. - r}\sum_{i=1}^{r}\sum_{j=1}^{s_i} n_{ij}(\tilde{p}_{Aij} - \tilde{p}_{Ai\cdot})^2 \\
MSI &= \frac{2}{n.. - s.}\sum_{i=1}^{r}\sum_{j=1}^{s_i} n_{ij}[\tilde{p}_{Aij}(1 - \tilde{p}_{Aij}) - \frac{1}{4}\tilde{H}_{Aij}] \\
MSG &= \frac{1}{2n..}\sum_{i=1}^{r}\sum_{j=1}^{s_i} n_{ij}\tilde{H}_{Aij}
\end{aligned}
$$

The method of moments provides estimators for each of the three measures of differentiation:

$$
\begin{aligned}
\hat{F} &= 1 - \frac{R_4}{R_2} \\
\hat{\theta}_S &= \frac{R_3}{R_2} \\
\hat{\theta}_P &= \frac{R_1}{R_2}
\end{aligned}
$$

where

$$
\begin{aligned}
R_1 &= \frac{MSP - MSI}{2n_{c2}} - \frac{n_{c1}(MSS - MSI)}{2n_{c2}n_{c3}} \\
R_2 &= \frac{MSP - MSI}{2n_{c2}} + \frac{(n_{c2} - n_{c1})(MSS - MSI)}{2n_{c2}n_{c3}} + \frac{MSG - MSI}{2} \\
R_3 &= \frac{MSP - MSI}{2n_{c2}} + \frac{(n_{c2} - n_{c1})(MSS - MSI)}{2n_{c2}n_{c3}} \\
R_4 &= MSG
\end{aligned}
$$

The combination over alleles and loci proceeds as before, as does numerical resampling over loci or populations.

There is an alternative to using these algebraic expressions for the various mean squares and subsequent estimates of measures of differentiation. Every individual in the data set could be coded as 11 for AA homozygotes, 10 for heterozygotes for allele A, and 00 for all other genotypes. This is equivalent to introducing the indicator variable $x_{ijk\ell}$ for the ℓth gene ($\ell = 1, 2$), in the kth individual ($k = 1, 2, \ldots, n_{ij}$), from the jth subpopulation ($j = 1, 2, \ldots, s_i$), of the ith population ($i = 1, 2, \ldots, r$). The variable takes the value 1 if that gene is of type A and the value 0 otherwise. Table

5.4 can be produced by a statistical computer package as a nested analysis of variance for the x's, and such packages generally allow the estimation of components of variance, even for unbalanced data. With the components of variance estimated, it is not difficult to recover the measures of differentiation.

Four-Level Hierarchy

As a final example, consider the case of individuals (indexed by ℓ) sampled from subsubpopulations (indexed by k) within subpopulations (indexed by j), within populations (indexed by i). There are r populations, s_i subpopulations from the ith population, t_{ij} subsubpopulations from the jth subpopulation of the ith population, and u_{ijk} individuals sampled from that subsubpopulation. The analysis of variance layout is shown in Table 5.5.

The variance components are for populations

$$\sigma^2_{\mathrm{P}} = p_A(1-p_A)\theta_{\mathrm{P}}$$

for subpopulations within populations

$$\sigma^2_{\mathrm{S}} = p_A(1-p_A)(\theta_{\mathrm{S}} - \theta_{\mathrm{P}})$$

for subsubpopulations within subpopulations

$$\sigma^2_{\mathrm{SS}} = p_A(1-p_A)(\theta_{\mathrm{SS}} - \theta_{\mathrm{S}})$$

for individuals within subsubpopulations

$$\sigma^2_{\mathrm{I}} = p_A(1-p_A)(F - \theta_{\mathrm{SS}})$$

and for genes within individuals

$$\sigma^2_{\mathrm{G}} = p_A(1-p_A)(1-F)$$

The coefficients for these components in the mean squares are

$$\begin{aligned}
n_{c1} &= (C_2 - C_3)/(r-1) \\
n_{c2} &= (C_4 - C_5)/(r-1) \\
n_{c3} &= (u_{\cdots} - C_6)/(r-1) \\
n_{c4} &= (C_1 - C_2)/(s_{.} - r) \\
n_{c5} &= (u_{\cdots} - C_5)/(s_{.} - r) \\
n_{c6} &= (u_{\cdots} - C_1)/(t_{..} - s_{.})
\end{aligned}$$

Table 5.5 Analysis of variance layout for a four-level sampling hierarchy in random populations.

Source	d.f.*	Mean Square	Expected Mean Square†
Between populations	$r-1$	*MSP*	$(1-F)+2(F-\theta_{SS})$ $+2n_{c1}(\theta_{SS}-\theta_S)+2n_{c2}(\theta_S-\theta_P)$ $+2n_{c3}\theta_P$
Subpopulations in populations	$s.-r$	*MSS*	$(1-F)+2(F-\theta_{SS})$ $+2n_{c4}(\theta_{SS}-\theta_S)+2n_{c5}(\theta_S-\theta_P)$
Subsubpopulations in subpopulations	$t_{..}-s.$	*MSSS*	$(1-F)+2(F-\theta_{SS})$ $+2n_{c6}(\theta_{SS}-\theta_S)$
Individuals in subsubpopulations	$u_{...}-t_{..}$	*MSI*	$(1-F)+2(F-\theta_{SS})$
Genes in individuals	$u_{...}$	*MSG*	$(1-F)$

$^*s. = \sum_{i=1}^{r} s_i,\ t_{..} = \sum_{i=1}^{r}\sum_{j=1}^{s_i} t_{ij},\ u_{...} = \sum_{i=1}^{r}\sum_{j=1}^{s_i}\sum_{k=1}^{t_{ij}} u_{ijk}$
†to be multiplied by $p_A(1-p_A)$

where

$$C_1 = \sum_{i=1}^{r}\sum_{j=1}^{s_i}\left[(\sum_{k=1}^{t_{ij}} u_{ijk}^2)/u_{ij.}\right]$$

$$C_2 = \sum_{i=1}^{r}\left[(\sum_{j=1}^{s_i}\sum_{k=1}^{t_{ij}} u_{ijk}^2)/u_{i..}\right]$$

$$C_3 = (\sum_{i=1}^{r}\sum_{j=1}^{s_i}\sum_{k=1}^{t_{ij}} u_{ijk}^2)/u_{...}$$

$$C_4 = \sum_{i=1}^{r}\left[(\sum_{j=1}^{s_i} u_{ij.}^2)/u_{i..}\right]$$

$$C_5 = (\sum_{i=1}^{r}\sum_{j=1}^{s_i} u_{ij\cdot}^2)/u_{\cdots}$$

$$C_6 = (\sum_{i=1}^{r} u_{i\cdot\cdot}^2)/u_{\cdots}$$

and dots indicate summation over subscripts for s, t, u.

It is informative to express the mean squares in a way that reveals the nature of the corresponding sums of squares:

$$\begin{aligned} MSP &= (S_1 - C)/(r-1) \\ MSS &= (S_2 - S_1)/(s_. - r) \\ MSSS &= (S_3 - S_2)/(t_{..} - s_.) \\ MSI &= (S_4 - S_3)/(u_{\cdots} - t_{..}) \\ MSG &= (S_5 - S_4)/(u_{\cdots}) \end{aligned}$$

The analysis may be formed in terms of indicator variables x_{ijklm} for the mth gene in the ℓth individual of the kth subsubpopulation from the jth subpopulation of the ith population. As previously, these variables take the value 1 if that gene is allele A, and have the value 0 otherwise. A statistical computer package will produce the structure of Table 5.5 if the genotypes are recoded as 11 for AA, 10 for A heterozygotes, and 00 for all other genotypes at that locus. Alternatively, a program can be written that uses the indicator variables, and for which the above sums of squares are

$$C = x_{\cdots\cdots}^2/2u_{\cdots}$$

$$S_1 = \sum_{i=1}^{r} \left(x_{i\cdots\cdot}^2/2u_{i\cdot\cdot}\right)$$

$$S_2 = \sum_{i=1}^{r}\sum_{j=1}^{s_i} \left(x_{ij\cdots}^2/2u_{ij\cdot}\right)$$

$$S_3 = \sum_{i=1}^{r}\sum_{j=1}^{s_i}\sum_{k=1}^{t_{ij}} \left(x_{ijk\cdot\cdot}^2/2u_{ijk}\right)$$

$$S_4 = \sum_{i=1}^{r}\sum_{j=1}^{s_i}\sum_{k=1}^{t_{ij}}\sum_{\ell=1}^{u_{ijk}} \left(x_{ijk\ell\cdot}^2/2\right)$$

$$S_5 = \sum_{i=1}^{r}\sum_{j=1}^{s_i}\sum_{k=1}^{s_{ij}}\sum_{\ell=1}^{u_{ijk}}\sum_{m=1}^{2} x_{ijk\ell m}^2$$

It is also possible to express the sums of squares in terms of average gene frequencies in various levels of the hierarchy. Within one subsubpopulation, the sample frequencies of A and heterozygotes for A are $\tilde{p}_{Aijk}$ and $\tilde{H}_{Aijk}$. Within subpopulations, within populations, or over the whole data set, the average sample allelic frequencies are

$$\begin{aligned}
\tilde{p}_{Aij\cdot} &= \sum_{k=1}^{t_{ij}} u_{ijk}\tilde{p}_{Aijk} / \sum_{k=1}^{t_{ij}} u_{ijk} \\
\tilde{p}_{Ai\cdot\cdot} &= \sum_{j=1}^{s_i} t_{ij}\tilde{p}_{Aij\cdot} / \sum_{j=1}^{s_i} t_{ij} \\
\tilde{p}_{A\cdots} &= \sum_{i=1}^{r} s_i\tilde{p}_{Ai\cdot\cdot} / \sum_{i=1}^{r} s_i
\end{aligned}$$

The sums of squares become

$$\begin{aligned}
S_1 - C &= 2\sum_{i=1}^{r} s_i(\tilde{p}_{Ai\cdot\cdot} - \tilde{p}_{A\cdots})^2 \\
S_2 - S_1 &= 2\sum_{i=1}^{r}\sum_{j=1}^{s_i} t_{ij}(\tilde{p}_{Aij\cdot} - \tilde{p}_{Ai\cdot\cdot})^2 \\
S_3 - S_2 &= 2\sum_{i=1}^{r}\sum_{j=1}^{s_i}\sum_{k=1}^{t_{ij}} u_{ijk}(\tilde{p}_{Aijk} - \tilde{p}_{Aij\cdot})^2 \\
S_4 - S_3 &= 2\sum_{i=1}^{r}\sum_{j=1}^{s_i}\sum_{k=1}^{t_{ij}} u_{ijk}[\tilde{p}_{Aijk}(1 - \tilde{p}_{Aijk}) - \frac{1}{4}\tilde{H}_{Aijk}] \\
S_5 - S_4 &= \frac{1}{2}\sum_{i=1}^{r}\sum_{j=1}^{s_i}\sum_{k=1}^{t_{ij}} u_{ijk}\tilde{H}_{Aijk}
\end{aligned}$$

GENETIC DISTANCE

When genetic data are available from several populations, it is natural to ask how genetically similar are the populations. Generally, genetic distance is thought of as being related to the time since the populations being compared diverged from a single ancestral population. This, in turn, requires a genetic model specifying the processes such as mutation and drift causing the populations to diverge. Genetic distances must therefore satisfy criteria different than those for geometric distances, but the latter kind will be reviewed first.

The GEOMETRIC DISTANCE d_{PQ} between points P and Q in Euclidean space is a quantity that satisfies three axioms:

- $d_{PQ} \geq 0, d_{PP} = 0$

  distances are not negative

- $d_{PQ} = d_{QP}$

  distances are symmetric

- $d_{PQ} + d_{QR} \geq d_{PR}$

  distances satisfy the triangle inequality

The most common distance satisfying these axioms is the EUCLIDEAN DISTANCE, defined in terms of the coordinates $\{p_i\}$, $\{q_i\}$ of points $\mathbf{P}$ and $\mathbf{Q}$ in n-dimensional space as

$$d_{PQ} = \sqrt{\sum_{i=1}^{n}(p_i - q_i)^2}$$

GENETIC DISTANCES are designed to express the genetic difference between two populations as a single number (Smith 1977). If there are no differences, the distance could be set to zero, whereas if the populations have no alleles in common at any locus the distance may be set equal to its maximum value, say 1. Distances may be regarded simply as data reduction devices, or as a means of comparing pairs of extant populations, or as the basis for constructing evolutionary histories for the populations. These different goals require different distances and, especially for inferring phylogenies, require genetic models. A review was given by Reynolds (1981), and some of the genetic distances that have been proposed are now considered.

Geometric Distances

Simply as a means of data reduction, a population may be represented as a point in v-dimensional space on the basis of frequencies of the v alleles at a locus. All populations lie on the hyperplane defined by $\sum_{u=1}^{v} p_u = 1$. This is illustrated in two dimensions in Figure 5.1.

The Euclidean distance between the populations in Figure 5.1 is expressed in terms of the allelic frequencies. For a locus with two alleles, if

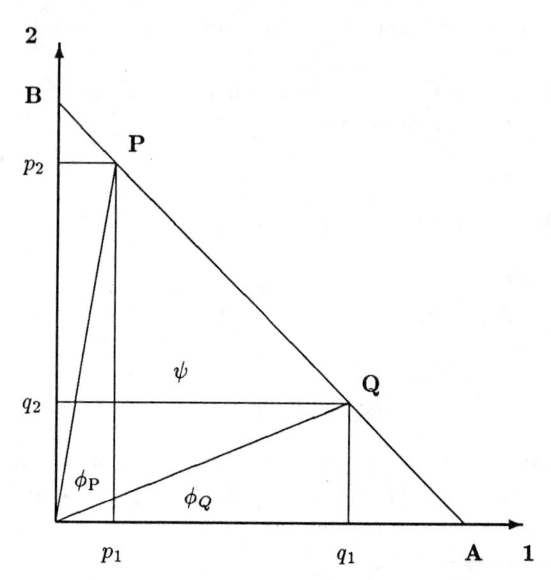

Figure 5.1 Representation of two populations in gene frequency space. Each population is represented by a point on the line from **A**:(1,0) to **B**:(0,1).

population P has allelic frequencies p_1 and $p_2 = 1 - p_2$ and population Q has frequencies q_1, q_2, then the Euclidean distance between the populations, based on that locus, is

$$d_{PQ} = \sqrt{[(p_1 - q_1)^2 + (p_2 - q_2)^2]}$$

or, for v alleles,

$$d_{PQ} = \sqrt{[\sum_{u=1}^{v}(p_u - q_u)^2]}$$

An alternative measure is based on the angle between the lines joining the origin to the points representing the populations. If ϕ_P and ϕ_Q are the angles between the allele 1 axis and these two lines, the angle ψ between the two lines is $(\phi_P - \phi_Q)$ and has cosine

$$\begin{aligned}\cos(\psi) &= \cos(\phi_P - \phi_Q) \\ &= \cos(\phi_P)\cos(\phi_Q) + \sin(\phi_P)\sin(\phi_Q)\end{aligned}$$

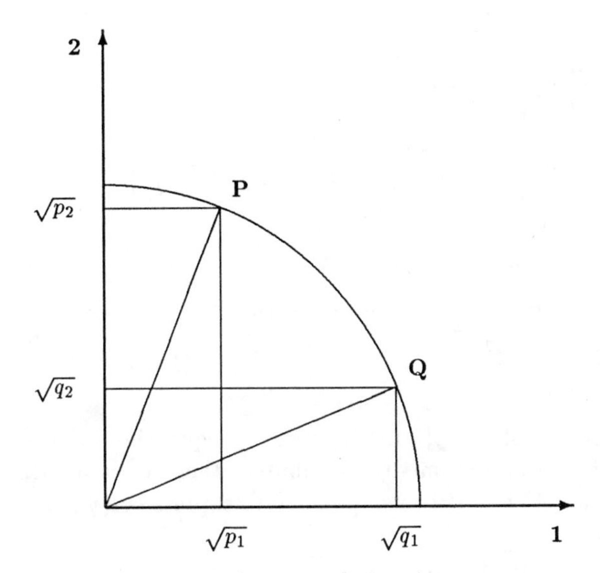

Figure 5.2 Representation of two populations in square root gene frequency space. Each population is represented by a point on a quarter-circle centered on the origin.

$$= \frac{\sum_{u=1}^{2} p_u q_u}{\sqrt{\sum_{u=1}^{2} p_u^2}\sqrt{\sum_{u=1}^{2} q_u^2}}$$

A more appropriately scaled measure would be $1 - \cos(\psi)$, which takes the value 0 when **P** and **Q** have the same frequencies, $\mathbf{P} \equiv \mathbf{Q}$, and the value 1 when there are no alleles in common so that $\sum_u p_u q_u = 0$.

These geometric distances can also be established when the square roots of gene frequencies are used as coordinates. Populations now lie on the surface of a hypersphere with radius 1 as illustrated in Figure 5.2 for two alleles. The cosine of ψ with this scaling is

$$\cos(\psi) = \sum_u \sqrt{(p_u q_u)}$$

and $1 - \cos(\psi)$ may again be used as a distance measure, as may ψ itself, or as may the chord length $\sqrt{[2 - 2\cos(\psi)]}$ between points **P** and **Q**. Such distance measures were used by Cavalli-Sforza and Bodmer (1971).

If a sample of size n, drawn from a population with frequencies p_u, has frequencies $\tilde{p}_u$, the geometric distance between sample and population can be based on

$$\begin{aligned}
\cos(\psi) &= \sum_u \sqrt{p_u \tilde{p}_u} \\
&= \sum_i p_u \sqrt{1 + \frac{\tilde{p}_u - p_u}{p_u}} \\
&\approx \sum_u p_u \left[1 + \frac{1}{2}\left(\frac{\tilde{p}_u - p_u}{p_u}\right) - \frac{1}{8}\left(\frac{\tilde{p}_u - p_u}{p_u}\right)^2 \right] \\
&= 1 - \frac{1}{8}\frac{X^2}{n}
\end{aligned}$$

where X^2 is the chi-square goodness-of-fit statistic

$$X^2 = \sum_u \frac{(n\tilde{p}_u - np_u)^2}{np_u}$$

for testing agreement of sample to population frequencies. If the sample allelic counts $n\tilde{p}_u$ are multinomially distributed, then X^2 is distributed as chi-square with $(v-1)$ d.f. It is also possible to express $\cos(\psi)$ as

$$\begin{aligned}
\cos(\psi) &= 1 - 2\sin^2\left(\frac{\psi}{2}\right) \\
&\approx 1 - 2\left(\frac{\psi}{2}\right)^2
\end{aligned}$$

to establish

$$\psi^2 \approx \frac{1}{4n} X^2$$

A similar argument holds for the distance between two populations:

$$\begin{aligned}
\cos(\psi) &= \sum_u \sqrt{p_u q_u} \\
&= \frac{1}{2}\sum_u (p_u + q_u) \sqrt{\left[1 - \left(\frac{p_u - q_u}{p_u + q_u}\right)^2\right]}
\end{aligned}$$

so that

$$\psi^2 \approx \frac{1}{2}\sum_u \left[\frac{(p_u - q_u)^2}{p_u + q_u}\right]$$

Population frequencies are replaced by sample frequencies to provide sample distances. This distance takes the value of 0 for identical populations, and the value 1 for completely distinct populations.

None of these geometric distances involves any genetic concepts.

Coancestry as Distance

For a finite population of size N mating at random, including a random amount of selfing, the inbreeding and coancestry coefficients are equal, $F = \theta$, and these values change over time according to

$$\theta_{t+1} = \frac{1}{2N} + \left(1 - \frac{1}{2N}\right)\theta_t$$

If the initial value θ_0 is zero

$$\theta_t = 1 - \left(1 - \frac{1}{2N}\right)^t$$

This shows that an estimate of θ from two or more populations of the same size and same history that have been diverging because of drift will give information about t, the time since the populations diverged. Specifically

$$\begin{aligned} d &= -\ln(1-\theta) \\ &= t\ln\left(1 - \frac{1}{2N}\right) \\ &\approx \frac{t}{2N} \end{aligned}$$

is an appropriate distance for divergence due to drift. Estimation can proceed by the variance component method discussed earlier in this chapter. The analysis shown in Table 5.2 is used since it is being assumed $F = \theta$. For equal-sized samples from two populations in which the frequencies of allele A_u at locus ℓ are $\tilde{p}_{\ell u_1}$ and $\tilde{p}_{\ell u_2}$ the estimator becomes

$$\hat{\theta} = \frac{\sum_\ell \left\{\frac{1}{2}\sum_u(\tilde{p}_{\ell u_1} - \tilde{p}_{\ell u_2})^2 - \frac{1}{2(2n-1)}\left[2 - \sum_u(\tilde{p}^2_{\ell u_1} + \tilde{p}^2_{\ell u_2})\right]\right\}}{\sum_\ell(1 - \sum_u \tilde{p}_{\ell u_1}\tilde{p}_{\ell u_2})} \qquad (5.9)$$

This corrects a typographical error in the corresponding expression of Reynolds et al. (1983).

If only one locus is scored, and the two populations are fixed for different alleles, each sample has only one nonzero frequency, but for different u values. The estimated distance is then equal to 1. The populations cannot diverge any further. If the two samples have no alleles in common but each has more than one allele, the estimated distance is less than 1, reflecting the fact that drift has not yet caused fixation for one allele per locus. When each of the two samples has only the same single allele at a single scored locus, the estimator is undefined, since Equation 5.9 becomes 0/0. This is because it cannot be determined from observing two samples fixed for the same allele whether the two populations have just become homozygous or have been homozygous for some time that may have extended back to the ancestral population. It is unlikely that this will cause a problem in practice, since more than one locus will be used in constructing the estimator and it is unlikely that all loci will be invariant for the same allele in both samples. If they were, however, there is no information in the data about genetic distance between those populations.

Nei's Genetic Distance

The most widely used measure of genetic distance was proposed by Nei (1972). With the same notation as in the previous section, a "genetic identity" I is defined between the two populations as

$$I = \frac{\sum_\ell \sum_u \tilde{p}_{\ell u_1} \tilde{p}_{\ell u_2}}{\sqrt{\sum_\ell \sum_u \tilde{p}^2_{\ell u_1} \sum_\ell \sum_u \tilde{p}^2_{\ell u_2}}}$$

or, with a bias correction (Nei 1978),

$$I = \frac{(2n-1) \sum_\ell \sum_u \tilde{p}_{\ell u_1} \tilde{p}_{\ell u_2}}{\sqrt{\sum_\ell (2n \sum_u \tilde{p}^2_{\ell u_1} - 1) \sum_\ell (2n \sum_u \tilde{p}^2_{\ell u_2} - 1)}} \tag{5.10}$$

and Nei's standard genetic distance D is

$$D = -\ln(I)$$

The quantity I is a ratio of the proportions of genes that are alike between and within populations, and Equation 5.10 can be also be written in terms $\tilde{Q}$ and $\tilde{q}$ of Equations 5.1 and 5.2

$$I = \frac{\tilde{q}_{12}}{\sqrt{\tilde{Q}_1 \tilde{Q}_2}}$$

Cockerham (1984) derived transition equations for the corresponding parameters Q and q. For a random mating population of finite size N, and for loci at which there is mutation away from any allele at rate μ per generation and mutation to any other specific allele at rate ν ($\nu = 0$ for the INFINITE ALLELE model), in generation t

$$Q_t = Q^* - (Q^* - Q_0)\lambda_Q^t$$
$$q_t = q^* - (q^* - q_0)\lambda_q^t$$

where $Q_0 = q_0$ are the values at time zero, before the two populations diverged, and Q^*, q^* are the final, or equilibrium, values given by

$$Q^* = \frac{1 + 4N\nu}{1 + 4N\mu + 4N\nu}, \; q^* = \frac{\nu}{\mu + \nu}$$

The two rates of change are

$$\lambda_Q = 1 - 2\mu - 2\nu - \frac{1}{2N}, \; \lambda_q = 1 - 2\mu - 2\nu$$

all supposing that squares and products of μ, ν, and $1/2N$ and smaller terms can be ignored. The statistics $\tilde{Q}$ and $\tilde{q}$ are unbiased for the parameters Q and q. To see what genetic parameters are being estimated by Nei's distance, approximate the expectation of the ratio in Equation 5.10 by

$$\mathcal{E}(I) = \frac{q}{Q} = \frac{q^* - (q^* - Q_0)\lambda_q^t}{Q^* - (Q^* - Q_0)\lambda_Q^t}$$

It is reasonable to assume the ancestral population was in equilibrium for the joint forces of drift and mutation, and that each population separately remains in equilibrium over time, $Q_t = Q^* = Q_0$. Then

$$\mathcal{E}(I) = \frac{q^*}{Q^*} + \left(1 - \frac{q^*}{Q^*}\right)\lambda_q^t$$

Although this has a finite final value, $-\ln I$ does increase approximately linearly with time until t becomes moderately large (Weir and Basten 1990). There is greater simplification with the infinite alleles model for then $q^* = 0$ and

$$\mathcal{E}(I) = \lambda_q^t$$

so that Nei's distance has an expected value of

$$\mathcal{E}(D) \approx 2\mu t$$

which, as desired, is linearly related to the time since divergence.

This discussion of Nei's distance and the coancestry distance has included explicit statements of the underlying genetic models. Nei's distance is appropriate for long-term evolution when populations diverge because of drift and mutation. The distance is proportional to the time since divergence in the special case of the infinite alleles mutation model and equilibrium in the ancestral population. The coancestry distance is appropriate for divergence due to drift only, and no assumptions need be made about the ancestral population. Neither distance should be used in the situation for which it was not intended.

When there is no mutation, the proportions of like genes change over time according to

$$Q_t = Q_0 - (1 - Q_0)\left(\frac{2N-1}{2N}\right)^t, \; q_t = Q_0$$

with Q_0 equal to the sum of squares of gene frequencies in the ancestral population. In this drift situation Nei's distance is estimating

$$\mathcal{E}(D) \approx \ln\left[1 - \left(\frac{1}{Q_0} - 1\right)\left(1 - \frac{1}{2N}\right)^t\right]$$

which is confounded by the unknown Q_0. This dependence on initial gene frequencies is also a feature of Rogers' (1972) distance

$$D_R = \frac{1}{m}\sum_{\ell}\left[\frac{1}{2}\sum_{u}(\tilde{p}_{1\ell u} - \tilde{p}_{2\ell u})^2\right]^{1/2}$$

Goodman (1972) pointed out that equal geometric distance may not indicate equal biological divergence, and a similar lack of correspondence between distance and degree of divergence can occur if mutation-based distances are applied to drift situations.

An alternative estimate of θ for the pure drift situation is provided by one of the geometric measures, as proposed by Balakrishnan and Sanghvi (1968). If locus ℓ has v_ℓ alleles, their measure G^2 is

$$G^2 = \frac{1}{\sum_\ell (v_\ell - 1)}\sum_{\ell}\sum_{u}\left[\frac{(\tilde{p}_{\ell u_1} - \tilde{p}_{\ell u_2})^2}{\tilde{p}_{\ell u_1} + \tilde{p}_{\ell u_2}}\right]$$

Taking expectations for each allele and summing over all alleles in the case of equal sample sizes leads to

$$\mathcal{E}(G^2) = \frac{1}{2n} + \frac{2n-1}{2n}\theta$$

so that $-\ln(1 - G^2)$ will estimate θ for large n.

Variance of Distance Estimates

All of the distance estimators discussed here are ratios of quadratic functions of gene frequencies and the methods of numerical resampling appear to be the best for providing estimates of sampling variances. From the discussion on page 151, resampling over loci is expected to mimic the genetic sampling process, and so will lead to appropriate variances when distances are being used in a random population framework.

If only the variances among repeated samples from the same populations are wanted, then the delta method discussed in Chapter 2 could be used.

Because the distances are not homogeneous functions of degree zero in the genotypic counts, Fisher's variance formula cannot be used. Of course, numerical resampling over individuals (or genes) within the two populations is also appropriate in this fixed population framework.

SUMMARY

When observations are collected from several populations, it is possible to estimate measures of population structure – specifically, measures of association between pairs of genes in various levels of a hierarchy. When data are available from only a single population it is not possible to estimate the effects of genetic sampling between populations unless different loci can be regarded as being independent. If inferences are to be made on only the population(s) sampled, it is not necessary to invoke the variance-based measures represented by F-statistics. Analyses of mean gene frequencies are appropriate.

A useful quantification of the structure in a population is provided by genetic distances between components of the population. Again, care is needed to match the distance to the intended scope of inference.

EXERCISES

Exercise 5.1

A locus with two alleles, A, a, has been scored in four populations. The four sets of allelic frequencies are listed below. Calculate the Euclidean distance, Nei's standard distance and the coancestry-based distance between populations 1,2 and 3,4. Why do these three pairs of distances differ?

Population	A	a	Population	A	a
1	0.55	0.45	3	0.95	0.05
2	0.45	0.55	4	0.85	0.15

Exercise 5.2

The two sets of data on the next page are the 11-locus genotypes for two populations (Ghana and N'goye) of the mosquito *Aedes aegypti* (J. Powell, personal communication). Using all the data, estimate the three components of variance (between populations, between individuals within populations, and between genes within individuals) for gene frequencies, and hence estimate the F-statistics.

Ghana

1	2	3	4	5	6	7	8	9	10	11
11	11	11	12	11	34	12	11	11	11	11
11	11	11	12	11	22	12	11	11	11	11
11	11	11	12	11	23	11	11	11	11	11
11	11	11	11	11	11	11	11	11	11	11
11	11	11	12	11	34	12	11	11	11	11
11	11	11	11	11	22	11	11	11	11	11
11	11	11	11	11	11	11	11	11	11	11
11	11	11	12	11	24	12	11	11	11	11
11	11	11	11	12	11	11	11	11	11	11
11	11	11	11	11	12	11	11	11	11	11
11	11	11	12	11	24	11	11	11	11	11
11	11	11	12	11	24	12	11	11	11	11
11	11	11	12	11	34	11	11	11	11	11
11	11	11	11	11	11	11	11	11	11	11
11	11	11	12	11	22	12	11	11	11	11
11	11	11	12	11	34	11	11	11	11	11
11	11	11	12	11	23	12	11	11	11	11
11	11	11	12	11	24	11	11	11	11	11
11	11	11	12	11	22	11	11	11	11	11
11	11	11	12	11	34	11	11	11	11	11
11	11	11	11	12	11	11	11	11	11	11
11	11	11	12	11	23	12	11	11	11	11
11	11	11	12	11	34	11	11	11	11	11
11	11	11	11	11	11	11	11	11	11	11
11	11	11	11	11	34	11	11	11	11	11
11	11	11	11	11	24	12	11	11	11	11
11	11	11	11	11	11	11	11	11	11	11
11	11	11	11	11	34	12	11	11	11	11
11	11	11	11	11	34	12	11	11	11	11
11	11	11	11	11	34	11	11	11	11	11
11	11	11	11	11	23	11	11	11	11	11
11	11	11	11	11	22	11	11	11	11	11
11	11	11	11	11	23	11	11	11	11	11
11	11	11	11	11	11	11	11	11	11	11
11	11	11	11	11	34	12	11	11	11	11
11	11	11	11	11	34	11	11	11	11	11
11	11	11	11	11	34	11	11	11	11	11
11	11	11	11	11	23	12	11	11	11	11
11	11	11	11	11	11	11	11	11	11	11
11	11	11	11	11	23	11	11	11	11	11

N'goye

1	2	3	4	5	6	7	8	9	10	11
11	11	11	11	11	11	13	11	11	11	11
11	11	11	11	11	11	11	11	11	11	11
11	11	11	11	11	33	11	11	11	11	11
11	11	11	11	11	24	11	11	11	11	11
11	11	11	11	12	11	13	11	11	11	11
11	11	11	11	11	11	11	11	11	11	11
11	11	11	11	11	14	12	11	11	11	11
11	11	11	11	11	12	13	11	11	11	11
11	11	11	12	11	13	23	11	11	11	11
11	11	11	11	11	14	12	11	11	11	11
11	11	11	11	11	11	11	11	11	11	11
11	11	11	11	11	11	33	11	11	11	11
11	11	11	11	11	11	12	11	11	11	11
11	11	11	11	11	24	13	11	11	11	11
11	11	11	11	11	11	11	11	11	11	11
11	11	11	11	11	12	11	11	11	11	11
11	11	11	11	11	11	11	11	11	11	11
11	11	11	11	11	12	11	11	11	11	11
11	11	11	11	11	11	33	11	11	11	11
11	11	11	11	11	13	12	11	11	11	11
11	11	11	11	11	12	11	11	11	11	11
11	11	11	11	11	11	22	11	11	11	11
11	11	11	11	11	11	13	11	11	11	11
11	11	11	11	11	14	13	11	11	11	11
11	11	11	11	11	12	13	11	11	11	11
11	11	11	11	11	11	13	11	11	11	11
11	11	11	11	11	14	11	11	11	11	11
11	11	11	11	11	11	12	11	11	11	11
11	11	11	11	11	14	12	11	11	11	11
11	11	11	12	11	12	12	11	11	11	11
11	11	11	11	11	11	13	11	11	11	11
11	11	11	12	11	12	11	11	11	11	11
11	11	11	11	12	12	11	11	11	11	11
11	11	11	12	11	11	11	11	11	11	11
11	11	11	11	11	12	11	11	11	11	11
11	11	11	11	11	11	11	11	11	11	11
11	11	11	11	11	12	11	11	11	11	11
11	11	11	11	11	12	11	11	11	11	11
11	11	11	11	11	11	11	11	11	11	11
11	11	11	11	11	11	33	11	11	11	11
11	11	11	11	11	24	11	11	11	11	11
11	11	11	11	11	12	11	11	11	11	11
11	11	11	11	11	11	23	11	11	11	11
11	11	11	11	11	12	13	11	11	11	11
11	11	11	11	11	11	11	11	11	11	11
11	11	11	11	11	14	11	11	11	11	11
11	11	11	12	11	34	12	11	11	11	11
11	11	11	22	11	23	11	11	11	11	11
11	11	11	11	11	11	33	11	11	11	11
11	11	11	11	11	12	11	11	11	11	11

Chapter 6

Analyses between Generations

When genetic data are collected from individuals in successive generations, inferences can be made on the processes concerned with the transmission of genetic material from parent to offspring. Attention is paid in this chapter to several such processes. Outcrossing rates are estimated for plant populations and the exclusion or determination of paternity determined for human populations. Methods for estimating selection coefficients and recombination fractions are also considered.

ESTIMATION OF OUTCROSSING

Plant data are often obtained from seed collected from a set of maternal parent plants, leaving open the question of paternal parentage. In the simplest approach, estimates of the proportion of offspring produced by outcrossing as opposed to selfing are based on the proportion of heterozygous offspring carrying one allele that could not be of maternal origin. Various methods have been used to estimate outcrossing depending on whether the maternal genotype is known or not, whether single or multiple loci are used, and whether one or more paternal parents are possible for offspring from a single female parent. Maximum likelihood estimation is generally used, although this may require numerical methods.

Estimation from Homozygous Female Parents

Outcrossing rates can be estimated from progeny arrays derived from female parents used because they are known to be homozygous. Suppose there is a constant probability, t, that each offspring of any maternal plant A_uA_u is an outcross, and probability $1-t$ that it is a self. An offspring is heterozygous only if it is both an outcross and it received an allele different from that carried by the mother. This probability is

$$\sum_{v \neq u} tp_v = t(1-p_u)$$

where p_u is the frequency of A_u alleles in the pollen pool. The heterozygote numbers $\tilde{H}_u$ in a sample of size N_u from A_uA_u mothers are binomially distributed

$$\tilde{H}_u \sim B[N_u, t(1-p_u)]$$

Combining over maternal genotypes gives the likelihood of t

$$L(t) \propto \prod_u [t(1-p_u)]^{\tilde{H}_u}[1-t(1-p_u)]^{N_u-\tilde{H}_u}$$

so that the score (Chapter 2) for t is

$$\begin{aligned} S(t) &= \frac{\partial \ln L(t)}{\partial t} \\ &= \frac{1}{t}\sum_u \tilde{H}_u - \sum_u \left[\frac{(N_u-\tilde{H}_u)(1-p_u)}{1-t(1-p_u)}\right] \end{aligned}$$

Setting the score to zero shows that the MLE $\hat{t}$ for t must satisfy an equation that requires numerical solution

$$\frac{1}{\hat{t}}\sum_u H_u = \sum_u \frac{(N_u-\tilde{H}_u)(1-p_u)}{1-\hat{t}(1-p_u)} \tag{6.1}$$

Note that this equation assumes the allelic frequencies p_u are known.

When only one maternal genotype, say AA, is used, the estimate can be found explicitly as

$$\hat{t} = \frac{\tilde{H}_A}{N_A(1-p_A)}$$

if the N_A offspring contain $\tilde{H}_A$ heterozygous offspring.

The data shown in Table 6.1 are for esterase frequencies used by Allard et al. (1972) to estimate the outcrossing rate in a barley population. From Equation 6.1, $\hat{t}$ must satisfy

$$\frac{10}{\hat{t}} = \frac{277.6608}{1-0.8928\hat{t}} + \frac{357.8382}{1-0.6518\hat{t}} + \frac{446.5235}{1-0.4723\hat{t}} + \frac{58.0029}{1-0.9831\hat{t}}$$

Table 6.1 Outcrossing data for an esterase locus in a barley population. Source: R. W. Allard (personal communication).

Maternal Class i	No. of Offspring N_u	Heterozygotes H_u	Gene Frequencies p_u
1	315	4	0.1072
2	554	5	0.3482
3	946	1	0.5277
4	59	0	0.0169
Totals	1874	10	1.0000

The easiest approach to solving such equations is to use iterations. An initial guess is used on the right-hand side of the equation, and the value found from the left-hand side is regarded as the next iterate. In this particular example, solving the quartic equation directly produces only one valid root in the range between 0 and 1, and that is $\hat{t} = 0.0087$, reinforcing the impression from Table 6.1 that there is very little outcrossing in the population.

For such small outcrossing values, the products $\hat{t}(1 - p_u)$ can be ignored as being much smaller than 1, and Equation 6.1 simplifies to

$$\hat{t} = \frac{\sum_u \tilde{H}_u}{\sum_u (N_u - \tilde{H}_u)(1 - p_u)} \tag{6.2}$$

To find the variance of the MLE $\hat{t}$, notice that it is a function of the counts $\tilde{H}_u$ and the pollen frequencies p_u. In the simple case considered by Allard et al. (1972) the allelic frequencies were estimated from different samples, and so were independent of the heterozygote counts. Sample values $\tilde{p}_u$ replaced parameters p_u in Equation 6.2 and the delta method was used to calculate the variance

$$\begin{aligned} \mathrm{Var}(\hat{t}) &= \sum_u \left(\frac{\partial \hat{t}}{\partial \tilde{H}_u}\right)^2 \mathrm{Var}(\tilde{H}_u) + \sum_u \left(\frac{\partial \hat{t}}{\partial \tilde{p}_u}\right)^2 \mathrm{Var}(\tilde{p}_u) \\ &\quad + \sum_u \sum_{v \neq u} \left[\frac{\partial \hat{t}}{\partial \tilde{p}_u} \frac{\partial \hat{t}}{\partial \tilde{p}_v} \mathrm{Cov}(\tilde{p}_u, \tilde{p}_v)\right] \end{aligned}$$

Derivatives of $\hat{t}$ can be found from Equations 6.1 or 6.2.

Estimation from Equilibrium Population

A population with a constant proportion t of outcrossing and a proportion $1-t$ of selfing eventually reaches an equilibrium in which genotypic frequencies remain constant over time for neutral loci. For a locus with alleles A_u these frequencies are

$$\begin{aligned} P_{uu} &= p_u^2 + p_u(1-p_u)\frac{1-t}{1+t}, \quad \text{for } A_uA_u \text{ homozygotes} \\ P_{uv} &= 2p_up_v\frac{2t}{1+t}, \quad \text{for } A_uA_v \text{ heterozygotes} \end{aligned}$$

which suggested to Fyfe and Bailey (1951) that observed genotypic frequencies could provide estimates of both gene frequencies p_u and outcrossing rate t. The procedures are the same as were considered in Chapter 2, where gene frequencies and the inbreeding coefficient f were estimated. Outcrossing and inbreeding coefficients are related by

$$f = \frac{1-t}{1+t} \tag{6.3}$$

It is a simple matter to invert Equation 6.3 to provide an estimate and its variance for t

$$\begin{aligned} \hat{t} &= \frac{1-\hat{f}}{1+\hat{f}} \\ \text{Var}(\hat{t}) &= \left(\frac{d\hat{t}}{d\hat{f}}\right)^2 \text{Var}(\hat{f}) \\ &= \frac{4}{(1+f)^4}\text{Var}(\hat{f}) \end{aligned}$$

with the variance of $\hat{f}$ following from Exercise 2.4.

Estimation from Offspring of Arbitrary Female Parents

It is not necessary to assume that the population is in equilibrium to estimate outcrossing rates and allelic frequencies. The approach of Brown et al. (1975) assumes that sufficient offspring are available per parent that the maternal genotype can be inferred, and here the *EM* algorithm method given by Cheliak et al. (1983) will be treated.

The *EM* algorithm accommodates the fact that some offspring genotypes may arise from either selfing or outcrossing, and it is not known which event occurred. The E step consists of separating offspring numbers into these two classes. Cheliak et al. (1983) required maternal genotypes to be known,

Table 6.2 Notation for estimating selfed and crossed offspring frequencies with the *EM* algorithm.

Maternal Genotype	Offspring Genotype	Observed Count	Probability	Expected Count
$A_u^*A_u^*$	$A_u^*A_u^*$	$_{uu}x_{uu}$	$1-t$	$_{uu}s_{uu}$
	$A_u^*A_u$		tp_u	$_{uu}c_{uu}$
	$A_u^*A_v$	$_{uu}x_{uv}$	tp_v	$_{uu}c_{uv}$
$A_u^*A_v^*$	$A_u^*A_u^*$	$_{uv}x_{uu}$	$(1-t)/4$	$_{uv}s_{uu}$
	$A_u^*A_u$		$tp_u/2$	$_{uv}c_{uu}$
	$A_v^*A_v^*$	$_{uv}x_{vv}$	$(1-t)/4$	$_{uv}s_{vv}$
	$A_v^*A_v$		$tp_v/2$	$_{uv}c_{vv}$
	$A_u^*A_v^*$	$_{uv}x_{uv}$	$(1-t)/2$	$_{uv}s_{uv}$
	$A_u^*A_v$		$tp_v/2$	$_{uv}c_{uv}$
	$A_v^*A_u$		$tp_u/2$	$_{uv}c_{vu}$
	$A_u^*A_w$	$_{uv}x_{uw}$	$tp_w/2$	$_{uv}c_{uw}$
	$A_v^*A_w$	$_{uv}x_{vw}$	$tp_w/2$	$_{uv}c_{vw}$

and appropriate notation is shown in Table 6.2. In this table, $_{ab}x_{cd}$ indicates the observed count of A_cA_d genotypes from A_aA_b female parents; $_{ab}s_{cd}$ is for selfed and $_{ab}c_{cd}$ is for crossed offspring, respectively. An asterisk denotes a maternal gene. The E step of the *EM* algorithm starts by assuming initial values of the unknowns t and p_u and then the numbers of crossed and outcrossed offspring can be estimated:

$$_{uu}s_{uu} = \left[\frac{1-t}{(1-t)+tp_u}\right] {}_{uu}x_{uu} \qquad _{uu}c_{uu} = \left[\frac{tp_u}{(1-t)+tp_u}\right] {}_{uu}x_{uu}$$

$$_{uv}s_{uu} = \left[\frac{1-t}{(1-t)+2tp_u}\right] {}_{uv}x_{uu} \qquad _{uv}c_{uu} = \left[\frac{2tp_u}{(1-t)+2tp_u}\right] {}_{uv}x_{uu}$$

$$_{uv}s_{vv} = \left[\frac{1-t}{(1-t)+2tp_v}\right] {}_{uv}x_{vv} \qquad _{uv}c_{vv} = \left[\frac{2tp_v}{(1-t)+2tp_v}\right] {}_{uv}x_{vv}$$

$$_{uv}s_{uv} = \left[\frac{1-t}{(1-t)+t(p_u+p_v)}\right] {}_{uv}x_{uv} \qquad _{uv}c_{uv} = \left[\frac{tp_v}{(1-t)+t(p_u+p_v)}\right] {}_{uv}x_{uv}$$

With all offspring genotypic counts estimated, gene frequencies in the pollen pool can be estimated straightforwardly as the proportion of outcrossed offspring that receives pollen carrying allele A_u

$$\hat{p}_u = \frac{_{..}c_{.u}}{_{..}c_{..}}$$

A dot indicates summation. Similarly, the outcrossing rate is estimated as the proportion of crossed offspring among all offspring

$$\hat{t} = \frac{_{..}c_{..}}{_{..}s_{..} + {}_{..}c_{..}}$$

These estimates are used in another E step, and the *EM* algorithm continues until the estimates stabilize.

Cheliak et al. (1983) applied the *EM* method to data on three isozyme loci in the white lupin, *Lupinus albus*, collected by Green at al. (1980). The appropriate version of Table 6.2 for the *Pgm*-1 locus is given as Table 6.3, and the data are shown in Table 6.4. Progress of the iterations for each of the three loci is shown in Table 6.5. The same end- points were obtained for other initial values of t and p_F, unless $t = 1$, in which case it remained at 1. The three estimates of t are not significantly different, although the allelic frequencies estimated for the pollen pool are different from those found in the maternal population (Green et al. 1980).

It may be that the actual numbers of offspring per parent are not large enough to infer the maternal genotype. It is then necessary to assign probabilities to each possible maternal genotype, as explained by Clegg et al. (1978), and their methodology now follows. Offspring data are collected from a series of maternal plants, and the probability π_{my} that the yth plant has genotype m needs to be estimated. In the population, the probability that an individual has genotype m is P_m.

Table 6.3 Notation for applying the *EM* algorithm to *Pgm*-1 in ***Lupinus albus*** data of Green et al. (1980).

Maternal Genotype	Offspring Genotype	Observed Count	Probability	Expected Count
F^*F^*	F^*F^*	120	$1-t$	${}_{FF}s_{FF}$
	F^*F		tp_F	${}_{FF}c_{FF}$
	F^*S	5	tp_S	${}_{FF}c_{FS}$
S^*S^*	S^*S^*	104	$1-t$	${}_{SS}s_{SS}$
	S^*S		tp_S	${}_{SS}c_{SS}$
	S^*F	9	tp_F	${}_{SS}c_{SF}$
S^*F^*	S^*S^*	16	$(1-t)/4$	${}_{SF}s_{SS}$
	S^*S		$tp_S/2$	${}_{SF}c_{SS}$
	F^*F^*	11	$(1-t)/4$	${}_{SF}s_{FF}$
	F^*F		$tp_F/2$	${}_{SF}c_{FF}$
	S^*F^*	28	$(1-t)/2$	${}_{SF}s_{SF}$
	S^*F		$tp_F/2$	${}_{SF}c_{SF}$
	F^*S		$tp_S/2$	${}_{SF}c_{FS}$

Let θ_{mo} be the probability an offspring from female parent of type m is of type o, as shown in Table 6.6 for a single locus. The multinomial probability

Table 6.4 Parent-offspring genotype combinations for *Lupinus albus* data of Green et al. (1980).

Maternal Genotype	Offspring Genotype	Locus		
		Pgm-1	*6Pgd-2*	*Aat-2*
FF	*FF*	120	108	80
	FS	5	4	7
FS	*FF*	11	17	14
	FS	28	35	30
	SS	16	14	16
SS	*FS*	9	16	4
	SS	104	263	283

of observing offspring array $\{n_{yo}\}$, where n_{yo} is the number of type o from parent y, assuming parent y is of genotype m, is

$$\Pr[\{n_{yo}\}|m] = (\sum_o n_{yo})! \prod_o \frac{(\theta_{mo})^{n_{yo}}}{n_{yo}!} \tag{6.4}$$

The joint probability of maternal type m and offspring array $\{n_{yo}\}$ is

$$\Pr(\{n_{yo}\}, m) = P_m \Pr(\{n_{yo}\}|m)$$

Then the probability of maternal type m given the offspring array follows from Bayes' Theorem as

$$\begin{aligned} \pi_{my} &= \Pr(m|\{n_{yo}\}) \\ &= \frac{\Pr(\{n_{yo}\}, m)}{\Pr(\{n_{yo}\})} \\ &= \frac{P_m \Pr(\{n_{yo}\}|m)}{\sum_m P_m \Pr(\{n_{yo}\}|m)} \end{aligned}$$

The estimation process is begun by assigning values to t, p_u, and P_m and using them to calculate the π_{my} values. These, in turn, lead to revised estimates of the genotypic frequencies from

$$P_m = \frac{1}{M} \sum_{y=1}^{M} \pi_{my}$$

Table 6.5 Outcrossing estimates for *Lupinus albus* data of Green et al. (1980). Initial iterates were 0.5 for t and p_F.

Iterate	*Pgm-1*		*6Pgd-2*		*Aat-2*	
	t	p_F	t	p_F	t	p_F
1	0.118	0.704	0.089	0.702	0.068	0.468
2	0.117	0.678	0.088	0.674	0.065	0.300
3	0.118	0.663	0.089	0.658	0.069	0.228
4	0.118	0.655	0.089	0.648	0.073	0.197
5	0.118	0.650	0.090	0.641	0.077	0.182
6	0.118	0.648	0.091	0.636	0.081	0.173
7	0.118	0.646	0.091	0.633	0.083	0.166
8	0.118	0.646	0.092	0.630	0.086	0.162
9	0.118	0.645	0.092	0.628	0.087	0.159
10	0.118	0.645	0.092	0.627	0.089	0.156
20	0.118	0.644	0.093	0.623	0.094	0.148
30	0.118	0.644	0.093	0.623	0.094	0.148

and iteration continues until the P_m values stabilize. Summation over y values means summing over all the M parents available. Equation 6.4 is then used to provide maximum likelihood estimates of the quantities t and

Table 6.6 Offspring genotypes and probabilities for partial outcrossing for a single locus with arbitrary number of alleles.

Maternal Genotype m	Offspring Genotype o	Probability θ_{mo}	
A_uA_u	A_uA_u	$(1-t)+tp_u$	
	A_uA_v	tp_v	
A_uA_v	A_uA_u	$(1-t)/4+tp_u/2$	
$(u \neq v)$	A_vA_v	$(1-t)/4+tp_v/2$	
	A_uA_v	$(1-t)/2+t(p_u+p_v)/2$	
	A_uA_w	$tp_w/2$	$w \neq u, v$
	A_vA_w	$tp_w/2$	$w \neq u, v$

Table 6.7 Identification of outcross events with multiple loci in hypothetical data.

Offspring	Loci				
Number	A	B	C	D	E
	Maternal genotypes				
	11	22	12	13	23
	Offspring genotypes				
1	11	22	12	13	13*
2	11	22	22	11	33
3	11	12*	12	23*	13*

* A discriminatory offspring genotype.

p_u, with numerical methods being needed. At this point the whole process needs to be repeated. Maternal genotypic frequencies are estimated and then allelic frequencies and the outcrossing rate is reestimated until all estimates have stabilized.

Multilocus Estimates

As data are usually collected at several loci for the same individual, and as all loci are subject to the same outcrossing process, it seems appropriate to base estimation procedures on all these loci. The immediate benefit of multiple loci is that it is easier to identify outcross individuals, as shown by Shaw et al. (1981). Their hypothetical data are shown in Table 6.7, illustrating that, with more loci, it is easier to detect outcross events. For offspring 1 in that table, locus E shows that the offspring is an outcross even though the other four loci were nondiscriminatory.

Shaw et al. (1981) distinguish between outcrosses and ambiguous outcrosses that cannot be identified as such on the basis of their genotypes. The probability t of outcrossing can be modified to $t(1-\alpha)$, where α is the probability that an outcross will not be discerned. The number n of discernible outcrosses in a sample of size N can be taken to be binomially distributed

$$n \sim B[N, t(1-\alpha)]$$

so that t can be estimated as

$$\hat{t} = \frac{n}{N(1-\alpha)}$$

if a value for α is available.

A discernible outcross is one in which at least one locus carries an allele that is different from both maternal alleles at that locus. The probability of an outcross being discernible is 1 minus the probability that outcrosses cannot be detected at all loci. If, at locus ℓ, the probability that outcrossing cannot be detected is β_ℓ

$$1 - \alpha = 1 - \prod_{\ell} \beta_\ell$$

or

$$\alpha = \prod_{\ell} \beta_\ell$$

Note that β_ℓ refers to an outcross that has occurred but is not discernible. Shaw et al. (1981) give β_ℓ as

$$\beta_\ell = \sum_u P_{\ell u,\ell u} p_{\ell u} + \sum_u \sum_{v \neq u} P_{\ell u,\ell v}(p_{\ell u} + p_{\ell v})$$

and they weight the maternal genotype frequencies $P_{\ell u,\ell u}$ and $P_{\ell u,\ell v}$ by the number of offspring contributed by that type. The allelic frequencies $p_{\ell u}$ refer to the pollen pool. This equation points out a potential problem. The probability of not detecting an outcross varies with maternal genotype, but Shaw et al. (1981) apply the same average value α to all offspring in the sample, regardless of maternal genotype. For small values of t this may not be too much of a problem.

Yeh and Morgan (1987) extended the *EM* approach to the multilocus case. Based on all loci, offspring are classified as discernible outcrosses or as ambiguous. Table 6.2 needs additional lines for the discernible outcrosses. The numbers of selfs and outcrosses among the ambiguous types are estimated, with the estimates now needing multilocus gametic frequencies instead of single-locus gene frequencies. The possibility of gametic linkage disequilibrium among pollen gametes needs to be accommodated.

Estimating Number of Paternal Parents

As well as estimating the proportion of individuals that arises from outcrossing in otherwise selfing species, there is the related question of estimating the number of paternal parents for a family when outcrossing has occurred. Multiple paternity is found in natural populations of many plant and animal

Table 6.8 Probabilities of mother-father pairs and offspring arrays when there is a single father for a family.

Mother	Father	Pr(Parents)	Pr(Offspring)		
			AA	Aa	aa
AA	AA	p_A^4	1	0	0
	Aa	$2p_A^3(1-p_A)$	0.5	0.5	0
	aa	$p_A^2(1-p_A)^2$	0	1	0
Aa	AA	$2p_A^3(1-p_A)$	0.5	0.5	0
	Aa	$4p_A^2(1-p_A)^2$	0.25	0.5	0.25
	aa	$p_A(1-p_A)^3$	0	0.5	0.5
aa	AA	$p_A^2(1-p_A)^2$	0	1	0
	Aa	$2p_A(1-p_A)^3$	0	0.5	0.5
	aa	$(1-p_A)^4$	0	0	1

species, with a recent review being given by Williams and Evarts (1989). Their approach to distinguishing between one and two paternal parents will now be discussed.

Although a multiple-allele treatment is possible, Williams and Evarts (1989) worked with a two-allele model, obtained by focusing attention on one allele at a locus and combining all others into a single allelic class. Observations are made on a set of n offspring from a single maternal parent. If there is a single father for all the offspring, then the probabilities of observing a family with counts n_{AA}, n_{Aa}, n_{aa} for genotypes AA, Aa, aa are given in Table 6.8. The father is not observed, and the genotype of the mother is supposed not to be available, so Table 6.8 lists all nine possible mother-father combinations. Each combination has a probability that can be expressed in terms of the allele frequency p_A, assuming Hardy-Weinberg equilibrium. For the first row, the probability of seeing counts n_{AA}, n_{Aa}, n_{aa} is

$$p_A^4 \frac{n!}{n_{AA}!n_{Aa}!n_{aa}!}(1)^{n_{AA}}(0)^{n_{Aa}}(0)^{n_{aa}}$$

Similar expressions hold for the other rows, and summing over all nine rows in the table gives the overall probability $P_1(\mathbf{n}|p_A)$ of observing the offspring array $\mathbf{n} = n_{AA}, n_{Aa}, n_{aa}$ for a specific value of p_A.

If two males inseminated the mother of a family, there are 27 possible combinations of parental genotypes. These are listed in Table 6.9 along with the corresponding probabilities. If random mixing of male gametes takes place, so that each of the two males is equally likely to be the father of any of the offspring, the probabilities of each offspring genotype for the 27 parental trios are also displayed in the table. The overall probability $P_2(\mathbf{n}|p_A)$ of an offspring array given the allelic frequency is now the sum of 27 terms.

Now the data allow the estimation of the unknown allelic frequency as well as the probability ϕ that an offspring comes from a family with multiple paternity. Allowing for both one (P_1) or two (P_2) fathers, the total probability of an offspring array is

$$\Pr(\mathbf{n}|p_A, \phi) = (1 - \phi)P_1(\mathbf{n}|p_A) + \phi P_2(\mathbf{n}|p_A)$$

In practice, data will be available from several families, and if these are independent their probabilities may be multiplied. Maximum likelihood can be used to estimate the two parameters, p_A, ϕ, from the support function

$$\ln L(p_A, \phi) = \sum_{\mathbf{n}} \log[(1 - \phi)P_1(\mathbf{n}|p_A) + \phi P_2(\mathbf{n}|p_A)]$$

Numerical methods will generally be needed to find the MLEs.

INFERENCES ABOUT PATERNITY

In paternity disputes it is necessary to determine whether a particular man is the father of a particular child. Classical considerations of such questions were limited to excluding a man from paternity of a child that has neither of the man's alleles at some locus. The increasing availability of diagnostic loci, including DNA fingerprinting, has given rise to calculations of the probability that a man is the father, although these calculations are not without their critics.

Paternity Exclusion with Codominant Loci

For a single autosomal locus with codominant alleles, those males that can be excluded as possible fathers for all possible mother-child combinations are shown in Table 6.10. The table also shows the probabilities of the various genotypes for random mothers and random men. The offspring probabilities

Table 6.9 Probabilities of mother-male trios and offspring arrays when two males inseminate the mother.

Mother	Male 1	Male 2	Pr(Parents)	Pr(Offspring)		
				AA	*Aa*	*aa*
AA	*AA*	*AA*	p_A^6	1	0	0
		Aa	$2p_A^5(1-p_A)$	0.75	0.25	0
		aa	$p_A^4(1-p_A)^2$	0.5	0.5	0
	Aa	*AA*	$2p_A^5(1-p_A)$	0.75	0.25	0
		Aa	$4p_A^4(1-p_A)^2$	0.5	0.5	0
		aa	$2p_A^3(1-p_A)^3$	0.25	0.75	0
	aa	*AA*	$p_A^4(1-p_A)^2$	0.5	0.5	0
		Aa	$2p_A^3(1-p_A)^3$	0.25	0.75	0
		aa	$p_A^2(1-p_A)^4$	0	1	0
Aa	*AA*	*AA*	$2p_A^5(1-p_A)$	0.5	0.5	0
		Aa	$4p_A^4(1-p_A)^2$	0.375	0.5	0.125
		aa	$2p_A^3(1-p_A)^3$	0.25	0.5	0.25
	Aa	*AA*	$4p_A^4(1-p_A)^2$	0.375	0.5	0.125
		Aa	$8p_A^3(1-p_A)^3$	0.25	0.5	0.25
		aa	$4p_A^2(1-p_A)^4$	0.125	0.5	0.375
	aa	*AA*	$2p_A^3(1-p_A)^3$	0.25	0.5	0.25
		Aa	$4p_A^2(1-p_A)^4$	0.125	0.5	0.375
		aa	$2p_A(1-p_A)^5$	0	0.5	0.5
aa	*AA*	*AA*	$p_A^4(1-p_A)^2$	0	1	0
		Aa	$2p_A^3(1-p_A)^3$	0	0.75	0.25
		aa	$p_A^2(1-p_A)^4$	0	0.5	0.5
	Aa	*AA*	$2p_A^3(1-p_A)^3$	0	0.75	0.25
		Aa	$4p_A^2(1-p_A)^4$	0	0.5	0.5
		aa	$2p_A(1-p_A)^5$	0	0.25	0.75
	aa	*AA*	$p_A^2(1-p_A)^4$	0	0.5	0.5
		Aa	$2p_A(1-p_A)^5$	0	0.25	0.75
		aa	$(1-p_A)^6$	0	0	1

Table 6.10 Paternity exclusion configurations at one locus with an arbitrary number of alleles.

Mother		Child		Excluded Male	
Genotype	Prob.	Genotype	Probability	Genotype	Probability
A_uA_u	p_u^2	A_uA_u	p_u	$A_wA_x,\ w,x \neq u$	$(1-p_u)^2$
		A_uA_v	p_v	$A_wA_x,\ w,x \neq v$	$(1-p_v)^2$
A_uA_v	$2p_up_v$	A_uA_u	$p_u/2$	$A_wA_x,\ w,x \neq u$	$(1-p_u)^2$
$v \neq u$		A_vA_v	$p_v/2$	$A_wA_x,\ w,x \neq v$	$(1-p_v)^2$
		A_uA_v	$(p_u+p_v)/2$	$A_wA_x,\ w,x \neq u,v$	$(1-p_u-p_v)^2$
		A_uA_y	$p_y/2$	$A_wA_x,\ w,x \neq y$	$(1-p_y)^2$
		A_vA_y	$p_y/2$	$A_wA_x,\ w,x \neq y$	$(1-p_y)^2$

are conditional on those of the mother. An assessment of how good this locus is for being able to exclude a random man from paternity, assuming that he is not the father of the child, is given by summing the joint probabilities of all the mother-child-excluded man combinations shown in Table 6.10. The probability of an A_uA_u mother with an A_uA_u child is p_u^3, and this combination excludes all men that do not have an A_u allele. Such men have frequency $(1-p_u)^2$, so that the trio in the first line of Table 6.10 has a combined probability of $p_u^3(1-p_u)^2$. Adding such probabilities from all seven lines of the table gives the exclusion probability Q

$$Q = \sum_u p_u(1-p_u)^2 - \frac{1}{2}\sum_u \sum_{v \neq u} p_u^2 p_v^2 (4 - 3p_u - 3p_v)$$

and will be maximized when all V alleles at the locus are equally frequent

$$Q_{\max} = 1 - \frac{2V^3 + V^2 - 5V + 3}{V^4}$$

The utility of a locus increases with the number of alleles, although even with 10 alleles the value of $Q_{\max}$ is only 0.79. Larger number of alleles are being found with RFLP loci (e.g., Chimera et al. 1989), and $V = 30$ is not unusual. With 30 equally frequent alleles, the value of Q increases to 0.9324.

Exclusion probabilities are increased with the use of several loci, since it is sufficient to exclude at even one out of several loci. If Q_ℓ is the exclusion probability at locus ℓ, then the overall probability of exclusion for unlinked loci follows from being able to exclude from at least one locus. In other words, Q is one minus the probability that none of the loci allows exclusion.

$$Q = 1 - \prod_{\ell}(1 - Q_{\ell})$$

With two loci, each with 10 equally frequent alleles, the value of Q increases to $1 - (1 - 0.79)^2 = 0.96$.

Now a man either is or is not the father of a child, but the probability π of the alleged father being the actual father can be defined with reference to the population from which he was drawn. Such a probability needs to be calculated only when the man has not been excluded by a genetic test. Prior to testing, there may be some belief π_0 that the alleged father is the actual father. After testing, if the man has not been excluded from paternity there is a modified probability that follows from Bayes Theorem. If N is the event of nonexclusion, then the conditional probabilities are

$$\Pr(N|\text{man is father}) = 1$$

since the true father cannot be excluded by a genetic test, and

$$\Pr(N|\text{man is not father}) = 1 - Q$$

Bayes Theorem provides the posterior probability

$$\begin{aligned} \pi_1 &= \Pr(\text{man is father}|N) \\ &= \frac{\Pr(\text{man is father}, N)}{\Pr(N)} \\ &= \frac{\Pr(N|\text{man is father})\Pr(\text{man is father})}{\Pr(N)} \\ &= \frac{\pi_0}{\pi_0 + (1 - Q)(1 - \pi_0)} \end{aligned} \tag{6.5}$$

If nonexclusion is based on several loci, ℓ, then the quantity $1 - Q$ in Equation 6.5 is replaced by $\prod_{\ell}(1 - Q_{\ell})$. As the number of loci that results in nonexclusion increases, Q increases and so does the π_1. Higher probabilities of exclusion result in higher probabilities of paternity among those men not excluded. Li and Chakravarti (1988) emphasize that π_1 is a function of the test loci, rather than of the particular genotypes of the mother-child-alleged father trios.

Table 6.11 Paternity exclusion arrangements for the three-allele *ABO* system.

		Mother			
	Excluded Men	A	B	AB	O
Child	A	None	B,O	None	B
	B	A,O	None	None	A
	AB	A,O	B,O	O	Impossible
	O	AB	AB	Impossible	AB

Paternity Exclusion with Dominant Loci

Exclusion calculations can also be made for diagnostic loci that have dominant alleles, as illustrated in Table 6.11 for the *ABO* system. Adding over all mother-child-excluded man combinations, the probability of paternity exclusion from the *ABO* system is found to be

$$Q = p(1-p)^4 + q(1-q)^4 + pqr^2(3-r)$$

where p, q, and r are the population frequencies of the A, B, and O alleles.

Examples of exclusion probabilities Q were compiled by Chakraborty et al. (1974). They used a four-allele model for the *ABO* system (alleles O, A_1, A_2, B) and reported Q values of 0.1774, 0.1342, and 0.1917 for the Black, White and Japanese populations, respectively. Greater discrepancy was found between populations for the isozyme marker *acid phosphatase* with three alleles. The Q values were then 0.1588, 0.2323, and 0.1340 for Black, White, and Japanese populations.

Paternity Index

There have been several recent publications on the subject of paternity indicies, with a useful clarification of the issues being given by Elston (1986a,b). Before the collection of any genetic evidence, there may be some prior beliefs about π, and π_0 is the PRIOR PROBABILITY of paternity. The main concern is with the situation after the genotypes of mother, child, and alleged father are available. If G denotes the genetic evidence, Bayes' Theorem leads to

$$\frac{\Pr(\text{man is father}|G)}{\Pr(\text{man is not father}|G)} = \frac{\Pr(G|\text{man is father})}{\Pr(G|\text{man is not father})} \times \frac{\Pr(\text{man is father})}{\Pr(\text{man is not father})}$$

The left-hand side of this expression is the ratio of POSTERIOR PROBABILITIES, π_1 to $(1 - \pi_1)$, while the right-hand side is the product of the PATERNITY INDEX L and the ratio of prior probabilities π_0 to $(1 - \pi_0)$:

$$\frac{\pi_1}{1 - \pi_1} = L\frac{\pi_0}{1 - \pi_0}$$

Rearranging this equation gives the posterior probability π_1 as

$$\pi_1 = \frac{L\pi_0}{(1 - \pi_0) + L\pi_0}$$

When several loci are available, L may be formed as the product of the values for each locus. Note that π_1 is the probability of an alleged man with a particular genotype being the father when genetic evidence is available from the mother, child, and himself. It is a different way of assessing the probability of paternity from that expressed in Equation 6.5. A discussion of the two approaches is given by Li and Chakravarti (1988).

Discussion of posterior probabilities is quite valid, but debate begins when particular values are given to the prior probabilities. They could be set to 0.5, for instance, so that $p_1 = L/(L+1)$ and Gürtler (1956) suggested that the man be declared the father if p_1 exceeded 0.95 ($L > 19$) and be declared not the father if p_1 was less than 0.05 ($L < 0.053$).

Although there appears to be little justification for setting p_0 to 0.5, the paternity index L does offer some means of quantifying the weight of the genetic evidence. Both Elston (1986a) and Majumder and Nei (1983) have pointed out that, at a single locus, L can have quite a restricted range of values, and Elston showed that the expected value of the posterior probability over all mother-child-alleged father combinations in the population is just the true value π. The arguments put forward by Majumder and Nei (1983) and Elston (1986a) are summarized in Table 6.12. For a two-allele locus, the various mother-child-alleged father combinations are given, and paternity indices are shown.

Use of this table shows that the mean paternity index for fathers is always higher than that of nonfathers (Elston, 1986b). The paternity index, which reduces to

$$L = \frac{\Pr(C|M, F)}{\Pr(C|M)}$$

is a likelihood ratio.

Table 6.12 Paternity index calculations for a two-allele locus as given by Elston (1986a).

Mother	Child	Alleged Father	X^*	$Y^\dagger$	$L^\ddagger$
AA	AA	AA	p_A^4	p_A^5	$1/p_A$
		Aa	$p_A^3 p_a$	$2p_A^4 p_a$	$1/2p_A$
		aa	0	$p_A^3 p_a^2$	0
	Aa	AA	0	$p_A^4 p_a$	0
		Aa	$p_A^3 p_a$	$2p_A^3 p_a^2$	$1/2p_a$
		aa	$p_A^2 p_a^2$	$p_A^2 p_a^3$	$1/p_a$
Aa	AA	AA	$p_A^3 p_a$	$p_A^4 p_a$	$1/p_A$
		Aa	$p_A^2 p_a^2$	$2p_A^3 p_a^2$	$1/2p_A$
		aa	0	$p_A^2 p_A^3$	0
	Aa	AA	$p_A^3 p_a$	$p_A^3 p_a$	1
		Aa	$2p_A^2 p_a^2$	$2p_A^2 p_a^2$	1
		aa	$p_A p_a^3$	$p_A p_a^3$	1
	aa	AA	0	$p_A^3 p_a^2$	0
		Aa	$p_A^2 p_a^2$	$2p_A^2 p_a^3$	$1/2p_a$
		aa	$p_A p_a^3$	$p_A p_a^4$	$1/p_a$
aa	Aa	AA	$p_A^2 p_a^2$	$p_A^3 p_a^2$	$1/p_A$
		Aa	$p_A p_a^3$	$2p_A^2 p_a^3$	$1/2p_A$
		aa	0	$p_A p_a^4$	0
	aa	AA	0	$p_A^2 p_a^3$	0
		Aa	$p_A p_a^3$	$2p_A p_a^4$	$1/2p_a$
		aa	p_a^4	p_a^5	$1/p_a$

$^*X = \Pr(G|A = F) = \Pr(M)\Pr(A)\Pr(C|M, A = F)$
$^\dagger Y = \Pr(G|A \neq F) = \Pr(M)\Pr(A)\Pr(C|M)$
$^\ddagger L = X/Y$ = paternity index
M = mother , A = alleged father , F = father , C = child

DNA Fingerprinting

Traditional use of genetic markers in paternity testing has made use of genetic systems, such as blood groups, that have relatively few loci and few alleles per locus. Although exclusion can be established with such systems, it is difficult to show genetic identity because each of the various marker classes in a population is relatively common. At the ultimate level of the DNA sequence, every individual, apart from identical twins, is unique. Information at the sequence level would therefore seem to offer more help in paternity cases or other situations in which it is desirable to establish identity. Instead of going all the way to the sequence, great success has been achieved with patterns of restriction fragments that reveal variation at many loci, or at single loci with many alleles.

The term DNA FINGERPRINTS was coined by Jeffreys et al. (1985) to refer to the collection of restriction fragments produced by multiple loci and detected with probes for hypervariable minisatellite sequences. Because many loci are involved, and because restriction fragments of the same length (see Chapter 7) may be produced by different loci, it is not possible to identify the origins of any specific fragment or feature of the "fingerprint." So much of the genome is being studied, however, that the same pattern found in two different DNA samples almost certainly indicates that the samples are from the same individual, while differences in any aspect of two patterns almost certainly indicate that the samples are from different individuals. There are applications for such means of detecting identity or non-identity in paternity disputes, immigration cases, and forensic science. A brief review was given by King (1989).

A growing use of restriction fragments in this area concentrates instead on a small number of loci that has many alleles. These loci are generally of the VNTR (variable number of tandem repeats) type, where different alleles (fragment lengths) result from different numbers of repeated units. Calculations of probabilities of identity rest on population frequencies of these alleles. Some of the statistical issues in these calculations have been reviewed recently by Lander (1989) and his points are now discussed. The central problem, whether in forensic or paternity cases, is to determine whether two DNA samples are identical.

The first issue is in deciding whether two samples have the same size fragment — revealed by the same amount of migration on an electrophoretic gel. To accommodate the many sources of experimental error in determining the position of a band on a gel, it is usual to treat bands as being at the same position if their measured positions are within some specified number of standard deviations of each other. The standard deviations of the measurement process are estimated by running fragments known to be identical in different lanes of a gel. Lander (1989) stresses that whatever rule is used to decide identity of bands between two samples should also be used in the population surveys used to establish frequencies of the corresponding alleles in the population. Gjertson et al. (1988) formalized the effects of experimental error on calculations of paternity probabilities.

Once two samples have been shown to have collections of bands that are the same, the probability that such an event could have arisen by chance is calculated. Samples with different bands are regarded as establishing non-identity. Since DNA is collected at the genotypic level, it is necessary to perform calculations at this level also. The discussion in Chapter 5 on sampling from a fixed population is relevant here. For any one locus, genotypic frequencies P_{uv} in some target population must be known. The probability that two samples, taken at random from such a population, have the same marker genotype is then just $\sum_{u,v} P_{uv}^2$. This quantity will be small for systems with large numbers of alleles and small frequencies for any particular genotype. Lander (1989) points to the care needed to identify the target population and to ensure that it can be properly characterized by a single set of frequencies. In practice, use is made of the Hardy-Weinberg assumption to construct population genotypic frequencies from allelic frequencies. Problems of false assumptions of HWE would seem to be best avoided, if possible, by taking samples large enough to provide genotypic frequencies directly, although with 30 or more alleles per locus this would not be practical. Given that HWE is assumed, calculations may as well be done on the allelic frequencies on the basis of their (assumed) multinomial sampling distribution.

Several marker loci are usually used, and probabilities for each locus are combined by multiplication. This can be justified only if the absence of linkage disequilibrium among the markers in the population is first demonstrated. The techniques of Chapter 3 should therefore be applied to the reference data collected for each target population. Balasz et al. (1989) studied five hypervariable loci in three ethnic groups. They found from 30 to 80 distinct bands for each locus, and documented frequency differences among the three populations. They did not find evidence of linkage disequilibrium between loci.

Most Likely Paternal Plants

Paternity analyses have been extended to plant populations. An examination of the genotypes of maternal plant, offspring, and all available paternal plants may lead to a determination of which male is the most likely father. Such work is generally based strictly on the genotypes of mother, offspring, and male, without regard to the geographic distance between male and mother. Pollen dispersal distance effects are ignored. If R_1 is the event that male A

Table 6.13 One-locus, two-allele lod scores for possible paternal plants for one locus with two alleles.

Mother M	Offspring C	Male A	$\Pr(C\|M)$	$\Pr(C\|M,A)$	Ratio
Aa	AA	AA*	p_A	1	$1/p_A$
		Aa	p_A	1/2	$1/(2p_a)$
		aa	p_a	0	0

*Most likely father.

is the father F of offspring C from female parent M, and R_2 the event that the male is simply a random male from the population and not related to C, then the likelihoods

$$\begin{aligned} L(R_1) &= \Pr(C|M,A)\Pr(M)\Pr(A) \\ L(R_2) &= \Pr(C|M)\Pr(M)\Pr(A) \end{aligned}$$

are both functions of the genotypes of M and A and the probabilities of C conditional on the known parent(s). There is a direct similarity to the work on disputed paternity in humans. The logarithm of the ratio of these likelihoods, known as the LOD SCORE, is a function of quantities that can be determined by transmission probabilities for the loci scored

$$\text{lod}(R_1 : R_2) = \log\left[\frac{\Pr(C|M,A)}{\Pr(C|M)}\right]$$

These lods are calculated for each possible male A, and the male with the highest score, provided that score exceeds some preset value, is said to be the father of the offspring. Some simple cases are set out in Table 6.13 under the assumption of random mating. This methodology has been developed and used by Meagher (1986).

ESTIMATION OF SELECTION

There are several methods available for detecting the presence of natural selection operating within a population. Many of these depend on estimating genotypic frequencies in successive generations, and comparing the results to those expected on the basis of no selection. When specific selection models are formulated, the appropriate parameters may be estimated from the relation between successive frequencies.

Goodness-of-Fit Test for Selection

Lewontin and Cockerham (1959) showed that selection may produce genotypic frequencies that are indistinguishable from those that would be obtained in the absence of selection. They were concerned with attempts to demonstrate the presence of selection from genotypic frequencies within a single population.

For a locus **A** with alleles A, a, and allelic frequencies p_A, p_a among adults, the frequencies of genotypes in the offspring generation before the action of any selection are

$$\begin{array}{ccc} AA & Aa & aa \\ p_A^2 & 2p_Ap_a & p_a^2 \end{array}$$

If the relative viabilities of the AA and aa homozygotes with respect to the heterozygote Aa are w_{AA} and w_{aa}, then the genotypic frequencies after selection will be

$$\begin{array}{ccc} AA & Aa & aa \\ w_{AA}p_A^2/\bar{w} & 2p_Ap_a/\bar{w} & w_{aa}p_a^2/\bar{w} \end{array}$$

where the mean fitness $\bar{w}$ is

$$\bar{w} = w_{AA}p_A^2 + 2p_Ap_a + w_{aa}p_a^2$$

so that the allelic frequencies among adult offspring are

$$\begin{aligned} p_A' &= \frac{w_{AA}p_A^2 + p_Ap_a}{\bar{w}} \\ p_a' &= \frac{w_{aa}p_a^2 + p_Ap_a}{\bar{w}} \end{aligned}$$

If a sample is taken from the adult offspring generation, will the action of selection result in evidence for departures from Hardy-Weinberg equilibrium? If a sample of size n has genotypic numbers n_{AA}, n_{Aa}, and n_{aa}, the sample allelic frequencies

$$\tilde{p}_A = \frac{2n_{AA} + n_{Aa}}{2n}, \quad \tilde{p}_a = \frac{2n_{aa} + n_{Aa}}{2n}$$

provide expected values for the test of HWE:

$$\begin{array}{ccc} AA & Aa & aa \\ n\tilde{p}_A^2 & 2n\tilde{p}_A\tilde{p}_a & n\tilde{p}_a^2 \end{array}$$

Recall that the sample allelic frequencies $\tilde{p}_A, \tilde{p}_a$ estimate the offspring adult frequencies p'_A, p'_a. Hardy-Weinberg genotypic frequencies exist in the offspring population if

$$\begin{aligned} (p'_A)^2 &= \frac{w_{AA}p_A^2}{\bar{w}} \\ 2(p'_A)(p'_a) &= \frac{2p_Ap_a}{\bar{w}} \\ (p'_a)^2 &= \frac{w_{aa}p_a^2}{\bar{w}} \end{aligned}$$

Lewontin and Cockerham (1959) point out that this requires

$$w_{AA}w_{aa} = 1$$

so that selection schemes with this relation will produce populations that have Hardy-Weinberg proportions. Evidently then, neutrality is one of the sufficient conditions for HWE, but not a necessary condition. Li (1988) showed that HWE proportions can also be found for various nonrandom mating situations.

If the population is in gene-frequency equilibrium (but not Hardy-Weinberg equilibrium), meaning that frequencies do not change over time,

$$p'_A = p_A, \; p'_a = p_a$$

then it is possible to estimate the two relative viabilities w_{AA} and w_{aa}. In this situation

$$\bar{w} = w_{AA}p_A + p_a = w_{aa}p_a + p_A$$

so that there are only two independent parameters, p_A and w_{AA}. Applying Bailey's rule (Chapter 2) gives maximum likelihood estimates of

$$\begin{aligned} \hat{p}_A &= \frac{2n_{AA} + n_{Aa}}{2(n_{AA} + n_{Aa} + n_{aa})} \\ \hat{w}_{AA} &= \frac{2n_{AA}(2n_{aa} + n_{Aa})}{n_{Aa}(2n_{AA} + n_{Aa})} \end{aligned}$$

With gene-frequency equilibrium, Hardy-Weinberg deviations can be written in terms of the disequilibrium coefficient D_A

$$\begin{aligned} D_A &= \frac{w_{AA}p_A^2}{\bar{w}} - \left(\frac{w_{AA}p_A^2 + p_Ap_a}{\bar{w}}\right)^2 \\ &= \frac{p_A^2p_a^2(w_{AA}w_{aa} - 1)}{\bar{w}^2} \end{aligned}$$

When the disequilibrium is not zero, the chi-square goodness-of-fit statistic has noncentrality parameter ν

$$\begin{aligned}\nu &= n\left[\frac{p_A p_a(w_{AA}w_{aa}-1)}{\bar{w}}\right]^2 \\ &= n\left[\frac{p_A p_a(w_{AA}w_{aa}-1)}{(w_{AA}p_A+p_a)(w_{aa}p_a+p_A)}\right]^2\end{aligned}$$

to correct the expressions given in Lewontin and Cockerham (1959) and in Weir and Cockerham (1978). As in Chapter 3, this allows determination of sample size to detect specified departures of $w_{AA}w_{aa}$ from 1.

Estimation within One Generation

Another way of detecting selection is to compare frequencies of one class of individual with those of a standard type that has an equal expected frequency in the absence of selection. Suppose A, B are phenotypes with observed numbers a, b in a sample of size n. If the viability of B relative to that of A is v, the expected proportions of A and B are

$$A:\frac{1}{v+1},\quad B:\frac{v}{v+1}$$

A naïve estimate of v is b/a, but this can be quite biased. A better estimate (Haldane 1956) is

$$\hat{v}=\frac{b}{a+1}$$

and this has been used in several experiments on *Drosophila* by Mukai (e.g., Mukai et al. 1974). Properties of this estimate follow from the binomial distribution

$$b \sim B\left(n, \frac{v}{v+1}\right)$$

The expected value of the estimate is

$$\begin{aligned}\mathcal{E}(\hat{v}) &= \sum_{b=0}^{n}\hat{v}\,\Pr(b) \\ &= \sum_{b=0}^{n}\frac{b}{a+1}\frac{n!}{a!b!}\left(\frac{v}{v+1}\right)^b\left(\frac{1}{v+1}\right)^a \\ &= v\sum_{b=1}^{n}\frac{n!}{(b-1)![n-(b-1)]!}\frac{v^{(b-1)}}{(v+1)^n}\end{aligned}$$

$$
\begin{aligned}
&= v\left[1 - \left(\frac{v}{v+1}\right)^n\right] \\
&\approx v \text{ for large } n
\end{aligned}
$$

By contrast, the naïve estimate does not have a finite expectation since there is the possibility that a will be zero in a sample. This problem could be removed by using the distribution of b conditional on b being less than n (and a being greater than zero):

$$
\Pr(b|b < n) = \frac{\Pr(b)}{1 - \Pr(b = n)}
$$

To find the variance of $\hat{v}$ note that it is not a function to which Fisher's approximate formula (Chapter 2) may be applied. Instead, $\hat{v}$ can be expressed in terms of b and the delta method used:

$$
\begin{aligned}
\hat{v} &= \frac{b}{n + 1 - b} \\
\text{Var}(\hat{v}) &\approx \left(\frac{\partial \hat{v}}{\partial b}\right)^2 \text{Var}(b) \\
&= \frac{v(v+1)^2}{n}
\end{aligned}
$$

Components of Selection

Detecting selection from an analysis of frequencies in successive generations is complicated by the existence of several types of selection that act at different stages of the life cycle. A typical representation is

$$
\text{Mating} \longrightarrow \text{Zygote}_t \stackrel{\text{early}}{\longrightarrow} \text{Adult}_t \stackrel{\text{late}}{\longrightarrow} \text{Mating} \longrightarrow \text{Zygote}_{t+1} \stackrel{\text{early}}{\longrightarrow} \text{Adult}_{t+1}
$$

with viability (early) selection between zygotes and adults in each generation, and fertility (late) selection between adults in one generation and zygotes in the next.

A sample may be taken from the population at the same stage in successive generations with the aim of making inferences about selection. This procedure was discussed by Prout (1965). He considered a two-allele locus acted on by "early" (viability) and "late" (fertility) selection. Observations are taken between the early and late stages. Assuming random mating, the genotypic frequencies are given in Table 6.14. Divisors for mean fitnesses have been omitted in Table 6.14 as they drop out of the following ratios.

Table 6.14 Genotypic frequencies under Prout's model of early and late selection.

	AA	Aa	aa
GENERATION t			
Pre-selection	p_A^2	$2p_Ap_a$	p_a^2
Early selection	$p_A^2E_{AA}$	$2p_Ap_a$	$p_a^2E_{aa}$
Late selection	$p_A^2E_{AA}L_{AA}$	$2p_Ap_a$	$p_a^2E_{aa}L_{aa}$
GENERATION $t+1$			
Pre-selection	X_A^{2*}	$2X_AX_a$	X_a^2
Early selection	$X_A^2E_{AA}$	$2X_AX_a$	$X_a^2E_{aa}$

$^*X_A = p_A^2E_{AA}L_{AA} + p_Ap_a,\ \ X_a = p_a^2E_{aa}L_{aa} + p_ap_A$

Within each of the two generations, the ratios of the AA homozygote frequencies to the heterozygote frequencies are

$$R_{AA_t} = \frac{P_{AA_t}}{P_{Aa_t}} = \frac{p_A^2E_{AA}}{2p_Ap_a}$$

with a similar expression for R_{aa_t}, and

$$R_{AA_{t+1}} = E_{AA}\frac{p_A^2E_{AA}L_{AA} + p_Ap_a}{2(p_a^2E_{aa}L_{aa} + p_Ap_a)} = \frac{E_{AA}}{2}\frac{2R_{AA_t}L_{AA} + 1}{2R_{aa_t}L_{aa} + 1}$$

and similarly for $R_{aa_{t+1}}$. It is the ratios R that are the observable quantities.

The total fitness of a genotype may be defined as the product of early and late selection coefficients

$$w = EL$$

When there is no late selection, the expressions for the ratios in the second generation can be rearranged to provide estimators of fitness

$$w_{AA} = E_{AA} \ \hat{=}\ 2R_{AA_{t+1}}\frac{2R_{aa_t} + 1}{2R_{AA_t} + 1}$$
$$w_{aa} = E_{aa} \ \hat{=}\ 2R_{aa_{t+1}}\frac{2R_{AA_t} + 1}{2R_{aa_t} + 1} \qquad (6.6)$$

If late selection is present, however, these last expressions will not provide estimates of either early or total fitnesses. Even though observations are being made at the same point of the life cycle, Equations 6.6 do not lead to estimates of total fitness, as can be seen by including late fitness in those equations.

$$\begin{aligned}
\hat{w}_{AA} &= 2R_{AA_{t+1}}\frac{2R_{aa_t}+1}{2R_{AA_t}+1} \\
&= E_{AA}\frac{(2R_{AA_t}L_{AA}+1)(2R_{aa_t}+1)}{(2R_{aa_t}L_{aa}+1)(2R_{AA_t}+1)} \\
\hat{w}_{aa} &= 2R_{aa_{t+1}}\frac{2R_{AA_t}+1}{2R_{aa_t}+1} \\
&= E_{aa}\frac{(2R_{aa_t}L_{aa}+1)(2R_{AA_t}+1)}{(2R_{AA_t}L_{AA}+1)(2R_{aa_t}+1)}
\end{aligned}$$

and in order for

$$\begin{aligned}
\hat{w}_{AA} &= E_{AA}L_{AA} \\
\hat{w}_{aa} &= E_{aa}L_{aa}
\end{aligned}$$

it is necessary that

$$L_{AA}L_{aa} = 1$$

and

$$(4R_{AA_t}R_{aa_t} - 1)(L_{AA} - 1) = 0$$

This last expression requires either that $L_{AA} = 1$, implying the absence of late selection, or that $4R_{AA_t}R_{aa_t} = 1$, implying Hardy-Weinberg frequencies among the observed individuals. Prout (1965) explained that when these conditions are not met, the use of $\hat{w}_{AA}$ as an estimate of total fitness will give spurious indications of frequency-dependent selection. If selection has not been completed at the time observations are made, the estimated viabilities may not represent net fitness values. For samples taken before selection operates ($E_{AA} = E_{aa} = 1$), then $\hat{w}_{AA}\hat{w}_{aa} = 1$, giving a misleading suggestion of the heterozygote having fitness intermediate between those of the two homozygotes. Estimation of selection coefficients must take into account the mode of selection and the stage of the life cycle at which observations are made.

In a classical set of experiments on the fitnesses associated with *Drosophila* chromosomes Dobzhansky and Levene (1951) used information on successive

Table 6.15 Dobzhansky and Levene's selection model with one mode of selection.*

	AA	Aa	aa
Before selection	$x^2(1-p_A)^2$	$2x(1-p_A)^2$	$x(1-p_A)^2$
After selection	$w_{AA}x^2$	$2x$	w_{aa}

*$x = p_A/(1-p_A)$

pairs of observations, but pointed out that more than one pair is needed. Their notation is given in Table 6.15, showing that they wrote x for the ratio $p_A/(1-p_A)$ in one generation.

In the next generation

$$x' = x\frac{w_{AA}x+1}{x+w_{aa}}$$

and this can be arranged to give

$$w_{aa} = \left(\frac{x}{x'} - x\right) + x\left(\frac{x}{x'}\right)w_{AA}$$

In other words, the data provide only a linear relation between the two fitnesses, rather than separate estimates of each. At least two such data sets are needed to determine the two selection coefficients separately. The coefficients are found as the point of intersection of the two linear relations, one for each data set. Should the lines not intersect for positive w's, it may be concluded either that the model is wrong, or that there is a large sampling error.

Maximum Likelihood Estimation of Viability Selection

A general maximum likelihood method for estimating viability selection in experimental populations was given by DuMouchel and Anderson (1968). The method depends on setting up transition equations for allelic frequencies, under the assumption of random mating, and then comparing observed and predicted frequencies in generations subsequent to an initial generation that has known frequencies.

Consider a locus with alleles A_u that have frequencies p_{u_t} at the beginning of generation t. Under the random mating assumption, genotypic frequencies are $p_{u_t}^2$ for A_uA_u homozygotes and $2p_{u_t}p_{v_t}$ for A_uA_v heterozygotes.

Selection acts during the generation to modify these genotypic frequencies to $w_{uu}p_{u_t}^2/\bar{w}_t$ and $2w_{uv}p_{u_t}p_{v_t}/\bar{w}_t$, respectively, where the mean fitness is

$$\bar{w}_t = \sum_u \sum_v w_{uv} p_{u_t} p_{v_t}$$

Allelic frequencies at the beginning of generation $t+1$ are just those among the parents of generation t:

$$\begin{aligned} p_{u_{t+1}} &= \frac{w_{uu}p_{u_t}^2 + \sum_{v \neq u} w_{uv} p_{u_t} p_{v_t}}{\bar{w}_t} \\ &= p_{u_t} \frac{\sum_v w_{uv} p_{v_t}}{\bar{w}_t} \end{aligned}$$

DuMouchel and Anderson (1968) were able to use these recurrence equations by taking known allelic frequencies at the beginning of an experiment and predicting what the frequencies should be in a series of subsequent generations. At the beginning of generation t, still under the assumption of random mating, the numbers n_{u_t} of alleles in a sample are multinomially distributed. Therefore the likelihood of a set of selection coefficients is

$$L(\{w_{uv}\})_t \propto \prod_u (p_{u_t})^{n_{u_t}}$$

When data are available from several generations

$$L(\{w_{uv}\}) = \prod_t L(\{w_{uv}\})_t$$

Once again, numerical methods are needed to find the set of w_{uv} values that maximizes the likelihood.

Selection for Partial Selfers

Workman and Jain (1966) considered the estimation of selection coefficients in plant species that practice self-mating and random outcrossing. With the addition of selfing, the transition equations for allelic frequencies are not as simple as in the previous section. Workman and Jain repeated Prout's (1965) warning that estimation must take account of the stages of the life cycle at which selection acts and at which observations are made. They phrased their development in terms of the arrangement

Stage	zygotes	$\xrightarrow{1}$	mature adults	$\xrightarrow{2}$	zygotes
Generation	t		t		$t+1$

In their model I, both selection and observations occur at stage 1 with selection having occurred prior to scoring. This model seems to be appropriate for seed characters that are determined by the maternal genotype. Model II allows selection at either stages 1 or 2 but requires that scoring be done before the action of any selection. This model is used for allozyme loci scored on very young plants. Model III has only fecundity selection, meaning that selection is at stage 2, and scoring is prior to any selection. Each model therefore has selection acting during a prescribed stage and scoring not taking place at a partially selected stage.

Workman and Jain (1966) gave estimates for loci with two alleles under each of their three models. Generalizations to multiple alleles and two loci were given by Allard et al. (1972) and Weir et al. (1974). Under Model II, the genotypic transition equations for the frequencies P_{uu} of A_uA_u homozygotes and P_{uv} of A_uA_v heterozygotes are

$$\begin{aligned} P_{uu}^{t+1} &= s\left(w_{uu}P_{uu}^t + \frac{1}{4}\sum_{v\neq u} w_{uv}P_{uv}^t\right)/\bar{w} \\ &\quad + (1-s)\left(w_{uu}P_{uu}^t + \frac{1}{2}\sum_{v\neq u} w_{uv}P_{uv}^t\right)^2 /\bar{w}_t^2 \\ P_{uv}^{t+1} &= s\left(\frac{1}{2}w_{uv}P_{uv}^t\right)/\bar{w} + 2(1-s)\left(w_{uu}P_{uu}^t + \frac{1}{2}\sum_{x\neq u} w_{ux}P_{ux}\right) \\ &\quad \times \left(w_{vv}P_{vv}^t + \frac{1}{2}\sum_{x\neq v} w_{vx}P_{vx}\right)/\bar{w}_t^2, \quad v \neq u \end{aligned}$$

where s and $1-s$ are the selfing and outcrossing rates, respectively, and $\bar{w}_t$ is the mean fitness in generation t:

$$\bar{w}_t = \sum_u w_{uu}P_{uu}^t + \sum_u \sum_{v\neq u} w_{uv}P_{uv}^t$$

To simplify things, use is made of the fact that the allelic frequencies in generation $t+1$ are given by

$$p_u^{t+1} = (w_{uu}P_{uu}^t + \sum_{v\neq u} w_{uv}P_{uv}^t)/\bar{w}_t$$

This allows the transition equations to be rewritten as

$$P_{uu}^{t+1} - (p_u^{t+1})^2 = s(w_{uu}P_{uu}^t + \frac{1}{4}\sum_{v\neq u} w_{uv}P_{uv}^t)/\bar{w}_t$$

$$P_{uv}^{t+1} - 2p_u^{t+1}p_v^{t+1} = s(\frac{1}{2}w_{uv}P_{uv}^t)/\bar{w}_t$$

As only *relative* viabilities can be estimated from genotypic proportions, some condition needs to be placed on the coefficients w_{uv}. For example, genotype A_1A_1 could be assigned unit fitness and then the other w_{uv}'s will measure fitness relative to this genotype. With this convention, the method of moments applied to the transition equations leads to the following estimates as functions of the observed genotypic frequencies in generations t and $t+1$, and observed allelic frequencies in generation $t+1$:

$$\hat{w}_{uu} = \frac{\tilde{P}_{11}^t}{\tilde{P}_{uu}^t}\left\{\frac{2\tilde{P}_{uu}^{t+1} - \tilde{p}_u^{t+1}[s + 2(1-s)\tilde{p}_u^{t+1}]}{2\tilde{P}_{11}^{t+1} - \tilde{p}_1^{t+1}[s + 2(1-s)\tilde{p}_1^{t+1}]}\right\}$$

$$\hat{w}_{uv} = \frac{2\tilde{P}_{11}^t}{\tilde{P}_{uv}^t}\left\{\frac{\tilde{P}_{uv}^{t+1} - (1-s)\tilde{p}_u^{t+1}\tilde{p}_v^{t+1}}{\tilde{P}_{11}^{t+1} - \tilde{p}_1^{t+1}[s + 2(1-s)\tilde{p}_1^{t+1}]}\right\},\ v \neq u$$

Variances of these estimates can be found by the delta method, assuming independent multinomial sampling of genotypes within each of the two successive generations.

Under Workman and Jain's Model I, the transition equations are much simpler, since the selection coefficients enter linearly:

$$P_{uu}^{t+1} = \frac{w_{uu}}{\bar{w}_t}\left[s(P_{uu}^t + \frac{1}{4}\sum_{v\neq u}P_{uv}^t) + (1-s)(P_{uu}^t + \frac{1}{2}\sum_{v\neq u}P_{uv}^t)^2\right]$$

$$P_{uv}^{t+1} = \frac{w_{uv}}{\bar{w}_t}\left[s(\frac{1}{2}P_{uv}^t) + 2(1-s)(P_{uu}^t + \frac{1}{2}\sum_{x\neq u}P_{ux})(P_{vv}^t + \frac{1}{2}\sum_{x\neq v}P_{vx})\right]$$

With the same convention of unit fitness for A_1A_1, the method-of-moments estimators are

$$\hat{w}_{uu} = \frac{\tilde{P}_{11}^{t+1}}{\tilde{P}_{uu}^{t+1}}\left[\frac{s(\tilde{P}_{uu}^t + \frac{1}{4}\sum_{v\neq u}\tilde{P}_{uv}^t)}{s(\tilde{P}_{11}^t + \frac{1}{4}\sum_{v\neq 1}\tilde{P}_{1v}^t)}\right]$$

$$\hat{w}_{uv} = \frac{\tilde{P}_{uv}^{t+1}}{\tilde{P}_{uu}^{t+1}}\left[\frac{s(\frac{1}{2}\tilde{P}_{uv}^t) + 2(1-s)(\tilde{P}_{uu}^t + \frac{1}{2}\sum_{x\neq u}\tilde{P}_{ux})(\tilde{P}_{vv}^t + \frac{1}{2}\sum_{x\neq v}\tilde{P}_{vx})}{s(\tilde{P}_{11}^t + \frac{1}{4}\sum_{v\neq 1}\tilde{P}_{1v}^t)}\right]$$

$$v \neq u$$

For both Model I and II estimators it is necessary that independent estimates of the selfing rate s are available.

Use of Mother-Offspring Data

Another treatment of components of selection, with special reference to animal populations, was given by Christiansen and Frydenberg (1973). These authors considered the case in which data are collected on mothers and offspring, as may be the case in fish populations. They identified four selection components:

ZYGOTIC SELECTION Differential survival of genotypes from zygote to adult.

SEXUAL SELECTION Differential success of genotypes at mating.

FECUNDITY SELECTION Differential zygote production of matings.

GAMETIC SELECTION Distorted segregation in heterozygotes.

Disentangling these components requires observations at different stages of the life cycle. An ideal data set would consist of

1. Sample of the population of zygotes.

2. Sample of population of adults – males and females.

3. Sample of breeding population – males, females, and mate pairs.

4. Numbers of zygotes produced by mate pairs.

5. Sample of new population of zygotes.

Data of types 1 and 2 would allow estimation of the zygotic selection coefficients, types 2 and 3 estimation of sexual selection coefficients, type 4 estimation of fecundity selection coefficients, and types 3, 4, and 5 estimation of gametic selection coefficients. For species of plants or animals in which zygotes produced by individual females can be identified, all these data types may be available. In the fish populations studied by Christiansen and Frydenberg, the adult population was sampled when females were pregnant, so that samples consisted of males, and breeding and nonbreeding females. Such data differ from the ideal in that information about males is limited to male gametes transmitted to offspring. Gametic and sexual selection on males cannot be identified separately with such data.

The counts of each of the three offspring genotypes for a locus with two alleles from each of the maternal genotypes are shown in Table 6.16, in the notation of Christiansen and Frydenberg. The authors assumed genotypes were available for one randomly chosen fetal offspring per pregnant female. Counts for non-pregnant females and adult males are also shown. It can be seen from the table that there are seven offspring counts, three nonbreeding

Table 6.16 Notation for mother-offspring counts for the model of Christiansen and Frydenberg (1973).

		Offspring				Nonbreeding	Adult
		A_1A_1	A_1A_2	A_2A_2	Sums	Females	Males
Mother	A_1A_1	C_{11}	C_{12}	–	$F_1 = C_{11} + C_{12}$	S_1	M_1
	A_1A_2	C_{21}	C_{22}	C_{23}	$F_2 = C_{21} + C_{22} + C_{23}$	S_2	M_2
	A_2A_2	–	C_{32}	C_{33}	$F_3 = C_{32} + C_{33}$	S_3	M_3

$$\begin{aligned} \text{Male gametes} : G_1 &= C_{11} + C_{21} + C_{32}, \\ G_2 &= C_{12} + C_{23} + C_{33} \end{aligned}$$

female counts and three adult male counts, giving a total of $6 + 2 + 2 = 10$ d.f. for estimating parameters and testing hypotheses.

With no assumptions about the structure of the data, the expected proportions could be defined as in Table 6.17. Within each of the three types, MLE's for the parameters follow simply as the observed proportions as shown in Chapter 2:

$$\hat{\gamma}_{uv} = \frac{C_{uv}}{C_{..}}, \quad \hat{\sigma}_u = \frac{S_u}{S.}, \quad \hat{\alpha}_u = \frac{A_u}{A.}$$

There is an overall test available for random mating and no selection, when the genotypic counts would be in Hardy-Weinberg proportions. If the allelic frequency of A_1 is written as π_1, the various genotypic classes have the proportions shown in Table 6.18. Writing $\pi_2 = 1 - \pi_1$, the likelihood for the 10 counts under the Hardy-Weinberg hypothesis is

$$\begin{aligned} L(\pi_1) &\propto \pi_1^{3C_{11}}[\pi_1^2(1-\pi_1)]^{(C_{12}+C_{21})}[\pi_1(1-\pi_1)]^{C_{22}}[\pi_1(1-\pi_1)^2]^{(C_{23}+C_{32})} \\ &\quad \times (1-\pi_1)^{3C_{23}}\pi_1^{2(S_1+M_1)}[2\pi_1(1-\pi_1)]^{(S_2+M_2)}(1-\pi_1)^{2(S_3+M_3)} \end{aligned}$$

so that

$$\ln L(\pi_1) \propto X \ln \pi_1 + Y \ln(1 - \pi_1)$$

and

Table 6.17 Expected proportions for mother-offspring data for the model of Christiansen and Frydenberg (1973).

Category	Expected Proportion	Observed Count
Offspring u, Mother v	γ_{uv}	C_{uv}
Mother u	ϕ_u	F_u
Nonbreeding female u	σ_u	S_u
Adult male u	α_u	M_u

$$\hat{\pi}_1 = \frac{X}{X+Y}$$

where X is the number of A_1 alleles and $X+Y$ is the total number of alleles in the sample

$$\begin{aligned} X &= 3C_{11} + 2(C_{12} + C_{21}) + C_{22} + C_{23} \\ &\quad + C_{32} + 2(S_1 + M_1) + S_2 + M_2 \\ X + Y &= 3C_{..} - C_{22} + 2S_{.} + 2M_{.} \end{aligned}$$

Dots in this notation indicate summation over those subscripts. The estimation of π_1 accounts for 1 d.f., and the remaining 9 can be accounted for with a goodness-of-fit chi-square on the 10 categories:

Table 6.18 Hardy-Weinberg proportions for mother-offspring data for the model of Christiansen and Frydenberg (1973).

		Offspring			Nonbreeding Females and Adult Males
		A_1A_1	A_1A_2	A_2A_2	
Mother	A_1A_1	π_1^3	$\pi_1^2\pi_2$	–	π_1^2
	A_1A_2	$\pi_1^2\pi_2$	$\pi_1\pi_2$	$\pi_1\pi_2^2$	$2\pi_1\pi_2$
	A_2A_2	–	$\pi_1\pi_2^2$	π_2^3	π_2^2

Table 6.19 Hypotheses for selection components with mother-offspring data. From Christiansen and Frydenberg (1973).

	Hypothesis	Component if Hypothesis Rejected
H_1	Half offspring from A_1A_2 mothers are heterozygotes.	Gametic selection in A_1A_2 females.
H_2	Frequency of transmitted male gametes independent of the genotype of the mother.	Nonrandom mating in the breeding population, and female specific selection of male gametes.
H_3	Frequency of transmitted male gametes equals the gene frequency in adult males.	Differential male mating success, and gametic selection in males.
H_4	Equal genotypic frequencies among mothers and nonbreeding females.	Differential female mating success.
H_5	Equal genotypic frequencies among adult females and adult males.	Unequal zygotic selection in males and females.
H_6	Adult population equals estimated zygote population.	Zygotic selection.

$$X^2 = \frac{(C_{11} - \hat{\pi}_1^3)^2}{\hat{\pi}_1^3} + \cdots + \frac{[M_3 - (1 - \hat{\pi}_1)^2]^2}{(1 - \hat{\pi}_1)^2}$$

An alternative procedure is to look at the three types of counts separately and set up a series of tests for various components of selection. Christiansen and Frydenberg identify six hypotheses, as listed in Table 6.19.

Only hypothesis H_1 will be treated in detail here. The hypothesis is of no selection among female gametes and it implies that heterozygous females contribute equal proportions of each of their two alleles to offspring and

$$\gamma_{22} = \frac{1}{2}\phi_2$$

This leads to

$$\gamma_{22} = \gamma_{21} + \gamma_{23}$$

The likelihood for the counts of mothers and offspring is

$$L \propto \prod_{u,v} \gamma_{uv}^{C_{uv}}$$

With no constraints, the MLE's are just the observed frequencies, as given above. When the likelihood is constrained by the hypothesis, γ_{22} is replaced by $(\gamma_{21} + \gamma_{23})$

$$\begin{aligned} L &\propto \gamma_{11}^{C_{11}} \gamma_{12}^{C_{12}} \gamma_{21}^{C_{21}} (\gamma_{21} + \gamma_{23})^{C_{22}} \gamma_{23}^{C_{23}} \gamma_{32}^{C_{32}} \\ & \quad \times (1 - \gamma_{11} - \gamma_{12} - 2\gamma_{21} - 2\gamma_{23} - \gamma_{32})^{C_{33}} \end{aligned}$$

the maximum likelihood estimates follow from equating the scores to zero:

$$\begin{aligned} \frac{\partial L}{\partial \gamma_{uv}} &= \frac{C_{uv}}{\gamma_{uv}} - K, \ (u,v) \neq (2,1),(2,3) \\ \frac{\partial L}{\partial \gamma_{21}} &= \frac{C_{21}}{\gamma_{21}} + \frac{C_{22}}{\gamma_{21} + \gamma_{23}} - K \\ \frac{\partial L}{\partial \gamma_{23}} &= \frac{C_{23}}{\gamma_{23}} + \frac{C_{22}}{\gamma_{21} + \gamma_{23}} - K \end{aligned}$$

where

$$K = \frac{C_{33}}{1 - \gamma_{11} - \gamma_{12} - 2\gamma_{21} - 2\gamma_{23} - \gamma_{32}}$$

The solutions for heterozygous offspring are

$$\begin{aligned} \hat{\gamma}_{21} &= \frac{C_{21}(C_{21} + C_{22} + C_{23})}{2C_{..}(C_{21} + C_{23})} \\ \hat{\gamma}_{23} &= \frac{C_{23}(C_{21} + C_{22} + C_{23})}{2C_{..}(C_{21} + C_{23})} \end{aligned}$$

which implies that

$$\hat{\gamma}_{22} = \frac{C_{21} + C_{22} + C_{23}}{2C_{..}}$$

The hypothesis can be tested with a chi-square test on heterozygous offspring

$$X^2 = \sum_{j=1}^{3} \frac{(C_{2j} - C_{2\cdot}\hat{\gamma}_{2j})^2}{C_{2\cdot}\hat{\gamma}_{2j}}$$

A more general approach to testing nested sets of hypotheses, including those of Christiansen and Frydenberg, is described in Williams et al. (1990).

ESTIMATION OF LINKAGE

With data available from successive generations there is the opportunity to study the degree of linkage between pairs of loci. It is most likely that an estimate of the strength of linkage will be used to indicate the usefulness of an easily scored gene as a marker for a gene affecting an economic trait in domestic species or a disease in humans. The study of linkage has proceeded in two directions. On one hand, plant and animal breeders can carry out specified crosses to simplify the detection of linkage and the estimation of its strength. Methods have been developed to make use of F_2 populations between parental lines and backcrosses to those lines. On the other hand, linkage studies in humans make use of family pedigrees where genes of interest are segregating. In both situations, heavy use is made of likelihood procedures.

Good sources for linkage studies are books by Mather (1951) and Bailey (1961), while a more recent account of procedures for human populations is contained in the book by Ott (1985).

Distances between Genes

Suppose loci **A** and **B**, with alleles A, a and B, b are located on the same arm of a chromosome so that they are not transmitted independently between generations. The RECOMBINATION FRACTION c_{AB} between them is the probability that the gamete transmitted by an individual is recombinant, having been derived by crossing over between the two parental gametes received by that individual. In the absence of any segregation distortion, a double heterozygote AB/ab that received gametes AB and ab is expected to transmit the four possible gametic types in the proportions

$$\frac{1-c_{AB}}{2}AB + \frac{c_{AB}}{2}Ab + \frac{c_{AB}}{2}aB + \frac{1-c_{AB}}{2}ab$$

The recombination fraction is expected to lie in the interval $[0, \frac{1}{2}]$, and is sometimes transformed to a LINKAGE PARAMETER λ by

$$\lambda_{AB} = 1 - 2c_{AB}$$

which ranges from 0 for unlinked loci (free recombination) to 1 for completely linked loci (no recombination).

Two related quantities need to be introduced. There is a PHYSICAL DISTANCE d_{AB} between any two regions on a chromosome. Although this may be difficult to define for a structural gene with many exons and introns, once some feature is specified there is a definite distance, usually measured in numbers of thousands of bases, kilobases (kb). A complete DNA sequence between two restriction sites, for example, allows this distance to be given precisely. In other cases, lengths can be estimated from migration distances of DNA fragments on an electrophoretic gel, as discussed in Chapter 7. The human genome, the complete set of DNA passed from one parent to a child, has a length of about 3.3×10^6 kb (Vogel and Motulsky 1986).

As a result of linkage and other studies it is possible to specify the relative positions of genes on the chromosomes, and to assign them a position on a chromosome map. The distance between two genes can therefore also be described by their MAP DISTANCE x_{AB}. As a unit of map distance, the MORGAN (M) is defined as that distance along which one crossing over is expected to occur per gamete per generation. The total length of the (sex-averaged) human map is about 33 M, or about 3.3×10^3 centiMorgans, cM (Ott 1985). The rule, therefore, is that in humans 1 cM $\approx$ 1,000 kb. Although map distance increases with physical distance, there is no simple relation between them. In a recent review of the kb/cM ratios for various species, Meagher et al. (1988) stress that the ratio can vary considerably within a genome to the point at which it may be quite misleading to work with single values. These authors described an assay that may allow better estimates of the relation between physical and recombination distances.

Now the recombination fraction is a dimensionless quantity. To estimate it from the kinds of data discussed below, it is necessary to identify recombinant gametes, which in turn result from an odd number of crossovers. The recombination fraction increases with map distance, so the tendency for crossing over can be quantified by map distance, but once again there is no universal or simple relation between them. When crossovers occur independently, Haldane (1919) showed that the recombination fraction c could be expressed in terms of the map distance x as

$$c_{AB} = \frac{1}{2}\left(1 - e^{-2x_{AB}}\right)$$

Note that c_{AB} increases from 0 to 0.5 as x_{AB} increases from 0. Other mapping functions (e.g., Kosambi 1944) allow for INTERFERENCE, whereby one crossover tends to prevent other crossovers in the same region. More complex functions have been found empirically (e.g., Rao et al. 1977).

This section is concerned with estimating the recombination fraction on the basis of observing individuals that must have received recombinant gametes.

Table 6.20 Two-locus counts for a double backcross. Each locus has two alleles, with one dominant to the other.

		Offspring				
	Genotype	AB/ab	Ab/ab	aB/ab	ab/ab	
	Phenotype	AB	Ab	aB	ab	Total
Observed		a	b	c	d	n
Expected		$\frac{n}{4}$	$\frac{n}{4}$	$\frac{n}{4}$	$\frac{n}{4}$	n

Double Backcross Method

From two lines homozygous for different alleles at each locus, say $AABB$ and $aabb$, the F_1 cross population is entirely doubly heterozygous AB/ab and the backcross $AB/ab \times aabb$ yields four classes of offspring. In the first place, these four categories allow a check to be made on the segregations at each locus and the joint segregation at the two loci. The notation employed by Bailey (1961) is shown in Table 6.20. The table also shows the phenotypes of the four offspring types when alleles A, B are dominant to a, b.

At locus **A** there are $a+b$ individuals that received allele A and $c+d$ that received a, so that a chi-square test statistic for independent segregation of alleles at this locus is given by

$$\begin{aligned} X_A^2 &= \frac{(a+b-n/2)^2}{n/2} + \frac{(c+d-n/2)^2}{n/2} \\ &= \frac{(a+b-c-d)^2}{n} \end{aligned} \tag{6.7}$$

with a similar statistic for locus **B**:

$$X_B^2 = \frac{(a+c-b-d)^2}{n} \tag{6.8}$$

For individuals that receive allele B, segregation at locus **A** is measured by $a-c$, whereas for individuals receiving b it is measured by $b-d$. Linkage between the loci causes alleles not to be transmitted independently, so the segregation at one locus is affected by the allele present at the other. For locus **A** this effect can be quantified by $(a-c)-(b-d)$. The same quantity would result from looking at the effect of A on segregation at locus **B**. Linkage is therefore detected by comparing $a+d$ to $b+c$. These terms should be the same for independent loci. An appropriate chi-square, by analogy to

Equations 6.7 and 6.8, is

$$X_{AB}^2 = \frac{(a+d-b-c)^2}{n} \tag{6.9}$$

As Bailey (1961) points out, these three chi-squares, each with 1 d.f., add up to the goodness-of-fit chi-square with 3 d.f. found by comparing the four counts to one-fourth of the sample size

$$\begin{aligned} X^2 &= \frac{(a-\frac{1}{4}n)^2}{\frac{1}{4}n} + \frac{(b-\frac{1}{4}n)^2}{\frac{1}{4}n} + \frac{(c-\frac{1}{4}n)^2}{\frac{1}{4}n} + \frac{(d-\frac{1}{4}n)^2}{\frac{1}{4}n} \\ &= X_A^2 + X_B^2 + X_{AB}^2 \end{aligned}$$

As an example, consider the maize data reported by Goodman et al. (1980) for enzymes *Mdh5* with alleles *E12, E15*, and *Got3* with alleles *U4, U8* in maize. The following testcross was set up

$$\frac{E12,U4}{E15,U8} \times \frac{E12,U4}{E12,U4}$$

and the resulting offspring array was

$\frac{E12,U4}{E12,U4}$	$\frac{E12,U8}{E12,U4}$	$\frac{E15,U4}{E12,U4}$	$\frac{E15,U8}{E12,U4}$
53	13	9	55

The chi-square test statistics for these data are

$$X_{\text{Mdh5}}^2 = \frac{(66-64)^2}{130} = 0.03$$

$$X_{\text{Got3}}^2 = \frac{(62-68)^2}{130} = 0.28$$

$$X_{\text{MDH5,Got3}}^2 = \frac{(108-22)^2}{130} = 56.89$$

$$X^2 = \frac{20.5^2 + 19.5^2 + 23.5^2 + 22.5^2}{32.5} = 57.20$$

Both loci are segregating normally, but there is strong evidence for linkage between the two loci.

Once evidence of linkage has been obtained by these goodness-of-fit tests, the recombination fraction can be estimated from the same data. For the double backcross $AB/ab \times ab/ab$, the expected values of the four offspring types when the loci are linked are shown in Table 6.21.

Table 6.21 Two-locus counts for a double backcross for linked loci.

	Genotype Phenotype	Offspring AB/ab AB	 Ab/ab Ab	 aB/ab aB	 ab/ab ab	Total
Observed		a	b	c	d	n
Expected		$\frac{n}{2}(1-c_{AB})$	$\frac{n}{2}c_{AB}$	$\frac{n}{2}c_{AB}$	$\frac{n}{2}(1-c_{AB})$	n

The likelihood of c_{AB} for these counts is

$$L(c_{AB}) \propto (1-c_{AB})^{a+d}(c_{AB})^{b+c}$$

so that the score and information (Chapter 2) are

$$S_{c_{AB}} = -\frac{a+d}{1-c_{AB}} + \frac{b+c}{c_{AB}}$$

$$I_{c_{AB}} = \frac{n}{c_{AB}(1-c_{AB})}$$

The MLE and its variance are therefore given by

$$\hat{c}_{AB} = \frac{b+c}{n}$$

$$\text{Var}(\hat{c}_{AB}) = \frac{c_{AB}(1-c_{AB})}{n}$$

F_2 Population Method

The other common situation in plant or animal breeding is where F_2 data are available. When the F_1 double heterozygotes AB/ab are crossed there are 10 possible genotypes among the offspring, and these fall into four recognizable classes when there is dominance at both loci. The data now have the expectations, under independent segregation within and between loci, shown in Table 6.22.

Repeating the arguments made above for the double backcross leads to the following three test statistics:

$$X_A^2 = \frac{(a+b-\frac{3}{4}n)^2}{\frac{3}{4}n} + \frac{(c+d-\frac{1}{4}n)^2}{\frac{1}{4}n}$$

Table 6.22 Two-locus counts for F_2 population. Each locus has two alleles with one dominant to the other.

	Offspring				
Genotype	AB/ab	Ab/ab	aB/ab	ab/ab	
Phenotype	AB	Ab	aB	ab	Total
Observed	a	b	c	d	n
Expected	$\frac{9n}{16}$	$\frac{3n}{16}$	$\frac{3n}{16}$	$\frac{n}{16}$	n

$$= \frac{(a+b-3b-3d)^2}{3n}$$

$$X_B^2 = \frac{(a+c-3b-3d)^2}{3n}$$

$$X_{AB}^2 = \frac{(a+9d-3b-3c)^2}{9n}$$

The expected segregation at each locus is now 3:1, instead of 1:1 as for the backcross.

To estimate the recombination fraction from F_2 data, the expectations are displayed in Table 6.23 using Mather's (1951) simplifying notation of $\theta = (1 - c_{AB})^2$. The likelihood is

$$L(\theta) \propto (2+\theta)^a(1-\theta)^{b+c}(\theta)^d$$

Table 6.23 Two-locus counts for F_2 population for linked loci.

	Offspring				
Genotype	AB/ab	Ab/ab	aB/ab	ab/ab	
Phenotype	AB	Ab	aB	ab	Total
Observed	a	b	c	d	n
Expected	$\frac{n}{4}(2+\theta)$	$\frac{n}{4}(1-\theta)$	$\frac{n}{4}(1-\theta)$	$\frac{n}{4}\theta$	n

The first two derivatives with respect to θ give the score and information

$$\begin{aligned} S(\theta) &= -\frac{n\theta^2 - (a - 2b - 2c - d)\theta - 2d}{(2+\theta)\theta(1-\theta)} \\ I(\theta) &= \frac{n(1+2\theta)}{2\theta(1-\theta)(2+\theta)} \end{aligned}$$

Recall that the inverse of the information provides the variance of a maximum likelihood estimate. Estimation of θ therefore requires the solution of a quadratic equation. Invoking the properties of MLE's allows the MLE of c_{AB} to then be found as

$$\hat{c}_{AB} = 1 - \sqrt{\hat{\theta}}$$

and the variance of this estimate is

$$\begin{aligned} \mathrm{Var}(\hat{c}_{AB}) &= \left(\frac{d\hat{c}_{AB}}{d\hat{\theta}}\right)^2 \mathrm{Var}(\hat{\theta}) \\ &= \frac{(1-\theta)(2+\theta)}{2n(1+2\theta)} \end{aligned}$$

A method of moments estimator has been suggested for the F_2 population situation (Fisher and Balmakund 1928, Immer 1930). The ratio $K = ad/bc$, with each term replaced by its expectation, has a simple form:

$$\frac{\mathcal{E}(a)\mathcal{E}(d)}{\mathcal{E}(b)\mathcal{E}(c)} = \frac{\theta(2+\theta)}{(1-\theta)^2}$$

suggesting that $\hat{\theta}$ be found as a solution to

$$(K-1)\theta^2 - 2(K+1)\theta + K = 0$$

and Immer (1930) constructed tables of $\hat{\theta}$ for $K = ad/bc$ values. Since the estimator is a homogeneous equation in multinomial counts, Fisher's variance formula can be used:

$$\begin{aligned} \mathrm{Var}(K) &= \left(\frac{\partial T}{\partial a}\right)^2 (2+\theta) + \left(\frac{\partial T}{\partial b}\right)^2 (1-\theta) + \left(\frac{\partial T}{\partial c}\right)^2 (1-\theta) \\ &\quad + \left(\frac{\partial T}{\partial d}\right)^2 \theta - \left(\frac{\partial T}{\partial n}\right)^2 \\ &= \frac{2\theta(2+\theta)(1+2\theta)}{n(1-\theta)^5} \end{aligned}$$

However,

$$\mathrm{Var}(K) = \left(\frac{dK}{d\hat{\theta}}\right)^2 \mathrm{Var}(\hat{\theta})$$

and

$$\frac{dK}{d\theta} = \frac{2(1+2\theta)}{(1-\theta)^3}$$

so that

$$\mathrm{Var}(\hat{\theta}) = \frac{2\theta(1-\theta)(2+\theta)}{n(1+2\theta)}$$

Now this result is the same as that obtained for the MLE of θ. This finding is unusual in that MLE's are known to be the most EFFICIENT ESTIMATORS – i.e., to have the smallest variances. It is rare for another estimator to be equally efficient.

Mather (1951) allowed a more general treatment of the F_2 method in that different recombination fractions can be assigned to males and females. If the fractions are c_{AB_m}, c_{AB_f} in males and females, then the above analysis holds with

$$\theta = (1 - c_{AB_m})(1 - c_{AB_f})$$

Direct Method

Inferences about recombination in humans need to be made from families, since crosses between homozygous lines cannot be arranged. In his review of methodology, Ott (1985) covered the DIRECT METHOD in which the recombination fraction is estimated as the observed number of recombinant gametes in a set of offspring. His example is shown in Figure 6.1. Individuals in this three-generation pedigree have been classified by *ABO* blood type, and by status of the autosomal dominant trait *CMT1* (hereditary motor and sensory neuropathy type 1). Normal individuals (empty symbols in Figure 6.1) are homozygous for the normal allele *t*. Affected individuals (symbols marked with **Aff.**) from normal parents are heterozygous *Tt*. The blood group of children often allows the *ABO* genotype of a parent to be inferred – individual 3.1 is known to be *AA* or *AO* on the basis of his blood group, but the occurrence of his *OO* children rules out *AA*. Following through the pedigree allows the two-locus genotype of 3.1 to be determined as *TA/tO*, and those of his four offspring to be *TA/tO*, *TO/tO*, *TO/tO*, and *TO/tO* respectively. Evidently individual 3.1 produced one nonrecombinant gamete *TA* and three recombinant *TO* gametes or HAPLOTYPES. No estimate is possible from the

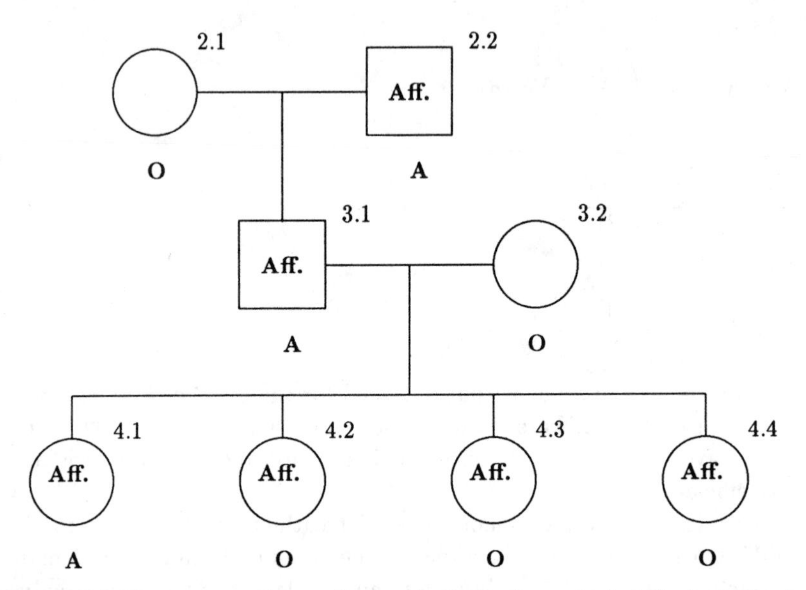

Figure 6.1 Three-generation pedigree. Individuals marked **Aff.** have the trait CMT1. Letters under each individual show *ABO* blood type. Source: Ott 1985.

other parent, 3.2, since she is double homozygous tO/tO. The estimated recombination fraction between the *CMT1* and *ABO* loci is therefore 3/4, and this analysis has been called the THREE-GENERATION METHOD.

Lod Scores with Phase Known

The only other method to be treated here is the likelihood method, introduced for human studies by Haldane and Smith (1947) in their estimation of the recombination fraction between the X-linked genes for color blindness and hemophilia. These authors sought to maximize the likelihood $L(c)$ for the recombination fraction, but did this by a variant of the grid-search method (Chapter 2) instead of using numerical iterations. To demonstrate that two loci were linked, they compared the likelihood of some c value with the likelihood of $c = 0.5$ (unlinked loci). The logarithm to base 10 of this ratio is termed the LOD SCORE (for log-odds) $Z(c)$ for that c value

$$Z(c) = \log \left[\frac{L(c)}{L(0.5)} \right]$$

For the pedigree in Figure 6.1, the likelihood calculated for the four children is that of obtaining three recombinants, each with probability c, and one nonrecombinant, with probability $1 - c$, among four independent gametes. The multinomial distribution is appropriate

$$L(c) \propto (1 - c)c^3$$

and the lod score is

$$Z(c) = \log[16(1 - c)c^3]$$

The maximum likelihood estimate of the recombination fraction is that value for which the lod score is maximized. When the MLE $\hat{c} < 0.5$ is more likely than $c = 0.5$, the lod score will be positive, and a conventional rule is to conclude that loci are linked whenever the lod score exceeds 3 ("odds of 1,000 to 1"). Conversely, if the MLE $\hat{c} < 0.5$ is less likely than $c = 0.5$, the lod score is negative, and the convention is to conclude a lack of linkage when $Z(c)$ is less than -2. For $\hat{c}$ values giving a lod score between these limits, Morton (1955) suggests the collection of more data. For the pedigree in Figure 6.1, the lod score is maximized, in the allowable range of c values, for $\hat{c} = 0.5$. As $Z(0.5) = 0$, however, no definite conclusions about linkage can be drawn from such a small amount of data.

A key feature of linkage analysis from human families is that information can be combined from independent families. The joint likelihood calculated from a set of independent families is just the product of the separate likelihoods, and the overall lod score is therefore the sum of the separate scores. The same lod score from the family with four children in Figure 6.1 could have been obtained from two independent families, of the same structure, one family with evidence of two recombination events and one family with one recombinant and one nonrecombinant.

Lod Scores with Phase Unknown

The probabilities of offspring combinations cannot be calculated simply from the binomial theorem when the parents are double heterozygotes of unknown phase. Suppose one parent is a double homozygote ab/ab but the other is a double heterozygote and no information in the available data indicates whether this parent has genotype AB/ab or Ab/aB. It is generally assumed in this situation that each phase is equally likely.

A child of genotype AB/ab or ab/ab is a nonrecombinant if the second parent was in the AB/ab phase, and is a recombinant if the parent was Ab/aB. Such a child is said by Ott (1985) to be of type 1. For dominant loci, this may be phrased as a child of phenotype AB or ab being of type 1. Other children are of type 2. The probability of obtaining a type 1 child is the probability of an AB/ab parent and no recombination plus the probability of an Ab/aB parent and recombination

$$\Pr(\text{Type 1 child}) = \frac{1}{2}(1-c) + \frac{1}{2}c = \frac{1}{2}$$

Similarly the probability of a type 2 child is 0.5. However, the probability of two type 1 children is

$$\Pr(\text{Two type 1 children}) = \frac{1}{2}(1-c)^2 + \frac{1}{2}c^2 \neq \frac{1}{4}$$

demonstrating that, although the recombination events in distinct gametes are independent, the genotypes of two children in the same family are not independent. A family with two children of type 2 (i.e., those that are recombinant if parental phase AB/ab is assumed and nonrecombinant if phase Ab/aB is assumed) gives a lod score of

$$Z(c) = \log[2(1-2c+2c^2)]$$

The lod score for two independent families, one with two type 1 children and one with one of each type of children, is the sum of the two separate lods

$$Z(c) = \log[2(1-2c+2c^2)] + \log[4c(1-c)]$$

Calculating likelihoods for pedigrees with more than two generations can be very difficult, but general algorithms have been developed. Reference should be made to Elston and Stewart (1971), Ott (1974), and Lange and Elston (1975).

SUMMARY

Data from successive generations allow many different genetic processes to be studied, from recombination between loci, to mating system, to selection. In each case a genetic model needs to be specified, and statements made about the expected behavior of statistics under such models. For each process considered in this chapter, inferences are generally based on likelihood methods. Although the underlying principles are straightforward, computational procedures can quickly become cumbersome.

EXERCISES

Exercise 6.1

Estimate the outcrossing rate for each locus in Table 6.3.

Exercise 6.2

Construct a table analogous to Table 6.11 for the *ABO* blood group system.

Exercise 6.3

Levine at al. (1980) gave some genotypic data for 10 larvae from each of 10 *Drosophila psedoobscura* females. (The genotypes refer to gene arrangements.) Use these data, and the methodology of Williams and Evarts (1989), to estimate the proportion of double paternity for the 10 families.

The data are

	Larva number									
Female	1	2	3	4	5	6	7	8	9	10
10	T/O	E/S	O/E	O/O	O/S	S/E	O/S	E/E	T/O	E/E
99	T/S	O/S	T/A	O/A	O/A	T/S	T/S	T/A	T/A	T/A
5	O/E	E/S	E/S	S/S	E/S	S/S	O/S	O/E	S/S	O/E
6	S/S	O/E	S/S	O/S	S/E	C/E	S/E	O/E	O/S	O/E
7	O/E	O/O	O/O	O/O	O/E	C/O	C/O	T/O	O/S	O/S
110	S/C	S/C	S/C	C/S	S/C	S/C	S/C	S/C	C/S	C/S
111	E/O	E/E	E/O	T/E	O/E	T/E	O/E	S/E	O/S	O/S
112	O/S	O/S	O/E	C/E	C/E	S/E	C/E	S/S	S/C	S/S
102	O/O	O/O	O/O	O/O	O/O	O/L	O/O	O/O	O/L	O/L
105	C/O	E/S	O/O	S/C	O/S	C/S	O/S	C/S	O/C	C/S

Chapter 7

Molecular Data

INTRODUCTION

A great amount of data on restriction site and sequence variation is now available for population genetic studies. Many of the analyses can be performed with the tools already developed in this book, but there are additional aspects that will be considered in this chapter.

The presence or absence of a recognition sequence for a restriction enzyme may be treated formally as two codominant alleles at a locus. In many studies, however, the restriction sites surveyed are quite close together so that assumptions of independence are unlikely to be valid. Great care needs to be taken with the sampling properties of sites within a few kilobases of each other. Once suitable models are formulated, restriction site variation can be used to infer the amount of underlying DNA sequence variation or to indicate the presence of a gene of interest.

DNA sequence data require several new analyses, if only because of their sheer magnitude. Analyses of several copies of a sequence of 1,000 bases or more are heavily dependent on computer algorithms. Many of the preliminary characterizations of sequence data are concerned with base composition, patterns of bases, and regions of similarity between sequences, and may be performed without reference to the biological processes that give rise to such features. A full analysis must also incorporate the effects of genetic forces such as drift, mutation, and recombination.

RESTRICTION SITE DATA

Estimating Fragment Lengths

The presence or absence of restriction sites is inferred from the numbers and sizes of fragments seen on an electrophoretic gel. Migration distance on a gel is inversely proportional to fragment length, and a set of standard fragments of known length run on a gel provides a means of calibrating that gel. If the standard fragments are used simply to identify other fragments that migrate the same distance there is little problem. If they are needed to estimate the length of fragments that migrate other distances, however, some method for carrying out the estimation is needed.

There is a physical limit to the resolution of bands on a gel that prevents a band from appearing as a line. Agard et al. (1981) describe a convolution of an "ideal profile" and a "smearing function" that results in observed bands. They establish a methodology for deconvoluting these two factors, by estimating the smearing function empirically. It is important to recognize that measurement error, beyond the control of the observer, limits the accuracy with which the positions of bands on a gel can be estimated. There is also the problem that very small bands may be lost off the gel.

The relation between migration distance and fragment length (or the logarithm of fragment length) is not linear, except over quite small ranges, so that simply drawing a calibration curve for the standard fragments will not be satisfactory. Nonlinear regression (Parker et al. 1977) of length on distance for the standard fragments may be used to predict the length of another fragment whose migration distance is measured. Another method makes use of a physical model that suggests migration distance m and fragment length L are related by

$$(m - m_0)(L - L_0) = c$$

where m_0, L_0, and c are constants to be estimated from the set of standard fragments. Schaffer (1983) suggested the following least-squares method for such estimation. Suppose the set of n standard fragments has lengths and distances $\{L_i, m_i\}$. Schaffer minimized the sum of squares

$$Q = \sum_i (c_i - \bar{c})^2$$

where the average value $\bar{c}$ follows from the model as

$$\begin{aligned}\bar{c} &= \frac{1}{n} c_i \\ &= \frac{1}{n} \sum_i m_i L_i - m_0 \bar{L} - L_0 \bar{m} + m_0 L_0\end{aligned}$$

with $\bar{m}$ and $\bar{L}$ being the means of the distances and lengths of the standard fragments. This allows Q to be written as

$$Q = \sum_i [m_i L_i - \frac{1}{n}\sum_j m_j L_j - m_0(L_i - \bar{L}) - L_0(m_i - \bar{m})]^2$$

Differentiating Q with respect to m_0 and L_0, and writing

$$S_{LL} = \sum_i (L_i - \bar{L})^2$$

$$S_{mL} = \sum_i (m_i - \bar{m})(L_i - \bar{L})$$

$$S_{mm} = \sum_i (m_i - \bar{m})^2$$

$$U_{mL} = \sum_i m_i L_i (L_i - \bar{L})$$

$$V_{mL} = \sum_i m_i L_i (m_i - \bar{m})$$

provides the least-squares estimates

$$\hat{m}_0 = \frac{S_{mm}U_{mL} - S_{mL}V_{mL}}{S_{LL}S_{mm} - S_{mL}^2}, \quad \hat{L}_0 = \frac{S_{LL}V_{mL} - S_{mL}U_{mL}}{S_{LL}S_{mm} - S_{mL}^2}$$

and

$$\hat{c}_i = (m_i - \hat{m}_0)(L_i - \hat{L}_0), \quad \bar{\hat{c}} = \frac{1}{n}\sum_i (m_i - \hat{m}_0)(L_i - \hat{L}_0)$$

For another fragment of length m, the estimated length $\hat{L}$ is

$$\hat{L} = L_0 + \frac{\bar{c}}{m - \hat{m}_0}$$

An example of the use of this procedure is shown in Table 7.1 for data supplied by A. H. D. Brown (personal communication). The standard fragments result from cutting the bacteriophage lambda with *Hind*III. From the equation given in the footnote to that table, a fragment with migration distance 30.5 mm on that gel has an estimated length of 4.7 kb.

Ordering fragments

Once the sizes of a series of fragments have been determined, the problem of restriction mapping is to infer the locations of the restriction sites that bound these fragments. The task is made easier by applying different restriction

Table 7.1 Estimating the sizes of *Hind*III restriction fragments for lambda.* Data supplied by A. H. D. Brown.

Distance (mm)		Length (kb)	
m_i	$(m_i - \bar{m})$	L_i	$(L_i - \bar{L})$
11.5	−22.04	23.1	15.14
17.25	−16.29	9.4	1.43
23.5	−10.04	6.6	−1.37
34	0.46	4.4	−3.57
55	21.46	2.3	−5.67
60	26.46	2.0	−5.97

*$\bar{m} = 33.54$ $\bar{L} = 7.97$
$S_{mm} = 2012.273$ $S_{mL} = -624.1879$ $S_{LL} = 313.5568$
$U_{mL} = 2074.8015$ $V_{mL} = -4096.0475$
$\hat{m}_0 = 6.7054$ $\hat{L}_0 = 0.0443$ $\bar{\hat{c}} = 108.5359$

$$\text{i.e., } \hat{L} = 0.0443 + \frac{108.5359}{m - 6.7054}$$

enzymes singly and in pairs. Ordering can often be done by a trial and error process by hand, but there have been attempts to construct algorithms to automate the process. The algorithm of Pearson (1982) is now illustrated.

Suppose that applying the restriction enzyme *Eco*RI to a 10-kb piece of DNA produces fragments of sizes 9 and 1 kb, while fragments of sizes 6 and 4 kb result from applying enzyme *Bam*HI. The relative positions of the single recognition sites for each enzyme are not known. As shown in Figure 7.1, there are two possible orders, with order A suggesting that restriction by both enzymes will produce fragments of lengths 6, 3, and 1 kb. The alternative order will produce lengths of 4, 5, and 1 kb, which is the order found in an actual double digest. Order B is taken to be the actual order.

Pearson's program begins by considering the fragment sets found from each of the first two enzymes applied singly. For each possible order of the recognition sites for the first enzyme, every possible ordering of sites for the second enzyme is investigated. Each arrangement is evaluated by comparing the fragment sizes from hypothetical and actual double digests. Arrangements are compared on the basis of the sums of squares of differences of fragment lengths between hypothetical and actual double digests.

1. Ordering of fragments.

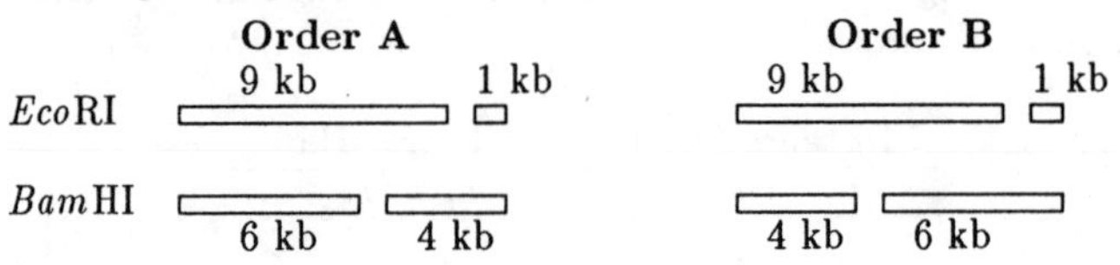

2. Calculation of hypothetical double digest fragments.

Order A

6 kb 3 kb 1 kb

Order B

4 kb 5 kb 1 kb

3. Comparison of hypothetical and actual double digest fragments.

Order A: 6 kb, 3 kb, 1 kb

Actual Order: 5 kb, 4 kb, 1 kb

Order B: 5 kb, 4 kb, 1 kb

4. Error calculation

Order A: $(6-5)^2+(3-4)^2+(1-1)^2=2$

Order B: $(5-5)^2+(4-4)^2+(1-1)^2=0$

Figure 7.1 Pearson's restriction site ordering algorithm illustrated for hypothetical data.

Estimating Location of Restriction Sites

Once the order of a set of restriction sites has been determined, there may still be a problem in determining the relative locations of the sites, since experimental error will prevent (estimated) fragment sizes adding exactly to the total length bteween the two outside sites. For a collection of r sites, there is a need to estimate $r-2$ coordinates for the interior sites.

If the coordinate of the ith site is written as X_i, and if the length of the fragment between sites i and j has been estimated to be d_{ij}, Schroeder and Blattner (1978) minimized the sum of squared relative errors to estimate coordinates. They chose X_i and X_j to minimize

$$Q = \sum_i \sum_j \left[\frac{d_{ij} - (X_j - X_i)}{d_{ij}} \right]^2$$

Longer lengths receive smaller weights in this expression. In general, minimization of Q will require numerical methods. Negative coordinates may indicate an incorrect ordering.

Inferring Nucleotide Variation

While a complete picture of variation at the nucleotide level depends on DNA sequencing, estimates of this variation can be provided by restriction site variation providing a suitable model is used.

A restriction site is a subsequence of particular bases, usually of length 4, 5, or 6. Mutation of any one of the bases to another type will cause the subsequence to cease being a recognition site, and the restriction enzyme will not cut the sequence at that point. Evidently the chance of mutation introducing a new recognition site will be much less than that of destroying a present site. Using standard population genetic arguments, Ewens (1983) considered a population of size N and restriction sites of length j that can be lost or gained by mutation or lost by genetic sampling (drift). For small mutation rates μ it may be supposed that at most one mutation will occur in a site per generation. The probability of a single nucleotide site being polymorphic in the population is

$$P = 4N\mu \ln 2N$$

and of being polymorphic in a sample of n

$$p = 4N\mu \ln 2n$$

Assuming independence of adjacent nucleotides, the probability of a restriction site being polymorphic in the sample is approximately jp, so that if k of the m observed restriction sites are found to be polymorphic, a naïve estimate of p would be

$$\hat{p} = \frac{k}{jm}$$

Ewens pointed out, however, that a restriction site must be observed at least once in the sample to be included in these calculations. A restriction site at any location is known to exist in a sample only if it is seen at least once in that sample. The probability of a restriction site appearing on x of n gametes in a sample is therefore

$$\begin{aligned}\Pr(x) &= \Pr(x|x > 0)\\ &= \frac{\frac{n!}{x!(n-x)!}p^x(1-p)^{n-x}}{1-(1-p)^n}\end{aligned}$$

Using such arguments, Ewens found that the conditional probability of polymorphism, given at least one occurrence, is $2jp$ and the correct estimate of p is

$$\hat{p} = \frac{k}{2jm}$$

with corresponding population value

$$\hat{P} = \frac{k \ln 2N}{2jm \ln 2n}$$

For the data on four restriction sites in Table 1.6, the sample size was $n = 17$. Eight enzymes of length $j = 6$ were used, and a total of $m = 24$ restriction sites found. Of these only $k = 4$ were polymorphic, meaning that 20 were present in every sample member. These figures provide an estimate of $\hat{p} = 0.014$. In other words, a level of polymorphism of (4/24=0.167) at the restriction site level translates to a level of polymorphism of 0.014 at the DNA level. The population size N is unknown.

Disease Associations

The growing abundance of RESTRICTION FRAGMENT LENGTH POLYMORPHISMS (RFLP's) along the whole lengths of chromosomes is enabling a map of the entire human genome to be constructed (Donis-Keller et al. 1987). These RFLPs are available to act as markers for genes of interest, such as disease genes. The methods of Chapter 6 can be employed to detect linkage between markers and disease genes from family pedigree studies, or among the offspring of specific mating types.

In the absence of family data, the proximity of a marker and a disease gene may be suggested by population associations. The simplest situation is when gametic, or haplotypic, data are available from the whole population and it is known whether the gamete has the normal (N) or defective (S) form of the disease gene. Associations between markers and disease genes are revealed by detection of linkage disequilibrium between marker and disease gene, with the data being set out as in Table 7.2.

A significant association is indicated by large values of the test statistic

$$X^2 = \frac{n_{..}(n_{N+} - n_{N.}n_{.+}/n_{..})^2}{n_{.+}n_{.-}n_{S.}n_{N.}}$$

For rare diseases, a sample taken randomly from the population may not contain any individuals with the disease. Associations are therefore sought from samples taken from the various disease categories in the population.

Table 7.2 Two-way table of haplotype frequencies for showing marker-disease associations.

Marker Status	Disease Status		
	Normal	Disease	Total
Restriction site present	n_{N+}	n_{S+}	$n_{.+}$
Restriction site absent	n_{N-}	n_{S-}	$n_{.-}$
Total	$n_{N.}$	$n_{S.}$	$n_{..}$

A procedure for the estimation of linkage disequilibrium in such conditional samples was given by Chakravarti et al. (1984), and will now be discussed in a modified form.

One of the first reports of an association between RFLP's and a disease was that between *Hpa*I fragments and sickle cell anemia. Data published by Kan and Dozy (1978) are shown in Table 7.3. These data are from three separate samples, one for each of the disease genotypes.

For a general discussion, suppose a disease locus has alleles N, S. In the whole population, a marker locus has two alleles M_1, M_2 and the allelic frequencies are p, q at the disease locus and m_1, m_2 at the marker locus. Allowing for linkage disequilibrium D between the loci the four gametes have frequencies

$$NM_1 : y_1 = pm_1 + D$$
$$NM_2 : y_2 = pm_2 - D$$

Table 7.3 Associations of RFLP's with sickle cell anemia in three separate samples. At the disease locus, N and S are the normal and defective alleles. Source: Kan and Dozy (1978).

Disease Genotype	Fragment Lengths			n	Number of M_1 Alleles
	7,7	7,13	13,13		
1. *NN*	39	4	0	$n_1 = 43$	$x_1 = 82$
2. *NS*	32	53	1	$n_2 = 86$	$x_2 = 117$
3. *SS*	4	24	30	$n_3 = 58$	$x_3 = 32$

$$
\begin{aligned}
SM_1 : y_3 &= qm_1 - D \\
SM_2 : y_4 &= qm_2 + D
\end{aligned}
$$

For a random-mating population, the conditional frequencies of marker allele M_1 in each of the three disease genotypic categories (i.e., in each of the three separate samples) are

$$
\begin{aligned}
a_1 = \Pr(M_1|NN) &= \frac{y_1^2 + y_1 y_2}{p^2} \\
&= m_1 + \frac{D}{p} \\
a_3 = \Pr(M_1|SS) &= \frac{y_3^2 + y_3 y_4}{q^2} \\
&= m_1 - \frac{D}{q} \\
a_2 = \Pr(M_1|NS) &= \frac{y_1 y_4 + y_2 y_3 + 2y_1 y_3}{2pq} \\
&= m_1 + \frac{D}{2p} - \frac{D}{2q} \\
&= \frac{1}{2}(a_1 + a_3)
\end{aligned}
$$

If the population frequencies p, q of the disease alleles are known, the two unknowns m_1, D can be estimated from the three separate samples. The two parameters a_1, a_3 are transformations of the parameters m_1, D. As random mating is assumed, the samples with homozygous disease genotypes are in Hardy-Weinberg equilibrium and the marker allele numbers follow a binomial distribution. For the heterozygous disease class, however, the marker genotypes have frequencies $a_1 a_3, a_1(1-a_3)+a_3(1-a_1), (1-a_1)(1-a_3)$ (Chakravarti et al. 1984) and HWE cannot be assumed unless there is no linkage disequilibrium. There are 4 d.f. in the data, ruling out use of Bailey's method for estimating the parameters. The likelihoods for the three samples are

$$
\begin{aligned}
L(a_1, a_3)_1 &\propto (a_1)^{x_1}(1 - a_1)^{2n_1 - x_1} \\
L(a_1, a_3)_2 &\propto (a_1 a_3)^{n_{21}}(a_1 + a_3 - 2a_1 a_3)^{n_{22}}(1 - a_1 - a_3 + a_1 a_3)^{n_{23}} \\
L(a_1, a_3)_3 &\propto (a_3)^{x_3}(1 - a_3)^{2n_3 - x_3}
\end{aligned}
$$

where x_i is the number of M_1 alleles among the $2n_i$ genes in sample $i, i = 1, 3$, and $n_{2j}, j = 1, 3$ are the three marker genotypic counts in sample 2. The total likelihood for the three independent samples is the product of the three separate likelihoods. Simple algebraic expressions for the maximum likelihood estimates of a_1, a_3 do not seem to be possible, but setting the derivatives of the logarithm of the likelihood with respect to each of them to zero, and rearranging leads to two equations suitable for iteration

$$a_1 = \frac{1}{2n_1 + n_{21} + n_{23}}\left[x_1 + n_{21} + \frac{n_{22}(1-2a_3)a_1(1-a_1)}{a_1 + a_3 - 2a_1a_3}\right]$$

$$a_3 = \frac{1}{2n_3 + n_{21} + n_{23}}\left[x_3 + n_{21} + \frac{n_{22}(1-2a_1)a_3(1-a_3)}{a_1 + a_3 - 2a_1a_3}\right]$$

Such equations have less trouble with nonconvergence than do the Newton-Raphson iterative equations (Chapter 2). Convergence is very rapid if the initial estimates of $a_1 = x_1/2n_1, a_3 = x_3/2n_3$ are used. Once the solutions are found, the estimates of interest and their variances can be obtained from

$$\begin{aligned} \hat{m}_1 &= p\hat{a}_1 + q\hat{a}_3 \\ \hat{D} &= pq(\hat{a}_1 - \hat{a}_3) \\ \text{Var}(\hat{m}_1) &= p^2\text{Var}(\hat{a}_1) + q^2\text{Var}(\hat{a}_3) + 2pq\text{Cov}(\hat{a}_1, \hat{a}_3) \\ \text{Var}(\hat{D}) &= p^2q^2\left[\text{Var}(\hat{a}_1) + \text{Var}(\hat{a}_3) - 2\text{Cov}(\hat{a}_1, \hat{a}_3)\right] \end{aligned}$$

and variances can be found from likelihood theory. For the data in Table 7.3 the estimates are $\hat{m}_1 = 0.8804, \hat{D} = 0.0237$ if it is assumed that $p = 0.95, q = 0.05$. The standard deviation for $\hat{m}_1$ is estimated as 0.0429, and for $\hat{D}$ as 0.0040.

A similar argument can be constructed for haplotypic, as opposed to genotypic data (Chakraborty 1986). Pedigrees studied for human diseases are usually ascertained through an affected child. For recessive diseases, this implies that each parent is heterozygous at the disease locus and each family can contribute these two (independent) heterozygotes for the analysis of association. In other words, only sample number 2 (with sample size n_2) of the situation illustrated in Table 7.3 is available. Maximizing the likelihood leads to the gene-counting (EM algorithm) estimate of a_2

$$\hat{a}_2 = \frac{2n_{21} + n_{22}}{2n_2}$$

since Bailey's method provides

$$\widehat{a_1a_3} = \frac{n_{21}}{n_2}, \quad \widehat{a_1 + a_3} = \frac{2n_{21} + n_{22}}{n_2}$$

Recovering estimates of m_1 and D from these requires the solution of quadratic equations.

Table 7.4 Observed and expected frequencies for four haplotypes when only heterozygotes at a disease locus are available.

	M_1		M_2		
	Exp.	Obs.	Exp.	Obs.	Total
N	$\frac{1}{2}a_1$	$\frac{1}{2n_2}(n_{21}+n_{22}^c)$	$\frac{1}{2}(1-a_1)$	$\frac{1}{2n_2}(n_{23}+n_{22}^r)$	$\frac{1}{2}$
S	$\frac{1}{2}a_3$	$\frac{1}{2n_2}(n_{21}+n_{22}^r)$	$\frac{1}{2}(1-a_3)$	$\frac{1}{2n_2}(n_{23}+n_{22}^c)$	$\frac{1}{2}$
Total	a_2	$\frac{1}{2n_2}(2n_{21}+n_{22})$	$1-a_2$	$\frac{1}{2n_2}(2n_{23}+n_{22})$	1

Family data often allow the phase of double heterozygotes to be determined, and haplotypic data to be recovered. From the sample of heterozygotes at the disease locus, the four haplotyes have the observed and expected frequencies shown in Table 7.4. There n_{22} double heterozygotes have been split into n_{22}^c of NM_1/SM_2 and n_{22}^r of NM_2/SM_1. The estimates of conditional allelic frequency a_2 and conditional linkage disequilibrium D_c are

$$\hat{a}_2 = \frac{2n_{21}+n_{22}}{2n_2}, \quad \hat{D}_c = \frac{n_{22}^c - n_{22}^r}{2n_2}$$

and it is such conditional values that were reported for some cystic fibrosis families by Weir (1989b). With knowledge of the population frequencies p, q of the disease alleles, the population estimates are

$$\hat{m}_1 = \hat{a}_2 - (q-p)\hat{D}_c, \quad \hat{D} = 2pq\hat{D}_c$$

DNA SEQUENCE DATA

Base Composition

An obvious first summary of a DNA sequence is just the distribution of the four base types. Although it would be convenient for mathematical modeling if the four bases were equally frequent, almost all empirical studies show an unequal distribution. The examples in Tables 7.5 and 7.6 show that bases have different frequencies within and between sequences. Table 7.5 contains

Table 7.5 Base compositions for nine complete DNA sequences, to illustrate different compositions between sequences.

Sequence	Locus Name*	Base A	C	G	T	Total
Bacteriophages						
Lambda	LAMCG	0.25	0.24	0.25	0.26	48502
T7	PT7	0.27	0.23	0.24	0.26	39936
ϕX174	PX1CG	0.24	0.22	0.31	0.23	5386
Viruses						
Cauliflower mosaic	MCACGDH	0.37	0.21	0.23	0.19	8016
Human papovirus BK	PVBMM	0.30	0.20	0.30	0.20	4963
Hepatitus B	HPBAYW	0.28	0.22	0.23	0.27	3182
Mitochondria						
Human	HUMMT	0.31	0.31	0.25	0.13	16569
Bovine	BOVMT	0.33	0.26	0.27	0.14	16338
Mouse	MUSMT	0.35	0.24	0.29	0.12	16295

* Names in GenBank database.

data from the sequences of nine entire DNA molecules, and Table 7.6 is for two fetal globin genes, ${}^G\gamma$, ${}^A\gamma$, each of which has three exons and two introns (Shen et al. 1981). An arbitrary 500-bp region on each side of each gene is called "flanking." The intergenic region is the remaining region between the two genes.

Dinucleotide Frequencies

One of the principal difficulties in analyzing DNA sequences is that the frequencies of neighboring bases are not independent. In particular, the frequencies of adjacent bases are generally different from the products of the single bases. Algebraically, if p_u is the frequency of base type u in the sequence, and p_{uv} is the frequency with which successive bases are of types u and v, then

$$p_{uv} \neq p_u p_v$$

This feature of dinucleotide frequencies has been studied empirically by Nussinov (1984), and some of her findings are shown in Table 7.7. These

Table 7.6 Base compositions for different regions of human fetal globin gene, to illustrate differences in composition within one sequence. Source: Locus HSGLBN in EMBL database.

Region	Length	A	C	G	T
5′ Flanking (2)	1000	0.33	0.23	0.22	0.22
3′ Flanking (2)	1000	0.29	0.15	0.26	0.30
Introns (4)	1996	0.27	0.17	0.27	0.29
Exons (6)	882	0.24	0.25	0.28	0.22
Intergenic (1)	2487	0.32	0.19	0.18	0.31

data are from 166 vertebrate sequences, totaling 136,731 bases. The ratios in the table are of the 16 dinucleotide frequencies divided by the products of the corresponding single base frequencies.

Table 7.7 Examples of dinucleotide frequencies in some vertebrate sequences. The frequency of each of the 16 dinucleotides is compared to the product of the corresponding pair of base frequencies. Source: Nussinov (1984).

Pair	Observed/Expected
TG	1.29
CT	1.26
CC	1.18
AG	1.16
AA	1.15
CA	1.15
GG	1.14
TT	1.07
GA	1.04
TC	1.00
GC	0.99
AT	0.85
AC	0.84
GT	0.82
TA	0.65
CG	0.42

```
  1 GTGCACTGGA CTGCTGAGGA GAAGCAGCTC ATCACCGGCC TCTGGGGCAA GGTCAATGTG
 61 GCCGAATGTG GGGCCGAAGC CCTGGCCAGG CTGCTGATCG TCTACCCCTG GACCCAGAGG
121 TTCTTTGCGT CCTTTGGGAA CCTCTCCAGC CCCACTGCCA TCCTTGGCAA CCCCATGGTC
181 CGCGCCCACG GCAAGAAAGT GCTCACCTCC TTTGGGGATG CTGTGAAGAA CCTGGACAAC
241 ATCAAGAACA CCTTCTCCCA ACTGTCCGAA CTGCATTGTG ACAAGCTGCA TGTGGACCCC
301 GAGAACTTCA GGCTCCTGGG TGACATCCTC ATCATTGTCC TGGCCGCCCA CTTCAGCAAG
361 GACTTCACTC CTGAATGCCA GGCTGCCTGG CAGAAGCTGG TCCGCGTGGT GGCCCATGCC
421 CTGGCTCGCA AGTACCAC
```

Figure 7.2 DNA sequence for the coding region of the chicken β-globin gene (GenBank database Locus CHKHBBM, Accession Number J00860).

As a specific example, consider the sequence for the chicken hemoglobin β chain mRNA shown in Figure 7.2. This figure, reading row by row, gives the 438 bases in that coding region. The four base and the 16 dinucleotide counts are displayed in Table 7.8. Regarding this table as a 4×4 contingency table and calculating the chi-square statistic for independence of rows and columns gives $X^2 = 59.3$, showing strong evidence for associations between successive nucleotide pairs.

In coding regions, there will be constraints on the DNA sequence to preserve the sequence of amino acids, and these constraints at the codon level are related to constraints at the dinucleotide level (Santibánez-Koref and Reich 1986, Nussinov 1987). In Table 7.9 the nuclear genetic code is shown, along with the numbers of the various codons in the sequence in Figure 7.2. Although the numbers are quite small, and strong statistical

Table 7.8 Dinucleotide counts for chicken β-globin sequence in Figure 7.2.

		Second Base				
		A	C	G	T	Total
First base	A	23	26	23	15	87
	C	37	51	14	41	143
	G	25	38	36	19	118
	T	2	29	44	14	89
Total		87	144	117	89	437

Table 7.9 The genetic code. All 64 possible base triplets (codons) and the corresponding amino acids are shown. The numbers are those for the sequence in Figure 7.2.

UUU Phe 3	UCU Ser 0	UAU Tyr 0	UGU Cys 2
UUC Phe 5	UCC Ser 5	UAC Tyr 2	UGC Cys 1
UUA Leu 0	UCA Ser 0	UAA Stop 0	UGA Stop 0
UUG Leu 0	UCG Ser 0	UAG Stop 0	UGG Trp 4
CUU Leu 1	CCU Pro 1	CAU His 3	CGU Arg 0
CUC Leu 6	CCC Pro 4	CAC His 4	CGC Arg 3
CUA Leu 0	CCA Pro 0	CAA Gln 1	CGA Arg 0
CUG Leu 11	CCG Pro 0	CAG Gln 0	CGG Arg 0
AUU Ile 1	ACU Thr 3	AAU Asn 1	AGU Ser 0
AUC Ile 6	ACC Thr 4	AAC Asn 6	AGC Ser 2
AUA Ile 0	ACA Thr 0	AAA Lys 1	AGA Arg 0
AUG Met 1	ACG Thr 0	AAG Lys 9	AGG Arg 3
GUU Val 0	GCU Ala 4	GAU Asp 1	GGU Gly 1
GUC Val 5	GCC Ala 11	GAC Asp 5	GGC Gly 4
GUA Val 0	GCA Ala 0	GAA Glu 4	GGA Gly 0
GUG Val 7	GCG Ala 1	GAG Glu 3	GGG Gly 3

conclusions cannot be drawn, it does appear that the different (synonymous) codons that encode the same amino acid are not represented equally in the chicken β-globin sequence. Such codon bias is bound to be reflected in associations at the dinucleotide level. Table 7.9 also makes it clear that there is much less constraint on the base present in the third position of a codon than in the second since changing the third base often does not change the amino acid.

There are several other biological reasons why successive base frequencies may not be independent. The average rate of recombination between adjacent bases is extremely low, of the order of 10^{-8} per generation, and may be less in some regions. Associations are expected to remain in a population for a very long time. There are also features such as possible avoidance of the pair CG because it is a methylation site and so prevents cleavage by restriction enzymes (Lewin 1987). Other authors (Smith et al. 1983) have related dinucleotide frequencies to functional features of sequences. Whatever the reason, it is of interest to characterize the extent of associations.

Markov Chain Analysis

Associations between adjacent bases will lead to associations between more distant bases, and an estimate of how far the relations extend may be found from Markov chain theory (Tavaré and Giddings 1989). Without invoking any biological mechanism, a MARKOV CHAIN OF ORDER k supposes that the base present at a certain position in a sequence depends only on the bases present at the previous k positions. A chain of order 1 supposes that the probability of a particular base being present at position i depends only on the probabilities of the four bases present at position $i - 1$. A sequence composed of independent bases will correspond to a Markov chain of order 0. The order can be estimated by likelihood methods.

Markov chain analyses are of use at the genome level, rather than at the level of an individual gene (J. Cuticchia, personal communication). If the genome of *Escherichia coli* is being studied, for example, there would be little point in seeking the order of a Markov chain describing the sequence of a gene since this involves only about 1 kb which is not sufficient to demonstrate the presence of high-order chains (Tavaré and Giddings 1989). Although such a short sequence may indicate only a second-order chain, it is unlikely that the entire *E. coli* genome can be described in terms of single base and dinucleotide frequencies. Once a Markov chain model has been fitted to a genome, no biological mechansim is implied, but useful questions can be answered. The frequency of particular subsequences can be predicted, so that the expected number of fragments produced when a specific restriction enzyme is applied to the genome can be estimated (Bishop et al. 1983). Of more current interest (Arnold et al. 1988) is the ability to determine oligonucleotide frequencies in very long sequences. Arnold et al. (1988) looked at the frequency of sequences up to six bases in length in 392 kb of yeast DNA, but their methods apply to the longer lengthed synthetic oligonucleotides being used as primers for DNA amplification (J. Cuticchia, personal communication).

As a notational device, suppose that X_i is the nucleotide at position i in the sequence, and that it is of base type a_i. Bases can be labeled in alphabetical order: A=1, C=2, G=3, and T=4. In the short sequence CTATAATAG, for example, $a_3 = 1$. If the sequence can be regarded as the result of a Markov chain of order k, the nth base depends only on the previous k

bases:

$$\Pr(X_n = a_n | X_{n-1} = a_{n-1}, X_{n-2} = a_{n-2}, \ldots, X_0 = a_0)$$
$$= \Pr(X_n = a_n | X_{n-1} = a_{n-1}, X_{n-2} = a_{n-2}, \ldots, X_{n-k} = a_{n-k})$$

The probability of a transition from a run of, say, three particular bases to a fourth, $a_1 \to a_2 \to a_3 \to a_4$, is estimated as the observed number of such tetranucleotides divided by the observed number of trinucleotides, $a_1 \to a_2 \to a_3$. Symbolically

$$\Pr(a_4|a_1, a_2, a_3) \;\hat{=}\; \frac{n_{a_1,a_2,a_3,a_4}}{n_{a_1,a_2,a_3}}$$

so that, from the short sequence above the probability that an ATA is followed by an A is

$$\Pr(\mathtt{A}\,|\,\mathtt{ATA}) \;\hat{=}\; \frac{n_{\mathrm{ATAA}}}{n_{\mathrm{ATA}}} = \frac{1}{2}$$

Under the hypothesis of an order k chain, the likelihood of a sequence is found as the product of the probability of the first k bases and the probabilities of each successive group of $k+1$ bases. The likelihood of an order two chain describing the short sequence above, for example, is

$$\begin{aligned} L(2) &= \Pr(\mathtt{CT})\Pr(\mathtt{A}|\mathtt{CT})\Pr(\mathtt{T}|\mathtt{TA})\Pr(\mathtt{A}|\mathtt{AT}) \\ &\quad \times \Pr(\mathtt{A}|\mathtt{TA})\Pr(\mathtt{T}|\mathtt{AA})\Pr(\mathtt{A}|\mathtt{AT})\Pr(\mathtt{G}|\mathtt{TA}) \end{aligned}$$

where Pr(A|CT) means the probability that an A follows a CT pair. This probability is estimated as the number of CTA triples divided by the number of CT pairs. The likelihood of an order three chain describing that same sequence is

$$\begin{aligned} L(3) &= \Pr(\mathtt{CTA})\Pr(\mathtt{T}|\mathtt{CTA})\Pr(\mathtt{A}|\mathtt{TAT})\Pr(\mathtt{A}|\mathtt{ATA}) \\ &\quad \times \Pr(\mathtt{T}|\mathtt{TAA})\Pr(\mathtt{A}|\mathtt{AAT})\Pr(\mathtt{G}|\mathtt{ATA}) \end{aligned}$$

The general expression, if $\Pr(a_1, a_2, \ldots, a_k)_F$ refers to the first k bases in the sequence, is

$$L(k) = \Pr(a_1, a_2, \ldots, a_k)_F \prod [\Pr(a_{k+1}|a_1, a_2, \ldots, a_k)]^{n_{a_1,a_2,\ldots,a_{k+1}}}$$

where the product is over all combinations of $k+1$ successive bases that have an observed count greater than zero. The logarithms of the likelihoods for the first five orders for the sequence displayed in Figure 7.2 are shown in Table 7.10. As the order gets higher, more parameters are being fitted and better fits to the data are expected. For an order k chain, the number of independent parameters is 3×4^k. Differences in successive log-likelihoods correspond to likelihood ratios for testing the effectiveness of one order versus the previous one. Taking minus two times these differences provides test statistics that are approximately chi-square (under the hypotheses that successive orders fit equally well) with degrees of freedom equal to the difference in degrees of freedom for each of the two orders (Chapter 3). Such test statistics are shown in Table 7.10.

Approximate critical values for chi-squares with large degrees of freedom can be found from the normal distribution. If $\alpha\%$ of values of the standard normal variable (Appendix Table 1) lie beyond $\pm z_{\alpha/2}$ then $\alpha\%$ of the values of a chi-square variable with ν degrees of freedom lie above $\frac{1}{2}(z_{\alpha/2}+\sqrt{2\nu+1})$ (Rohlf and Sokal 1981). Applying this approximate expression to the values in Table 7.10 shows that there is a significant difference (at the 5% level) in fit only between chains of order 0 and order 1. Higher order chains seem to fit equally well.

In estimating the order k that best describes the sequence, account needs to be taken of the fact that higher order chains use more parameters. Katz (1981) showed the validity of penalizing for an increased number of fitted parameters by using the Bayesian Information Criterion (*BIC*) defined as

$$BIC(k) = \text{Constant} - 2 \ln L(k) + 3 \times 4^k \ln n_k$$

where n_k is the number of subsequences of length $k+1$ that are found in the sequence. That k for which *BIC(k)* is minimized is taken as the estimate. Bearing in mind that the sequence in Figure 7.2 is really too short to allow the detection of chains of high order, the *BIC* values for that sequence are shown in Table 7.10. The lowest *BIC* value is for order $k = 1$. This reinforces the conclusion from the likelihood ratio tests that an order 1 chain is appropriate.

Once the order of the chain is established, such chains can be used to estimate the frequency of specific subsequences, as was done for yeast DNA by Arnold et al. (1988). They worked with chains of order 3, and estimated the probability of the recognition sequence GCGGCCGC of the rare-cutting enzyme *Not*I as

$$\begin{aligned}\Pr(\texttt{GCGGCCGC}) &= \Pr(\texttt{GCG})\Pr(\texttt{G}|\texttt{GCG})\Pr(\texttt{C}|\texttt{CGG})\Pr(\texttt{C}|\texttt{GGC}) \\ &\quad \times \Pr(\texttt{G}|\texttt{GCC})\Pr(\texttt{C}|\texttt{CCG}) \\ &\doteq \frac{n_{\texttt{GCGG}}\, n_{\texttt{CGGC}}\, n_{\texttt{GGCC}}\, n_{\texttt{GCCG}}\, n_{\texttt{CCGC}}}{n_{\texttt{CGG}}\, n_{\texttt{GGC}}\, n_{\texttt{GCC}}\, n_{\texttt{CCG}}}\end{aligned}$$

This probability can be estimated as the ratio of the various tetranucleotides and trinucleotides seen in a subset of the sequence. Arnold et al. used

Table 7.10 Values of the Bayesian Information Criterion and the log-likelihoods for estimating the order of the Markov chain to represent the chicken β-globin sequence of Figure 7.2.

k	BIC(k)	Log-likelihood(k)	No. Parameters	Chi-square (d.f.)
0	1213.04	−597.40	3	
1	1196.19	−562.93	12	68.94* (9)
2	1365.70	−540.12	48	45.62 (36)
3	2068.46	−454.77	192	170.70 (142)
4	5234.38	−290.52	768	328.50 (576)

* Significant at 5%.

50 kb to provide estimates of tetranucleotide frequencies and estimated the probability of an *Not*I site as 2.0×10^{-6}. The length of the sequence between two points that have a probability of P of occurring is approximately $1/P$ for small P. In this case, the size of fragments produced by the *Not*I enzyme is expected to be $1/(2 \times 10^{-6}) = 500$ kb.

Patterns in a Single Sequence

Apart from characterizing the extent to which associations hold along sequences, there is often interest in looking for patterns such as direct repeats. An efficient algorithm for doing this was given by Karlin et al. (1983). The method requires the entire sequence to be searched just once, and uses words composed of particular sets of base letters. Each base is assigned a value, α, such as 0, 1, 2, 3 for A,C,G,T. Each different subsequence, or word, of k letters $X_1, X_2, \ldots, X_k$ can be given a unique word value $1 + \sum_{i=1}^{k} \alpha_i 4^{k-i}$. These values lie in the range from 1 to 4^k. The 5-word TGACC, for example, has value $1+3\times 4^4+2\times 4^3+0\times 4^2+1\times 4^1+1\times 4^0 = 459$. Some low value of k is taken, and then the values of the k-words that start at each position in the sequence are recorded. Only those positions at which repeated k-words are found need be considered further in searches for repeated words of length greater than k.

For the sequence TGGAAATAAAACGTAAGTAG, the initial positions and values of all 2-words are shown in Table 7.11. Searches for DIRECT REPEATS, or subsequences that are repeated exactly, of length greater than 2 can be confined to the initial positions of 2-words that are repeated. Only four 2-

Table 7.11 Values and positions of 2-words for the sequence TGGAAATAAAACGTAAGTAG following the algorithm of Karlin et al. (1983).

Value	Position	Value	Position
1	4, 5, 8, 9, 10, 15	9	3
2	11	10	–
3	16, 19	11	2
4	6	12	13, 17
5	–	13	7, 14, 18
6	–	14	–
7	12	15	1
8	–	16	–

words are repeated in this example. Words of value 1 are found at positions 4, 5, 8, 9, 10, and 15, for example. The 3-words that start with each of the repeated 2-words are shown in Table 7.12, where words of value 1, 45, and 49 are found to be repeated. An examination of the 4-words that start with each of the repeated 3-words fails to reveal any longer repeated values, so the longest direct repeats are AAA at positions 4, 8, and 9, GTA at positions 13 and 17, and TAA at positions 7 and 14. Applying the technique to the chicken β-globin sequence in Figure 7.2 by means of the program listed in the Appendix locates the longest direct repeats shown in Table 7.13.

Karlin et al. (1983) discuss methods for assessing the statistical signifi-

Table 7.12 Values and positions of 3-words that begin with repeated 2-words for the sequence TGGAAATAAAACGTAAGTAG.

Value	Position
1	4, 8, 9
2	10
3	15
4	5
45	13, 17
49	7, 14
51	18

Table 7.13 Repeated words of lengths eight or more in the chicken β-globin sequence of Figure 7.2.

Length	Word	Starting Positions
8	GCCCTGGC	79, 418
	GCCAGGCT	85, 377
	CCAGGCTG	86, 378
	CAGGCTGC	87, 379
	TCCTTTGG	130, 208
	CCTTTGGG	131, 209
	TGGTCCGC	176, 398
	GGTCCGCG	177, 399
9	GCCAGGCTG	85, 377
	CCAGGCTGC	86, 378
	TCCTTTGGG	130, 208
	TGGTCCGCG	176, 398
10	GCCAGGCTGC	85, 377

cance of the longest direct repeat found within a sequence. Under the assumption that nucleotide positions are independent (a Markov chain of order zero), they give asymptotic expressions for L_n, the longest direct repeat in a sequence of length n. The expected length is

$$\mu_L = \frac{0.6359 + 2\ln n + \ln(1-P)}{\ln(1/P)} - 1$$

which increases as the logarithm of sequence length, while the variance is independent of sequence length

$$\sigma_L^2 = \frac{1.645}{(\ln P)^2}$$

In these expressions, P is the sum of squares of base frequencies in the sequence

$$P = \sum_{i=1}^{4} p_i^2$$

Computing effort may be saved by starting the algorithm of Karlin et al. (1983) with a word length k as close as possible to the expected mean of the longest length.

An approximate test of the hypothesis that the sequence is a random collection of bases, with the longest direct repeat having mean and variance given by these last expressions, would be to compare the observed longest repeat with the sum of the mean and 1.645 times the standard deviation. By supposing the direct repeat lengths were normally distributed, this calculated value would be exceed by chance only 5% of the time if the hypothesis was true. For the chicken β-globin sequence, the four bases have counts 87, 144, 118, 89 for `A, C, G, T` so $P = 0.2614$ and the longest repeat has an expected length of 8.31 with an expected variance of 0.9138. With a probability of 95% normal distribution theory says that the longest direct repeat is expected to be no longer than $9.88 \approx 10$.

Shuffling Tests

A procedure that does not depend on asymptotic expressions, or assumptions of normality, is related to numerical resampling (Chapter 5). A SHUFFLING TEST can be performed by randomly permuting the elements of the sequence at random a number of times, and searching for direct repeats for each permuted sequence. In this way a distribution of direct repeat lengths is built up, under the supposition that the sequence is a random collection of bases, and the hypothesis can be rejected at the 5% level if the observed longest repeat lies in the upper 5% of this distribution.

The permutations for the shuffling test preserve the base frequencies in the original sequence, and an algorithm for doing this is contained in the Appendix. More sophisticated permutations can be constructed to preserve the original dinucleotide or codon frequencies, or any other feature of the original sequence.

One hundred shuffles were carried out on the chicken β-globin sequence of Figure 7.2, with the results displayed in Table 7.14. The observed longest direct repeat of 10 appears in the top 8% of those expected by chance alone. It would probably not be regarded as being "significant." These shuffles ignored the codon structure in this particular sequence. The mean and variance calculated from the data in Table 7.14 are close to those predicted by Karlin et al. (1983).

COMPARING SEQUENCES

Dot Plots

Comparison of different sequences can be accomplished in many ways, the most direct being visually, with DOT PLOTS. The sequences to be compared

Table 7.14 Results of a shuffling test of the significance of the longest direct repeat found in the chicken β-globin sequence of Figure 7.2.

Longest Length L	Frequency* f	fL	fL^2
7	22	154	1078
8	54	432	3456
9	16	144	1296
10	8	80	800
Total†	100	810	6630

* Among 100 shuffles.

† Totals lead to sample mean of $\bar{L} = 8.10$ and sample variance of $s_L^2 = 0.6970$.

are arranged along the margins of a table, and a "dot" is placed in the table at the intersection of those rows and columns that are headed by the same letter. A diagonal stretch of dots will indicate equal subsequences in the two sequences. To reduce the number of dots, which may represent matches that have arisen by chance, dot plots can be enhanced by filtering (Maizel and Lenk 1981). Dots are placed only when a certain proportion of a small group of successive bases match. In Figure 7.3, the dot plot is shown for a comparison of the chicken β-globin sequence in Figure 7.2 with itself. Groups of 9 successive bases were compared, for all possible starting positions of groups of this length. If 7 or more bases matched in the groups starting at sites i and j, a dot was placed at position $i+4, j+4$ in the plot. When a sequence is compared with itself, there must be a solid line of dots on the diagonal, but several runs of "near" matches can be seen in Figure 7.3. The small circle indicates the regions of 10 matches, starting at positions 85 and 377, noted in Table 7.13. The larger circle indicates the regions between positions 72 and 89, and 411 and 428 that meet the 7 out of 9 criterion.

A region of high similarity between two sequences appears as a diagonal in the dot plot with a large number of runs of k matching bases. When many comparisons are being made, as is the case when a query sequence is compared to all entries in a databank, it is important to know how likely a set of runs on a diagonal is to have arisen by chance alone. A methodology for performing statistical tests on observed similarities in this context is described by Mott et al. (1989).

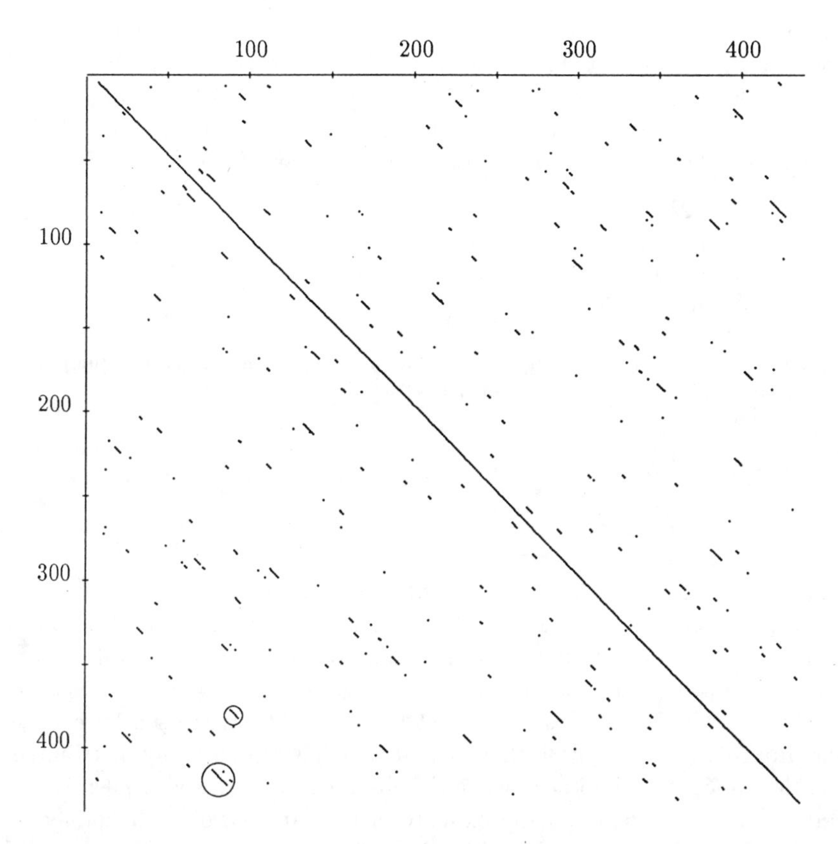

Figure 7.3 Dot-plot for comparison of sequence in Figure 7.2 with itself. Circles explained in text.

Exact Matches between Sequences

The simplest kind of similarity between two sequences is that of a region in one exactly matching a region in the other. As a basis for assessing the biological significance of detected similarities, a model is set up that supposes that a DNA sequence consists of a random array of nucleotides, each one with the same probability of having a particular base type. Under this random model, sequences with base frequencies a_i and b_i give a constant probability $P = \sum_{i=1}^{4} a_i b_i$ that any two positions, one in each sequence, carry the same base. A run of ℓ matches, terminated by a mismatch, therefore has probability $P^{\ell}(1 - P)$. A complete search along sequences of lengths n_1

and n_2 for matches of length ℓ will involve $(n_1 - \ell + 1)(n_2 - \ell + 1)$ starting positions, and there may be interest in the longest match, of length L, found in all of these comparisons. Approximate expressions can be found (Arratia and Waterman 1985) for the mean and variance of this longest match found from all starting positions in two random sequences of length n

$$\begin{aligned} \mu_L &= \frac{2 \ln n}{\ln(1/P)} \\ \sigma_L^2 &= \frac{1.645}{(\ln P)^2} + \frac{1}{12} \end{aligned}$$

These values are substantially larger than those for the length ℓ expected to be found from any particular pair of starting positions

$$\begin{aligned} \mu_\ell &= \frac{P}{1 - P} \\ \sigma_\ell^2 &= \frac{P}{(1 - P)^2} \end{aligned}$$

Two sequences of length 100 bases, with the four bases equally frequent in each, are expected to contain an exact match of 6.64 bases somewhere along their lengths. From any particular starting position, the expected length of an exact match is only 0.33 bases.

To test whether an observed exact match could have arisen by chance, one procedure is to suppose that longest matches are normally distributed, and then reject the hypothesis when the observed value exceeds $\mu_L + 1.645\sigma_L$. Once again, dependence on approximate theory can be avoided by numerical studies. The two sequences could be shuffled and each shuffled pair searched for exact matches. Alternatively, benchmark values could be established by simulation. For any given base composition, pairs of sequences of specified lengths could be generated at random, and matches sought in each such simulated pair. With many simulations a distribution of longest match lengths could be generated.

Queen and Korn Algorithm

Even if two sequences are known to be homologous, meaning that they have descended from a single ancestral sequence, it is unlikely that they will retain the same base composition or sequence length. Both features will change over time because of base substitution mutations or structural rearrangements. Searches for LOCAL SIMILARITIES between sequences should therefore accommodate some departures from exact matches. A variety of ad-hoc procedures for searching for less-than-perfect matches have been proposed, and that due to Queen and Korn (1980) will be treated here.

The algorithm requires that a number of parameters be set. In general, matches are sought from all possible starting positions between the two sequences providing that there is a match for at least a (e.g., $a = 2$) bases. Once the initial a bases are found to match, the region of local similarity can be extended in one of several ways. If there is matching between the bases in the next position of each sequence, these positions are added to the similarity. A mismatch can be added, however, if there is a specified level of matching following this position. For example, a mismatch can be allowed provided two of the next three bases match. Failing this test, a specified number of bases could be "looped out" of either sequence and the search for matching continued with this displacement. Such looping out of (say) one or two bases may be allowed provided two or three of the next three bases match. Finally, the addition of blocks of mismatches or loopouts followed by sets of bases with specified levels of matching can be added to the growing similarity only if the overall level of similarity in the extended region exceeds some specified level.

An illustration of a possible implementation of this algorithm is shown in Figure 7.4. A similarity can be extended by a single matching base or by blocks of four bases consisting of mismatches or loopouts followed by at least two of three matches. The overall level of matching is therefore constrained to be at least 50%. Applying the algorithm of Figure 7.4 to the two short sequences `CTATAATCCC` and `CTGTATC` provides a longest local similarity of

```
CTATAA
CTGTAT
```

showing an initial pair of matches `CT/CT`, a mismatch `A/G` followed by a block of two matches out of three `TAA/TAT`.

Although it is possible to calculate the probabilities of extending a local similarity in each possible way under the Queen and Korn algorithm (Brooks et al. 1988), in practice numerical methods provide a simpler way of assessing the statistical significance of an observed match. Each sequence can be shuffled and the algorithm applied to the resulting pairs of sequences.

Needleman-Wunsch Algorithm

The Queen and Korn algorithm represents the methods used to search for regions of local similarity between two sequences. No attempt is made to relate the sequences as a whole. Global searches, that take into account the total lengths of the sequences, are used to align sequences in such a way

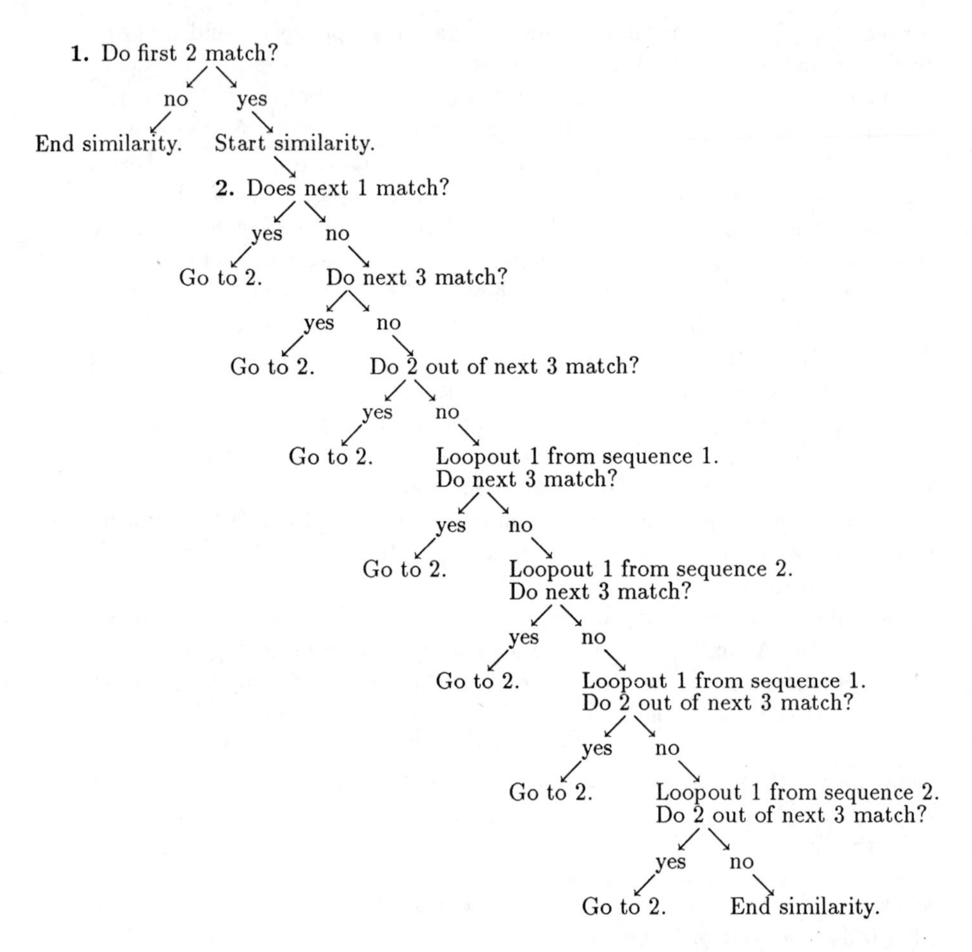

Figure 7.4 An implementation of the Queen and Korn algorithm that starts with two matches, and allows single mismatches or single loopouts that are followed by at least two matches in the next three positions.

as to maximize the degree of GLOBAL SIMILARITY. Alignment of sequences of unequal length necessarily requires the introduction of loopouts on one sequence, or gaps on the other. An example of such procedures is provided by the algorithm due to Needleman and Wunsch (1970). This algorithm was developed for amino acid sequences, but it can also be used for nucleotide sequences.

As originally introduced, the algorithm sought to minimize a distance between two sequences. Although the elements of such distances are specified in an ad-hoc manner, the algorithm has optimal properties in that it determines the minimum distance. It is an example of a DYNAMIC PROGRAMMING method, and the following description is based on that given by Kruskal (1983).

The two sequences to be aligned are placed along the axes of a two-way table. From any pair of bases, i.e., any cell in the table, an alignment can be extended in three possible ways: adding a base in each sequence, with a specified addition to the distance if the bases do not match, or adding a base in one sequence but a gap in the other, or vice versa. Introduction of a gap also contributes a specified amount to the distance. A cell in the table can therefore be reached from (at most) three neighboring cells, and the direction that results in the smallest distance to that cell from the upper left-hand corner is taken to be direction of extending the similarity. Equal distances imply that two directions are possible. These directions are recorded and when all cells have been investigated a path is traced from the bottom right-hand corner (the end of the two sequences) to the top left-hand corner (the beginning of the two sequences) along the recorded directions. The resulting path(s) give the alignments(s) of smallest distance.

The process is illustrated in Figure 7.5 for the two small sequences `CTGTATC` and `CTATAATCCC` and weights 1 for mismatched bases or 3 for introduced gaps. This figure shows that a boundary row and column are added to the table as initial conditions. In column 5 of row 3, corresponding to the second `T` in the shorter (the second) sequence and the first `T` in the longer (the first) sequence, the three possible increments to distance are shown. Adding the base `T` in each sequence contributes 0 to the distance. A horizontal movement, from column 4 of row 3, equivalent to adding the second sequence `T` but introducing a gap in sequence 1, adds a penalty of 3 in this example. A vertical movement from row 2 of column 5 corresponds to adding the first sequence `T` but introducing a gap in sequence 2, and this also has a penalty of 3 in this case. The additions that result in the smallest distance for the cell are the diagonal and horizontal ones, so arrows in those two directions are recorded on the table. Both result in a distance of 6 units to this cell from the top left-hand corner. Traversing the table along the arrows from bottom right to top left produces six possible alignments:

```
CTATAATCCC    CTATAATCCC    CTATAATCCC
CTGTA TC      CTGTA T C     CTGTA T  C

CTATAATCCC    CTATAATCCC    CTATAATCCC
CTGT ATC      CTGT AT C     CTGT AT  C
```

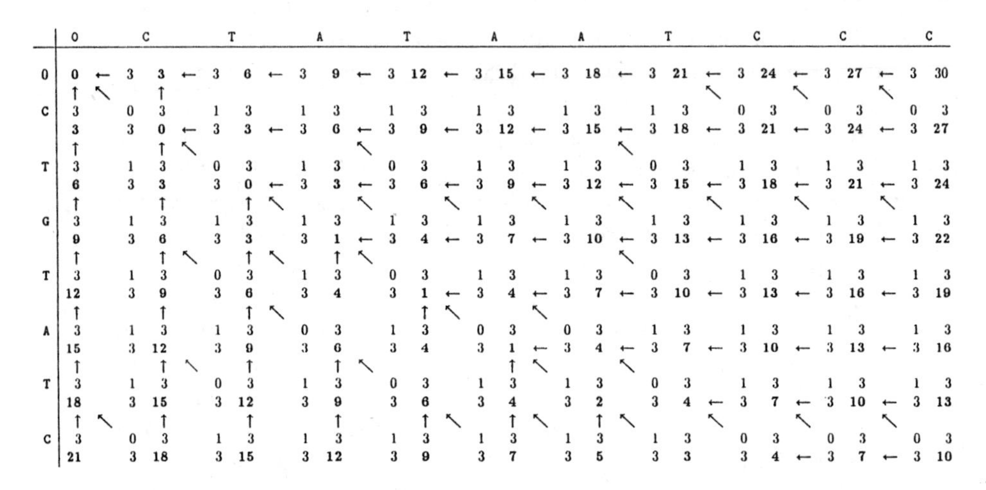

Figure 7.5 An implementation of the Needleman-Wunsch algorithm that weights mismatches by 1 and single deletions or insertions by 3.

In each case the distance is 10, corresponding to six matches, one mismatch, and three gaps in the shorter sequence.

Algebraically, the algorithm can be described for sequences **a** and **b** with bases a_i and b_j. The distance between the sequences is $d(\mathbf{a}, \mathbf{b})$ and is found recursively by evaluating the distances $d(\mathbf{a}^i, \mathbf{b}^j)$ that take into account the first i positions in **a** and the first j positions in **b**. If **a** and **b** are of lengths m and n, then the desired distance is $d(\mathbf{a}^m, \mathbf{b}^n)$. The introduced first row and column of the table are labeled with 0, corresponding to an empty sequence. Within cell (i, j), the additions to distance for the three possible events that lead to that cell are

1. Vertical movement from cell $(i-1, j)$ to cell (i, j) corresponds to extending the similarity by inserting a gap in sequence **b**. In other words **b** has resulted from the deletion of a_i in **a**, and the weight for this event is written as $w_-(a_j)$.

2. Diagonal movement from cell $(i-1, j-1)$ to cell (i, j) corresponds to extending the similarity by adding bases a_i and b_j. In other words, **b** has resulted from the substitution of a_i in **a** by b_j, and the weight for this event is written as $w(a_i, b_j)$.

3. Horizontal movement from cell $(i, j-1)$ to cell (i, j) corresponds to

extending the similarity by inserting a gap in sequence **b**. In other words, **b** has resulted from inserting b_j into **a**, and the weight for this event is written as $w_+(b_j)$.

The distance $d(\mathbf{a}^i, \mathbf{b}^j)$ associated with cell (i, j) is taken to be the minimum of the distances in each of the three neighboring cells plus the associated weights:

$$d(\mathbf{a}^i, \mathbf{b}^j) = \min \begin{cases} d(\mathbf{a}^{i-1}, \mathbf{b}^j) + w_-(a_i) \\ d(\mathbf{a}^{i-1}, \mathbf{b}^{j-1}) + w(a_i, b_j) \\ d(\mathbf{a}^i, \mathbf{b}^{j-1}) + w_+(b_j) \end{cases}$$

and there is a need for initial conditions

$$\begin{aligned} d(\mathbf{a}^0, \mathbf{b}^0) &= 0 \\ d(\mathbf{a}^0, \mathbf{b}^j) &= \sum_{k=1}^{j} w_+(b_k) \\ d(\mathbf{a}^i, \mathbf{b}^0) &= \sum_{k=1}^{i} w_-(a_k) \end{aligned}$$

In the example of Figure 7.5,

$$\begin{aligned} w_-(a_i) &= 3 \text{ for every } i \\ w(a_i, b_j) &= \begin{cases} 0 \text{ for every } i = j \\ 1 \text{ for every } i \neq j \end{cases} \\ w_+(b_j) &= 3 \text{ for every } j \end{aligned}$$

When two sequences have been aligned, an assessment of the minimum distance between them could be gained by applying the algorithm to shuffled versions of the sequences.

PROTEIN SEQUENCE DATA

Many of the algorithms applied to DNA sequences were originally derived for protein sequences of amino acids. Because proteins are more likely to be the target of natural selection, it could be argued that analyses of these sequences have more biological relevance than do analyses of DNA sequences. Protein analyses certainly avoid problems of degeneracy in the genetic code, whereby several triplets may encose the same amino acid. As protein sequences can be written in a 20-letter alphabet (Table 1.4), there is also less chance that two proteins will have the same letter (amino acid) by chance alone at any position than is the case for DNA sequences.

Table 7.15 Numbers of substitutions between amino acids observed in 1572 exchanges. Each entry has been multiplied by 10. Fractional exchanges occur when ancestral sequences are ambiguous.

	A	R	N	D	C	Q	E	G	H	I	L	K	M	F	P	S	T	W	Y
R	30																		
N	109	17																	
D	154	0	532																
C	33	10	0	0															
Q	93	120	50	76	0														
E	266	0	94	831	0	422													
G	579	10	156	162	10	30	112												
H	21	103	226	43	10	243	23	10											
I	66	30	36	13	17	8	35	0	3										
L	95	17	37	0	0	75	15	17	40	253									
K	57	477	322	85	0	147	104	60	23	43	39								
M	29	17	0	0	0	20	7	7	0	57	207	90							
F	20	7	7	0	0	0	0	17	20	90	167	0	17						
P	345	67	27	10	10	93	40	49	50	7	43	43	4	7					
S	772	137	432	98	117	47	86	450	26	20	32	168	20	40	269				
T	590	20	169	57	10	37	31	50	14	129	52	200	28	10	73	696			
W	0	27	3	0	0	0	0	0	3	0	13	0	0	10	0	17	0		
Y	20	3	36	0	30	0	10	0	40	13	23	10	0	260	0	22	23	6	
V	365	20	13	17	33	27	37	97	30	661	303	17	77	10	50	43	186	0	17

Source: Dayhoff et al. (1979).

There is a need for more refined analyses that reflect varying degrees of similarity between amino acids, and consequent varying likelihoods of one being subsituted for another during the course of evolution. It may be convenient, for example, to classify amino acids into such groups as neutral and hydrophobic (G, A, V, L, I, F, P, M), neutral and polar (S, T, Y, W, N, E, C), basic (K, R, H), or acidic (D, E) (Lewin 1987) in the one-letter notation of Table 1.4. Other classifications have been given (e.g., Dayhoff et al. 1979 use hydrophobic, aromatic, basic, acid or acid-amide, cysteine, other hydrophylic).

Alignment of protein sequences can take account of the differential rates at which amino acids substitute for each other, and the following discussion is based on Dayhoff (1979), Dayhoff et al. (1979), and Schwartz and Dayhoff (1979). On the basis of comparisons among many pairs of protein sequences, Dayhoff and her colleagues constructed a MUTATION PROBABILITY MATRIX **M**. They began by comparing many pairs of protein sequences to determine the empirical frequencies with which one amino acid is replaced by others during evolution. On the basis of 1572 observed substitutions, they found the "accepted point mutation matrix" **A** shown in Table 7.15 (note that rounding errors have prevented the entries in that table adding exactly to 1572). The empirical frequency with which amino acid of type i is replaced by type j (or vice versa) is found from that table, and is written as A_{ij}.

For the mutation probability matrix, element M_{ij} in row i and column j is the (empirical) probability that the amino acid in column j will be replaced by that in row i after a specified time interval. Dayhoff et al. measured time in units of PERCENT ACCEPTED MUTATIONS (*PAM*s). Assuming that mutation does not affect the same site twice, 1 *PAM* is the time over which one substitution is expected to occur in a 100 amino acid polypeptide chain of average composition. Translation from the A_{ij} terms to the M_{ij}s depends on the number of *PAM*s for which the matrix **M** is appropriate, and a scaling constant λ is necessary. Dayhoff et al. (1979) give

$$M_{ij} = \frac{\lambda A_{ij}}{\sum_i A_{ij}}, \quad i \neq j$$

$$M_{ii} = 1 - \lambda m_i$$

so that

$$\sum_i M_{ij} = 1$$

The quantity m_i is the "mutability" of amino acid i. This is another empirical measure, and is for the relative rates of change of the amino acids. Dayhoff et al. (1979) gave the values shown in Table 7.16. The relative frequencies, f_i, of the 20 amino acids are also shown in Table 7.16, and these allow calculation of the number of amino acids in a chain of 100 that will remain unchanged over a specified time. For the 1 *PAM* matrix

$$100 - 1 = 1 - \sum_i f_i M_{ii}$$

so that

$$\lambda = \frac{1}{100 \sum_i m_i f_i}$$

Unlike matrix **A**, matrix **M** is not symmetric. Premultiplying the vector **f** of 20 frequencies f_i by the matrix **M** leaves it unchanged.

Schwartz and Dayhoff (1979) found that 250 *PAM*s was an appropriate time for studying evolutionary relationships among distantly related proteins. The resulting matrix is the 250th power of **M**, and is shown in Table 7.17. Rounding errors prevent Table 7.17 following exactly from Tables 7.15 and 7.16.

Finally, Dayhoff et al. (1979) defined a relatedness odds matrix **R** with elements $R_{ij} = M_{ij}/f_i$. This odds matrix is symmetric and gives the probabilities of replacement of one amino acid by another. This matrix, after logarithms have been taken of each element, is shown in Table 7.18 in a way

Table 7.16 Relative mutabilities m_i and frequencies f_i of amino acids in the accepted point mutation data. Source: Dayhoff et al. (1979).

	m_i	f_i		m_i	f_i
A	100	0.087	L	40	0.085
R	65	0.041	K	56	0.081
N	134	0.040	M	94	0.015
D	106	0.047	F	41	0.040
C	20	0.033	P	56	0.051
Q	93	0.038	S	120	0.070
E	102	0.050	T	97	0.058
G	49	0.089	W	18	0.010
H	66	0.034	Y	41	0.030
I	96	0.037	V	20	0.065

that groups amino acids that are more likely to be interchangeable. Once again, rounding errors prevent an exact matching of Tables 7.15 – 7.18. Elements of the matrix are used, for example, in a Needleman-Wunsch-like algorithm as the weights w_{ij} for substituting amino acid i by amino acid j, although Dayhoff et al. recommended adding a bias term B to each element

Table 7.17 Mutation probability matrix for the evolutionary time of 250 *PAM*s. Each element has been multiplied by 100.

	A	R	N	D	C	Q	E	G	H	I	L	K	M	F	P	S	T	W	Y	V
A	13	6	9	9	5	8	9	12	6	8	6	7	7	4	11	11	11	2	4	9
R	3	17	4	3	2	5	3	2	6	3	2	9	4	1	4	4	3	7	2	2
N	4	4	6	7	2	5	6	4	6	3	2	5	3	2	4	5	4	2	3	3
D	5	4	8	11	1	7	10	5	6	3	2	5	3	1	4	5	5	1	2	3
C	2	1	1	1	52	1	1	2	2	2	1	1	1	1	2	3	2	1	4	2
Q	3	5	5	6	1	10	7	3	7	2	3	5	3	1	4	3	3	1	2	3
E	5	4	7	11	1	9	12	5	6	3	2	5	3	1	4	5	5	1	2	3
G	12	5	10	10	4	7	9	27	5	5	4	6	5	3	8	11	9	2	3	7
H	2	5	5	4	2	7	4	2	15	2	2	3	2	2	3	3	2	2	3	2
I	3	2	2	2	2	2	2	2	2	10	6	2	6	5	2	3	4	1	3	9
L	6	4	4	3	2	6	4	3	5	15	34	4	20	13	5	4	6	6	7	13
K	6	18	10	8	2	10	8	5	8	5	4	24	9	2	6	8	8	4	3	5
M	1	1	1	1	0	1	1	1	1	2	3	2	6	2	1	1	1	1	1	2
F	2	1	2	1	1	1	1	1	3	5	6	1	4	32	1	2	2	4	20	3
P	7	5	5	4	3	5	4	5	5	3	3	4	3	2	20	6	5	1	2	4
S	9	6	8	7	7	6	7	9	6	5	4	7	5	3	9	10	9	4	4	6
T	8	5	6	6	4	5	5	6	4	6	4	6	5	3	6	8	11	2	3	6
W	0	2	0	0	0	0	0	0	1	0	1	0	0	1	0	1	0	55	1	0
Y	1	1	2	1	3	1	1	1	3	2	2	1	2	15	1	2	2	3	31	2
V	7	4	4	4	4	4	4	5	4	15	10	4	10	5	5	5	7	2	4	17

Source: Dayhoff et al. (1979).

Table 7.18 Log-odds matrix for 250 *PAM*s. Elements are shown multiplied by 10.

	C	S	T	P	A	G	N	D	E	Q	H	R	K	M	I	L	V	F	Y	W
C	12																			
S	0	2																		
T	-2	1	3																	
P	-3	1	0	6																
A	-2	1	1	1	2															
G	-3	1	0	-1	1	5														
N	-4	1	0	-1	0	0	2													
D	-5	0	0	-1	0	1	2	4												
E	-5	0	0	-1	0	0	1	3	4											
Q	-5	-1	-1	0	0	-1	1	2	2	4										
H	-3	-1	-1	0	-1	-2	2	1	1	3	6									
R	-4	0	-1	0	-2	-3	0	-1	-1	1	2	6								
K	-5	0	0	-1	-1	-2	1	0	0	1	0	3	5							
M	-5	-2	-1	-2	-1	-3	-2	-3	-2	-1	-2	0	0	6						
I	-2	-1	0	-2	-1	-3	-2	-2	-2	-2	-2	-2	-2	2	5					
L	-6	-3	-2	-3	-2	-4	-3	-4	-3	-2	-2	-3	-3	4	2	6				
V	-2	-1	0	-1	0	-1	-2	-2	-2	-2	-2	-2	-2	2	4	2	4			
F	-4	-3	-3	-5	-4	-5	-4	-6	-5	-5	-2	-4	-5	0	1	2	-1	9		
Y	0	-3	-3	-5	-3	-5	-2	-4	-4	-4	0	-4	-4	-2	-1	-1	-2	7	10	
W	-8	-2	-5	-6	-6	-7	-4	-7	-7	-5	-3	2	-3	-4	-5	-2	-6	0	0	17

Source: Dayhoff et al. 1979.

of the matrix. A penalty for an insertion or deletion also needs to be defined. Dayhoff et al. gave a constant gap penalty for gaps of any length. A typical value for the bias correction and the gap penalty is 6. The alignment of fish protamine sequences shown in Figure 1.2 followed from such an algorithm.

DISTANCE BETWEEN DNA SEQUENCES

The distance employed in the Needleman-Wunsch algorithm was meant to accommodate the mutational events of base substitution or insertion and deletion, but the relative weights of these events were arbitrary. Such distances would not be used in the construction of evolutionary trees. The earliest distance for sequences, introduced for amino acid sequences, was given by Jukes and Cantor (1969). It is based on the quantity q discussed in Chapter 4, and introduced in Equation 4.2.

Although the construction of evolutionary trees is taken up in Chapter 8, this is a good place to repeat the observation of Felsenstein (1988) that alignment of sequences and construction of trees are not really separate issues. They are being treated separately here by way of introduction only. Sankoff et al. (1973) presented alignments of some 5S rRNA sequences by taking explicit account of their evolutionary history.

When a sequence is available from each of two populations, q is the probability that homologous nucleotides have the same base. Note that there is an explicit assumption that the sequences are homologous, having a single ancestral sequence. Detecting similarities with the Queen and Korn algorithm carries no such assumption. The comparison of sequences to allow q to be estimated as the proportion of homologous bases that is the same requires a preliminary alignment of the sequences. As shown in Chapter 4, q decreases over time because of base substitution, and has a value in generation t since divergence from the ancestral sequence of

$$q_t = \frac{1}{4} + \frac{3}{4}\left(1 - \frac{8\mu}{3}\right)^t$$

where μ is the rate of base substitution. The quantity

$$K = \frac{3}{4}\ln\left(\frac{3}{4q-1}\right) \approx 2\mu t$$

can serve as a distance measure.

Distance K is appropriate for indicating the time since two sequences diverged from an ancestral sequence, and can lead to evolutionary trees between sequences (gene trees) as described in Chapter 8. Care does need to be taken when the time of divergence is very large. Even though K increases linearly over all time, there is a limit to how different sequences may become. Eventually, with equal base frequencies, there is just a probability of 0.25 that homologous bases are the same. The quantity q can then no longer discriminate between sequences, and any discrimination based on K may be misleading. The behavior of both q and K, as a function of $2\mu t$ is shown in Figure 7.6.

When several sequences are available from each of two species, it is possible to construct species trees allowing for the ancestral species being polymorphic. Two sequences, within or between species, may have descended from different sequences in that ancestral population. A measure of sequence similarity, Q, was discussed in Chapter 4 and it depends on the population size as well as the mutation rate. When the ancestral population is in equilibrium between drift and mutation, the quantity Q is expected to remain constant at $\hat{Q}$ over time and to furnish the initial value of q. Although this constant value is not known, it can be estimated from the variation observed in each of two extant species (Cockerham 1984, Weir and Basten 1990). A modified version of the Jukes-Cantor distance, taking account of within-species variation, is then

$$K_{\mathrm{W}} = \frac{3}{4}\ln\left(\frac{4\hat{Q}-1}{4q-1}\right)$$

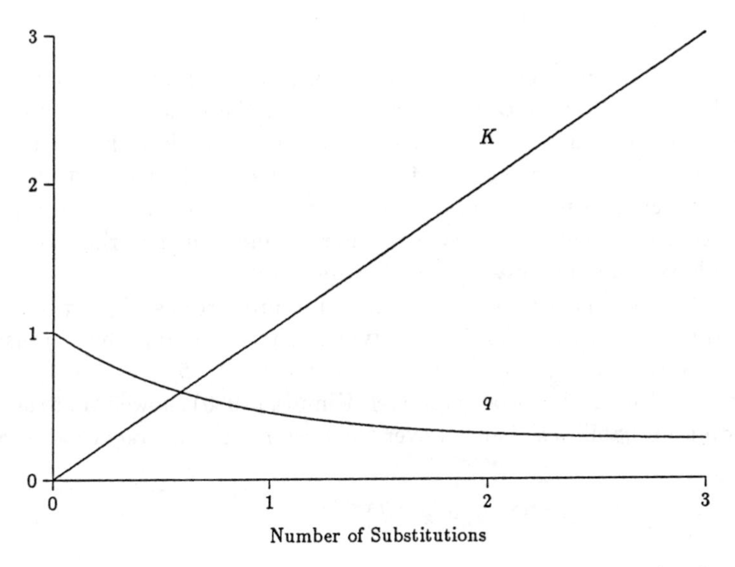

Figure 7.6 Probability q that homologous bases are the same, and Jukes-Cantor distance K, as a function of the number of base substitutions $2\mu t$.

This is still directly proportional to time, but time now refers to that between the extant and ancestral *populations* rather than sequences.

If n_i sequences are observed in population i, and $r_{ijj'}$ is the proportion of homologous bases that is the same between sequences j and j', then the sample similarity within that population is

$$\tilde{Q}_i = \frac{1}{n_i(n_i-1)} \sum_{j=1}^{n_i} \sum_{j' \neq j} r_{ijj'}$$

and $\hat{Q}$ could be estimated as

$$\tilde{Q} = \frac{1}{2}(\tilde{Q}_1 + \tilde{Q}_2)$$

If $s_{jj'}$ is the proportion of like bases between sequence j in population 1 and sequence j' in population 2, then the between-population similarity is estimated as

$$\tilde{q} = \frac{1}{n_1 n_2} \sum_{j=1}^{n_1} \sum_{j'=1}^{n_2} s_{jj'}$$

Putting these together

$$\tilde{K}_{\mathrm{W}} = \frac{3}{4} \ln \left(\frac{4\tilde{Q} - 1}{4\tilde{q} - 1} \right)$$

There have been several attempts to generalize such distances by allowing for less restrictive mutation models. Use of the Jukes-Cantor measure assumes that each base is equally likely to mutate to each of the three other bases with a constant rate $\mu/3$. Kimura (1980) allowed TRANSITIONS (mutation between two pyrimidines or two purines: $T \leftrightarrow C$, or $A \leftrightarrow G$) and TRANSVERSIONS (mutation between a pyrimidine and a purine: $T, C \leftrightarrow A, G$) to have different rates, α and β, respectively.

After the sequences have been aligned, the proportions of positions having bases that differ by a transition (a type I change), or differ by a transversion (a type II change), or are the same are denoted by $\tilde{p}_{\mathrm{I}}$, $\tilde{p}_{\mathrm{II}}$, or $1 - \tilde{p}_{\mathrm{I}} - \tilde{p}_{\mathrm{II}}$, respectively. Under his mutation model, Kimura (1980) showed that the corresponding probabilities changed over the time t since divergence according to

$$p_{\mathrm{I}_t} = \frac{1}{4}\left(1 - 2e^{-4(\alpha+\beta)t} + e^{-8\beta t}\right)$$

$$p_{\mathrm{II}_t} = \frac{1}{4}\left(1 - e^{-8\beta t}\right)$$

If $k = \alpha + 2\beta$ is the total rate of base substitution per unit time, an appropriate measure of distance for a sequence tree is

$$K = -\frac{1}{2} \ln \left[(1 - 2p_{\mathrm{I}} - p_{\mathrm{II}}) \sqrt{1 - 2p_{\mathrm{II}}} \right] \approx 2kt \qquad (7.1)$$

Such distances may be used in distance matrix methods of phylogeny construction and estimated with sample proportions substituted into Equation 7.1. J. Felsenstein (personal comuunication) points out that such estimates are not maximum likelihood, and that numerical methods will be need to find the ML estimate of K.

Kimura used β-globin sequences from rabbit and chicken as an example, and these sequences are displayed in Figure 7.7. The sequence lengths are 438 bp, and there are 58 type I changes and 63 type II changes. Therefore $\tilde{p}_{\mathrm{I}} = 0.1324$ and $\tilde{p}_{\mathrm{II}} = 0.1438$. Kimura's distance is 0.3513, which is not substantially different from the Jukes-Cantor distance of 0.3446 based only on the proportion $\tilde{q} = 0.7237$ of similar bases.

SUMMARY

Restriction site data may generally be analyzed with the techniques introduced in earlier chapters for such Mendelizing units as allozymes. When con-

```
type I        **     **            *    * *    *
Chicken  GTGCACTGGA CTGCTGAGGA GAAGCAGCTC ATCACCGGCC TCTGGGGCAA GGTCAATGTG  60
Rabbit   GTGCATCTGT CCAGTGAGGA GAAGTCTGCG GTCACTGCCC TGTGGGGCAA GGTGAATGTG
type II         * *    *            *** *        *    *            *

type I                  *  *                    * *
Chicken  GCCGAATGTG GGGCCGAAGC CCTGGCCAGG CTGCTGATCG TCTACCCCTG GACCCAGAGG 120
Rabbit   GAAGAAGTTG GTGGTGAGGC CCTGGGCAGG CTGCTGGTTG TCTACCCATG GACCCAGAGG
type II   **   **    * *            *                       *

type I        *             *           *         **  *   **    *  *
Chicken  TTCTTTGCGT CCTTTGGGAA CCTCTCCAGC CCCACTGCCA TCCTTGGCAA CCCCATGGTC 180
Rabbit   TTCTTCGAGT CCTTTGGGGA CCTGTCCTCT GCAAATGCTG TTATGAACAA TCCTAAGGTG
type II         *                 *   **  * * *        * *           *   *

type I    *   *  *         *       * *      **             *
Chicken  CGCGCCCACG GCAAGAAAGT GCTCACCTCC TTTGGGGATG CTGTGAAGAA CCTGGACAAC 240
Rabbit   AAGGCTCATG GCAAGAAGGT GCTGGCTGCC TTCAGTGAGG GTCTGAGTCA CCTGGACAAC
type II  * *                      *   *        *  *  * *    **

type I        ***       *  *   *     *         *                *     *  *
Chicken  ATCAAGAACA CCTTCTCCCA ACTGTCCGAA CTGCATTGTG ACAAGCTGCA TGTGGACCCC 300
Rabbit   CTCAAAGGCA CCTTTGCTAA GCTGAGTGAA CTGCACTGTG ACAAGCTGCA CGTGGATCCT
type II  *               *  *      **

type I                         **  *      * *            *  *   *  **    *
Chicken  GAGAACTTCA GGCTCCTGGG TGACATCCTC ATCATTGTCC TGGCCGCCCA CTTCAGCAAG 360
Rabbit   GAGAACTTCA GGCTCCTGGG CAACGTGCTG GTTATTGTGC TGTCTCATCA TTTTGGCAAA
type II                              *  *         *    *  **

type I                  *              *                 *
Chicken  GACTTCACTC CTGAATGCCA GGCTGCCTGG CAGAAGCTGG TCCGCGTGGT GGCCCATGCC 420
Rabbit   GAATTCACTC CTCAGGTGCA GGCTGCCTAT CAGAAGGTGG TGGCTGGTGT GGCCAATGCC
type II    *          *  ***            *       *     ***  **       *

type I          *    *
Chicken  CTGGCTCGCA AGTACCAC
Rabbit   CTGGCTCACA AATACCAC
type II
```

Figure 7.7 Chicken and rabbit β-globin sequences. Asterisks denote differences. Source: GenBank database loci CHKHBBM and RABHBB1A1, accession numbers J00860 and J00659.

clusions depend critically on knowing the lengths of restriction fragments, it must be remembered that there are large experimental errors in inferring lengths from gel migration distances. Use of restriction site markers in disease association studies generally requires attention to be paid to the sampling scheme by which normal and diseased individuals are chosen for study.

DNA sequence data analyses currently take one of two forms. In the first place, analyses proceed with little in the way of formal genetic models. The assessment of the significance of the length of longest direct repeats, for example, may rest on the assumption that nucleotides are independent and that bases have equal frequencies at all sites. Methods for finding local similarities or aligning whole sequences are often ad-hoc. Although there may be a biological basis for considering base substitutions and deletions as different events, there is little basis for the particular weights assigned to these events in the various algorithms. When coding regions are being studied, a greater degree of biological relevance is probably attained when amino acid sequences, rather than DNA sequences, are used. These protein sequences are more likely to reflect biological constraints. For evolutionary studies, it is essential that a genetic model be employed. Genetic distances between sequences should attempt to reflect an appropriate mutation model, for example.

EXERCISES

Exercise 7.1

Derive an expression for estimating the population value of linkage disequilibrium between a marker and a disease locus, when haplotype data are available from random samples within the two disease classes.

Exercise 7.2

Apply both the Queen and Korn and Needleman-Wunsch algorithms to the following two sequences: `CTTTGGGAACCTCTCCAGCC`, `CTTCGATCTGTCACA`. Try different penalty values for the Needleman-Wunsch algorithm.

Chapter 8

Phylogeny Construction

INTRODUCTION

Taxonomy is concerned with grouping organisms into a manageable number of groups whose characters are mainly constant. This enables the classification of species. For evolutionary studies, the classification also allows the construction of phylogenies. These activities may shed light on the question of whether observed patterns of species suggest anything about the nature of evolutionary forces. A fine summary of the process of constructing phylogenies is given by Nei (1987). Only a hint of the many methods and complexities is given in this chapter.

A distinction can be made between PHENETIC and CLADISTIC data. Phenetic relationships are defined by Sneath and Sokal (1973) as being similarities based on a set of phenotypic characters for the objects under study, whereas cladistic relationships contain information about ancestry and so are used to study evolutionary pathways. Both relationships are best represented as phylogenetic trees or DENDROGRAMS. The terms PHENOGRAM or CLADOGRAM have been used for trees based on phenetic or cladistic relationships, respectively. A cladogram displays evolutionary time between events or groups, but no concept of time is needed for a phenogram. In this chapter, no great attention will be paid to the distinction. As Nei (1987) points out, if measures of phenotypic similarity imply degrees of evolutionary similarity, then phenetic methodology can provide cladistic trees.

Phylogenetic trees represent the evolutionary pathways, and a distinction between SPECIES TREES and GENE TREES was made by Tateno et al. (1982). Branches in a species tree join extant species to an ancestral species and represent the times since those species diverged. Although such trees are of evolutionary interest, the data used to construct a tree are often from a

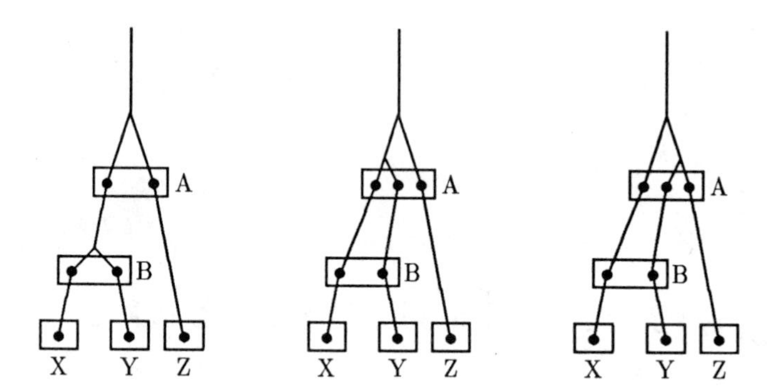

Figure 8.1 A species tree links species A, B, X, Y, and Z. Three possible gene trees, indicated by solid lines, link the genes indicated by solid circles. Source: Nei (1987).

single region of the genome of those species. A gene tree constructed from such data, for example, the DNA sequence of a short region, may not be the same as the species tree. Two species may carry genes that diverged prior to the species split, or introgression may have resulted in genes having diverged after the species split. Possible differences between species and gene trees are shown in Figure 8.1, taken from Nei (1987) (see also Pamilo and Nei 1988). For times of divergence between species that are long compared to the COALESCENT times between genes (i.e., the times at which two genes split from a single ancestral gene), the two trees will be similar.

Another distinction among trees is between ROOTED and UNROOTED trees. In Figure 8.2 two of the 15 possible rooted trees and one of the three possible unrooted trees for four species are shown. A rooted tree conveys the notion of temporal ordering of the species or genes on a tree, while an unrooted tree merely reflects distances between units with no notion of which was ancestral to which.

Three of the principal methods for ordering species in a phylogenetic history will be treated here. DISTANCE MATRIX methods are based on the set of distances calculated between each pair of species. The computations are generally quite straightforward, and the quality of the resulting trees depends on the quality of the distance measure. Whereas distances are generally based on genetic models, MAXIMUM PARSIMONY methods make less explicit mention of genetic assumptions. They proceed by seeking to minimize the

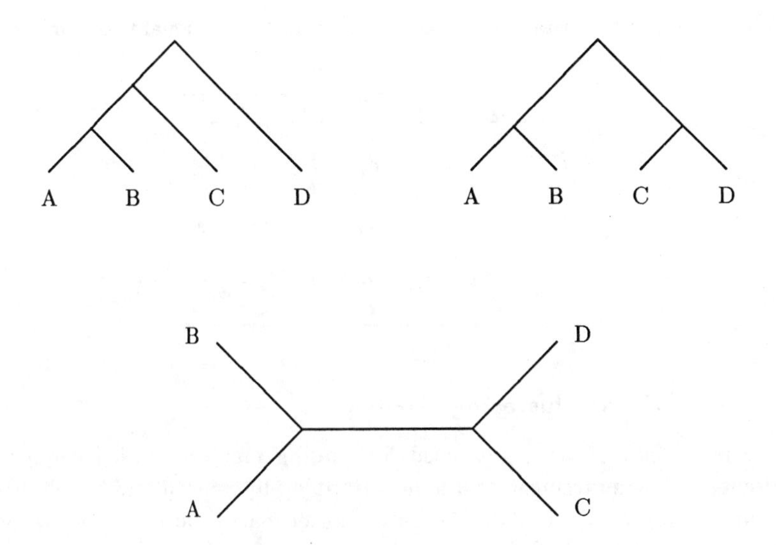

Figure 8.2 Two of the 15 rooted trees and one of the three unrooted trees for four species.

numbers of changes between species over the tree, although they may require approximately constant rates of change (Nei 1987). Heavy dependence on models is a feature of MAXIMUM LIKELIHOOD methods. Although this third class of construction is computationally demanding, it can provide a basis for statistical inference.

DISTANCE MATRIX METHODS

Phylogenetic trees may be based on distance matrices, here of genetic distances between all pairs of OPERATIONAL TAXONOMIC UNITS (OTU's). For t OTU's the matrix of distances between each pair is shown in Table 8.1.

Using these distances to group OTU's in a phenetic context may employ CLUSTERING, and possible approaches to clustering were given by Sneath and Sokal (1973). Clusters are characterized by having more OTU's per unit of, say, gene frequency space than do other areas, and the process of clustering consists of identifying these areas of higher density. Several methods of clustering can be used. When distances are used to construct additive trees, as with the Fitch-Margoliash algorithm described below, the process is said to use a pairwise method rather than clustering.

Table 8.1 Matrix of distances between all pairs of t operational taxonomic units.

		OTU's				
		1	2	3	...	t
OTU's	1	–	d_{12}	d_{13}	...	d_{1t}
	2	d_{21}	–	d_{23}	...	d_{2t}
	3	d_{31}	d_{32}	–	...	d_{3t}
	...	...	...	...	...	...
	t	d_{t1}	d_{t2}	d_{t3}	...	–

Average Linkage Clustering (UPGMA)

There is a class of strategies used for finding clusters, called sequential, agglomerative, hierarchical, and nonoverlapping by Sneath and Sokal (1973). The most widely used is UPGMA (unweighted pair-group method using an arithmetic average). It defines the intercluster distance as the average of all the pairwise distances for members of two clusters.

As an example, consider the mitochondrial DNA sequence data displayed in Figure 1.3. Jukes-Cantor distances (Chapter 7) between each pair of these sequences depend on the observed numbers of nucleotide differences between each pair. If a proportion $\tilde{q}$ of the bases in two sequences is the same, then the distance K is estimated as

$$\tilde{K} = \frac{3}{4}\ln\left(\frac{3}{4\tilde{q}-1}\right)$$

Differences and distances for the sequences are shown in Table 8.2.

Table 8.2 Numbers of differences (below diagonal) and Jukes-Cantor distances (above diagonal) for the five mitochondrial sequences of Figure 1.3.

	Human	Chimpanzee	Gorilla	Orangutan	Gibbon
Human	–	0.015	0.045	0.143	0.198
Chimpanzee	1	–	0.030	0.126	0.179
Gorilla	3	2	–	0.092	0.179
Orangutan	9	8	6	–	0.179
Gibbon	12	11	11	11	–

The closest pair is human and chimpanzee, and these are joined to form a cluster. The distance of each other sequence from this cluster is found as the average distance from the sequence to members of the cluster:

$$
\begin{aligned}
d_{(\text{hu-ch}),\text{go}} &= \frac{1}{2}(d_{\text{hu},\text{go}} + d_{\text{ch},\text{go}}) = 0.037 \\
d_{(\text{hu-ch}),\text{or}} &= \frac{1}{2}(d_{\text{hu},\text{or}} + d_{\text{ch},\text{or}}) = 0.135 \\
d_{(\text{hu-ch}),\text{gi}} &= \frac{1}{2}(d_{\text{hu},\text{gi}} + d_{\text{ch},\text{gi}}) = 0.189
\end{aligned}
$$

The distance matrix can be reduced to

	(hu-ch)	go	or	gi
(hu-ch)		0.037	0.135	0.189
go			0.092	0.179
or				0.179
gi				

and this has smallest distance between the human-chimpanzee cluster and gorilla. These are clustered, and the new distances required are

$$
\begin{aligned}
d_{(\text{hu-ch-go}),\text{or}} &= \frac{1}{3}(d_{\text{hu},\text{or}} + d_{\text{ch},\text{or}} + d_{\text{go},\text{or}}) = 0.121 \\
d_{(\text{hu-ch-go}),\text{gi}} &= \frac{1}{3}(d_{\text{hu},\text{gi}} + d_{\text{ch},\text{gi}} + d_{\text{go},\text{gi}}) = 0.185
\end{aligned}
$$

The next reduction in the distance matrix is

	(hu-ch-go)	or	gi
(hu-ch-go)		0.121	0.185
or			0.179
gi			

As the smallest distance is now that between the human-chimpanzee-gorilla cluster and orangutan, these sequences are clustered and the distance from this triple to the gibbon sequence is

$$
d_{(\text{hu-ch-go-or}),\text{gi}} = \frac{1}{4}(d_{\text{hu},\text{gi}} + d_{\text{ch},\text{gi}} + d_{\text{go},\text{gi}} + d_{\text{or},\text{gi}}) = 0.183
$$

The results of the clustering can be represented as the dendrogram shown in Figure 8.3. In constructing this dendrogram, branchpoints are placed

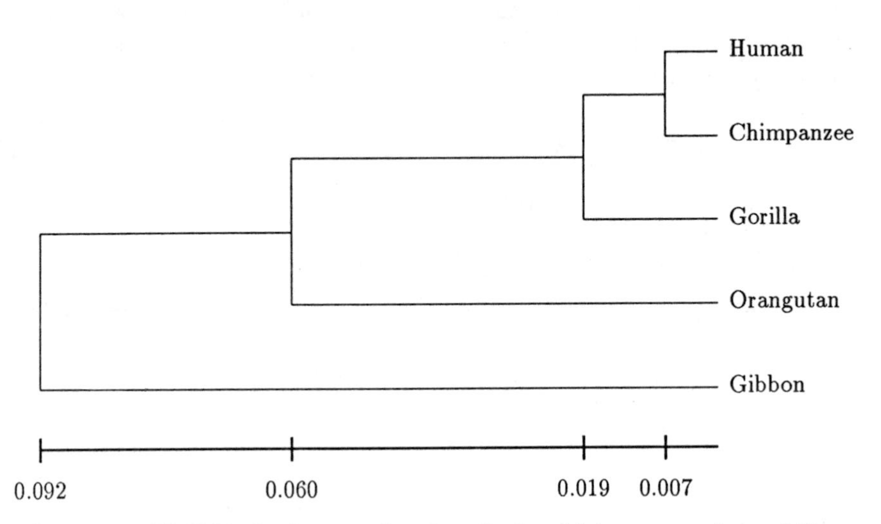

Figure 8.3 UPGMA dendrogram for the mitochondrial sequence data of Figure 1.3.

midway between two sequences, or clusters. The distance between a pair of sequences in the figure is the sum of the branch lengths.

The UPGMA method is widely used for matrices of distances. Degens (1983) has discussed conditions under which the estimated branch lengths are maximum likelihood. Nei et al. (1983) conducted simulation studies to compare different methods of constructing trees, and found that UPGMA generally performed well when the mutation rates were the same along all branches of the trees. It must be emphasized that the assumption of (nearly) equal rates is crucial for the UPGMA method to be appropriate. Other simulation studies (e.g., Kim and Burgman 1988) have demonstrated that the method is not satisfactory when the rates are unequal in different lineages. When rates are equal, a MOLECULAR CLOCK is said to be operating.

Fitch-Margoliash Algorithm

The UPGMA method implicitly assumes a constant rate of change along all branches in a tree. This assumption is removed by an algorithm developed by Fitch and Margoliash (1967). Their method proceeds by inserting "missing" OTU's as common ancestors of later OTU's and fits branch lengths to groups of three OTU's at a time. A similar additive approach was first discussed by

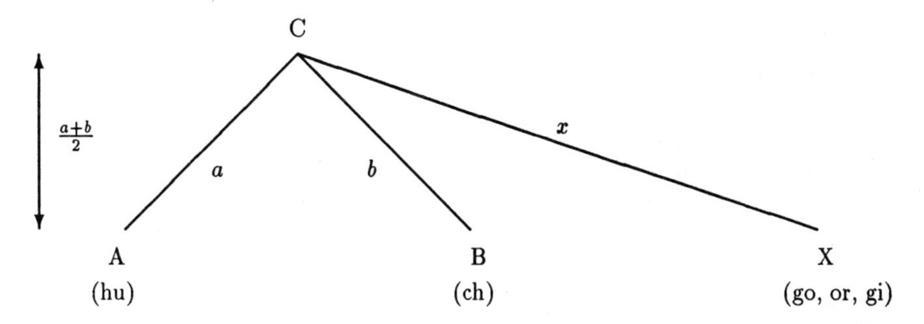

Figure 8.4 Initial step for Fitch-Margoliash algorithm applied to the mitochondrial data of Figure 1.3.

Cavalli-Sforza and Edwards (1967). The Fitch-Margoliash algorithm is now illustrated for the mitochondrial data of Figure 1.3.

The OTU's are divided into three groups: the closest pair, A = human and B = chimpanzee, and the remaining set, X = (gorilla, orangutan, gibbon). Node C, which is the immediate ancestor of A and B, is introduced. The branch lengths from C to A, B are a, b, while the branch from C to X is x (Figure 8.4). As a further notational device, d_{UV} means the distance from node U to node V, $d_{U\bar{V}}$ means the average distance from node U to all nodes other than node V, and d_{U*V} means the average distance from all tip nodes below U to V. An asterisked letter, U^*, denotes the set of tips descended from the node, U, of the same letter. The three pairwise distances between A, B, and C provide three equations in three unknowns:

$$\begin{aligned} a+x &= d_{AX} = d_{A\bar{B}} \\ &= \frac{1}{3}(0.045+0.143+0.198) = 0.129 \\ b+x &= d_{BX} = d_{B\bar{A}} \\ &= \frac{1}{3}(0.030+0.126+0.179) = 0.112 \\ a+b &= d_{AB} \\ &= 0.015 \end{aligned}$$

The first equation uses the average distance from A to each member of set X. Solving these three equations gives

$$a = 0.016, \; b = -0.001$$

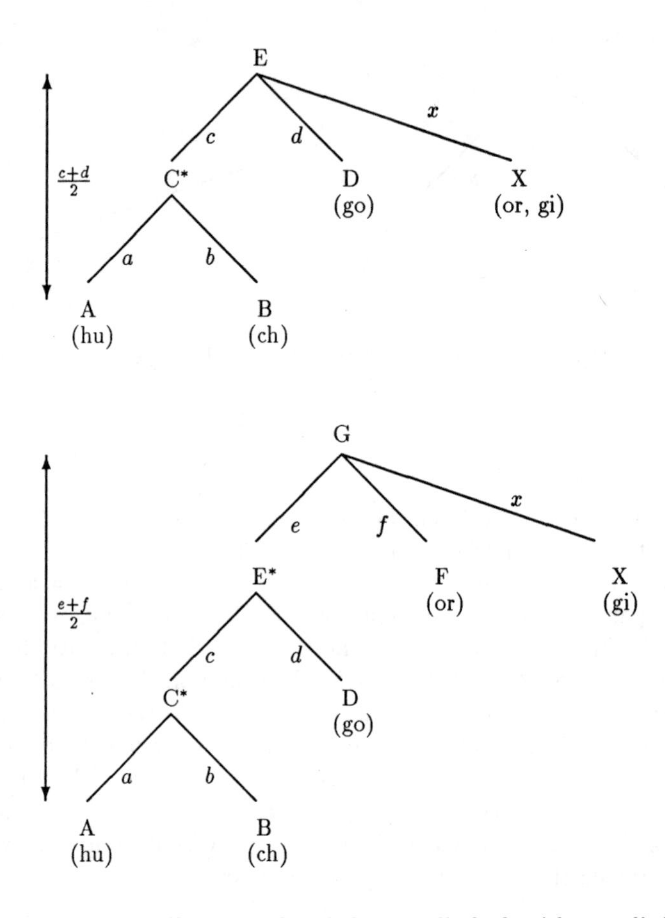

Figure 8.5 Intermediate steps for Fitch-Margoliash algorithm applied to mitochondrial sequence data of Figure 1.3.

By convention, negative estimates are set equal to zero so that $b = 0$ here. The average of a, b is called the height of node C. This value is 0.008. Replacing the pair A, B by C, and recalculating the distances as was done for the UPGMA case, gives the next closest pair of C and D = gorilla. Node E is introduced as the immediate ancestor of C and D. As shown in Figure 8.5, the branch lengths for C^* and E, D and E, and E and X are c, d and x, where the set X is now orangutan and gibbon. The three equations to be solved are

$$c + d = d_{C^*D}$$

$$\begin{aligned}
&= \frac{1}{2}(0.045 + 0.030) = 0.037 \\
c + x &= d_{C \bullet X} = d_{(AB)X} \\
&= \frac{1}{4}(0.143 + 0.198 + 0.126 + 0.179) = 0.162 \\
d + x &= d_{DX} \\
&= \frac{1}{2}(0.092 + 0.179) = 0.136
\end{aligned}$$

so that

$$c = 0.032, \; d = 0.006$$

The height of node E is $(c + d)/2 = 0.019$. The branch length c' between nodes C and E is given by c minus the height of node C, since c measures the distance from C to E *and* the average distance from A and B to C. In other words

$$c' = 0.032 - 0.008 = 0.024$$

With the OTU's reduced to E, orangutan and gorilla, the closest pair is E and F = orangutan. Node G is introduced as their immediate ancestor, and the remainder is now X = gibbon. The equations to be solved for branch lengths are

$$\begin{aligned}
e + f &= d_{E \bullet F} \\
&= \frac{1}{3}(0.143 + 0.126 + 0.092) = 0.121 \\
e + x &= d_{E \bullet X} \\
&= \frac{1}{3}(0.198 + 0.179 + 0.179) = 0.185 \\
f + x &= d_{FX} = 0.179
\end{aligned}$$

with solutions

$$e = 0.063, \; f = 0.057$$

The height of node G is $(e + f)/2 = 0.060$, and the branch length e' from E to G is the difference of e and the height of E, or $0.063 - 0.019 = 0.044$. The process could stop here, and the network is shown in Figure 8.6.

Without extra information, or without assuming the same rates of change along all branches, the Fitch-Margoliash algorithm can lead only to an unrooted tree. Additional information, from a distant sequence (an OUTGROUP), would allow the tree to be rooted. The same topology as provided

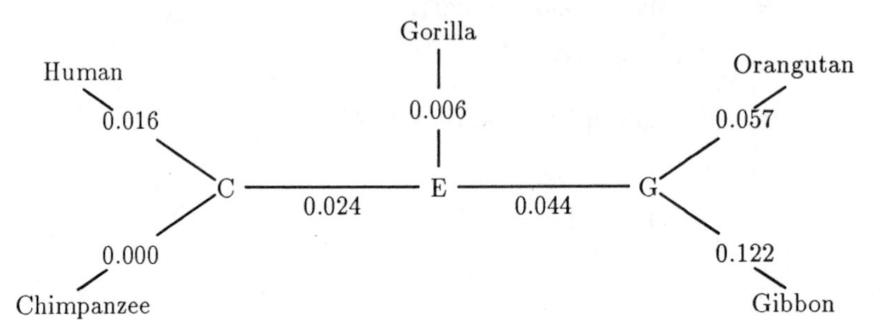

Figure 8.6 Fitch-Margoliash unrooted network for mitochondrial sequence data in Figure 1.3.

by the UPGMA method will result if the root I is placed so that the distances g, from G to I and h from H to I are equal (i.e., assuming equal rates of change in the two branches from I to the extant sequences). Since

$$\begin{aligned} g + h &= d_{G*H} \\ &= \frac{1}{4}(0.198 + 0.179 + 0.179 + 0.179) = 0.184 \end{aligned}$$

$g = h = 0.092$, and the distance g' from G to I is g minus the height of G, or 0.032. Putting all these branch lengths together produces the tree shown in Figure 8.7.

Fitch and Margoliash allow for the possibility that the topology found by their algorithm is incorrect, and they suggest that other topologies be examined. Different trees are compared on the basis of a measure of goodness of fit, called the "percent standard deviation" by Fitch and Margoliash (1967). The best tree will have the smallest percent standard deviation. If d_{ij} is the observed distance pair, i and j, of n OTU's and e_{ij} is the sum of the branch lengths between them on the tree,

$$s = \left\{ \frac{\sum_{i,j}[(d_{ij} - e_{ij})/d_{ij}]^2}{n(n-1)} \right\}^{1/2} \times 100$$

is the percent standard deviation. Note the additivity assumption, whereby the distance between any two nodes is the sum of the branch lengths between them. For the tree in Figure 8.6, the observed distances and branch lengths are shown in Table 8.3 and the percent standard deviation is 1.94. It may

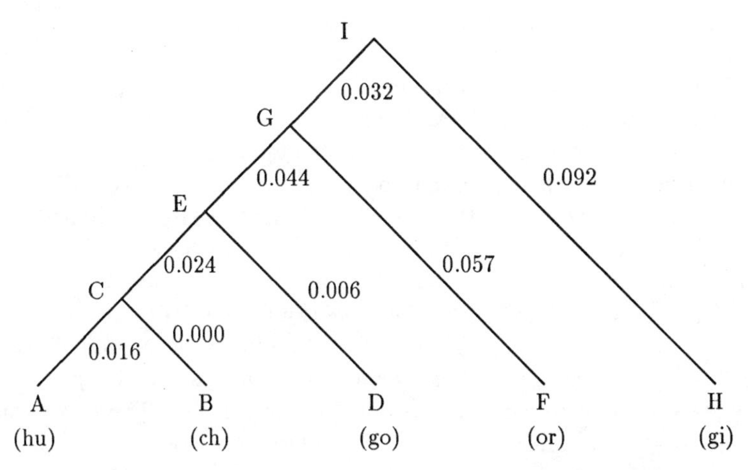

Figure 8.7 Fitch-Margoliash rooted tree for the mitochondrial sequence data in Figure 1.3.

also be possible to adjust branch lengths in the fitted tree to reduce the quantity s.

Choosing trees on the basis of the percent standard deviation should be referred to as using the Fitch-Margoliash criterion, and the best such tree may not be the same as found by the Fitch-Margoliash algorithm. Use of the criterion is expected to give similar results to the UPGMA method when there is a molecular clock. If there is not a clock, so that rates of change are different in different lineages, the Fitch-Margoliash criterion will do much

Table 8.3 Observed distances (above diagonal) and calculated distances (below diagonal) from Fitch-Margoliash algorithm for the five mitochondrial sequences of Figure 1.3.

	Human	Chimpanzee	Gorilla	Orangutan	Gibbon
Human	-	0.015	0.045	0.143	0.198
Chimpanzee	0.016	-	0.030	0.126	0.179
Gorilla	0.046	0.030	-	0.092	0.179
Orangutan	0.141	0.125	0.107	-	0.179
Gibbon	0.208	0.192	0.174	0.181	-

better than UPGMA. (J. Felsenstein, personal communication).

Other trees can be chosen for examination by selecting a different initial pair of OTU's. That tree with the lowest percent standard deviation is considered to be the best tree, and this criterion is the basis on which the Fitch-Margoliash algorithm operates. For example, human and gorilla could be grouped first, and then chimpanzee, orangutan, and gibbon added in that order. However, in that case the height of the second interior node E is less than the height of the first interior node C and the fit between observed and calculated distances is not nearly as good as for the first order.

For the data of Figure 1.3, the trees found by UPGMA and Fitch-Margoliash were very similar. If the correct topology is used, whereas Fitch-Margoliash uses a weighted least-squares approach, UPGMA gives least-squares estimates of branch lengths (Chakraborty 1977). The two methods are not expected to differ very much when there is a clock.

PARSIMONY METHODS

Parsimony methods take explicit notice of the character values observed for each species, rather than working with the distances between sequences that summarize differences between character values. The approach was introduced for gene frequency data by Edwards and Cavalli-Sforza (1963) under the name "Principle of Minimum Evolution." If sequences are available for a set of species then the most parsimonious topology linking them is sought. Branch lengths are generally not obtained.

For each possible topology, the sequences at each node are inferred to be those that require the least number of changes to give each of the two immediately descendant sequences. The total number of changes required to traverse the whole tree is then found, and that tree with the minimum total is the most parsimonious. To illustrate the procedure, consider the example given by Fitch (1971). Sequences from six species, A–F, are available, and at one particular position they have bases `C`, `T`, `G`, `T`, `A`, `A`, respectively. There are many possible topologies, and one of them is shown in Figure 8.7. Each of the nodes 1–5, starting with those nearest the extant sequences, is considered in turn. At each node the "parsimony" (W. M. Fitch, unpublished) of the two descendant sequences is written down. This operation, denoted here by $\diamond$, is a set operation defined as the intersection of two sets if this intersection is not empty, and the union of the two sets if their intersection is empty. For distinct sets (sequences) X, Y, Z, the usual set operations of union and

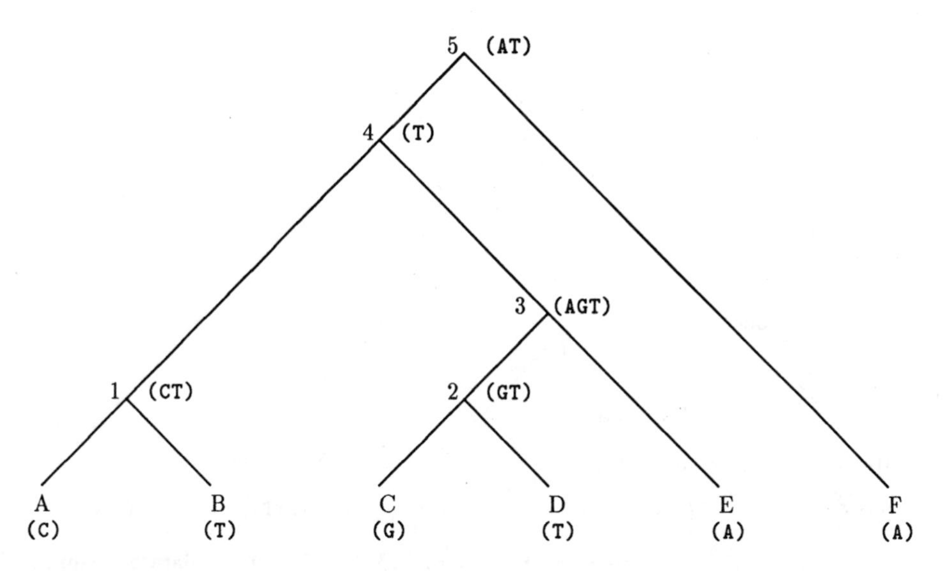

Figure 8.8 Illustration of procedure for finding maximum parsimony tree for one site in six sequences. Source: Fitch (1971).

intersection can be contrasted with parsimony:

intersection	$[X,Y] \cap [X,Z] = [X]$	$[X] \cap [Y] = \phi$
union	$[X,Y] \cup [X,Z] = [X,Y,Z]$	$[X] \cup [Y] = [X,Y]$
parsimony	$[X,Y] \diamond [X,Z] = [X]$	$[X] \diamond [Y] = [X,Y]$

If two sequences have the same base at a position, minimal changes will result if their immediate ancestor also has this base. If they have different bases, minimal changes require that their ancestor has either of those two. In Figure 8.8, nodes 1 and 2 are (CT) and (GT), meaning that either of the two bases shown will give the smallest number of changes. There are three possibilities for node 3, but only one for node 4, and then two for node 5. The smallest number of changes, four, for this topology results if nodes 1–5 all have base T. As Nei (1987) points out, however, the same minimum number results if each node has base A. There are a further nine possibilities: those in which the five nodes have bases TTTTA, TAAAA, CAAAA, AGAAA, ATAAA, CGAAA, CTAAA, TTAAA, TGAAA.

The process is repeated for other topologies, and that topology requiring the smallest number of changes is taken as the final tree. For maximum parsimony, it is necessary only to consider the INFORMATIVE SITES. For DNA sequences, these are sites at which there are at least two different

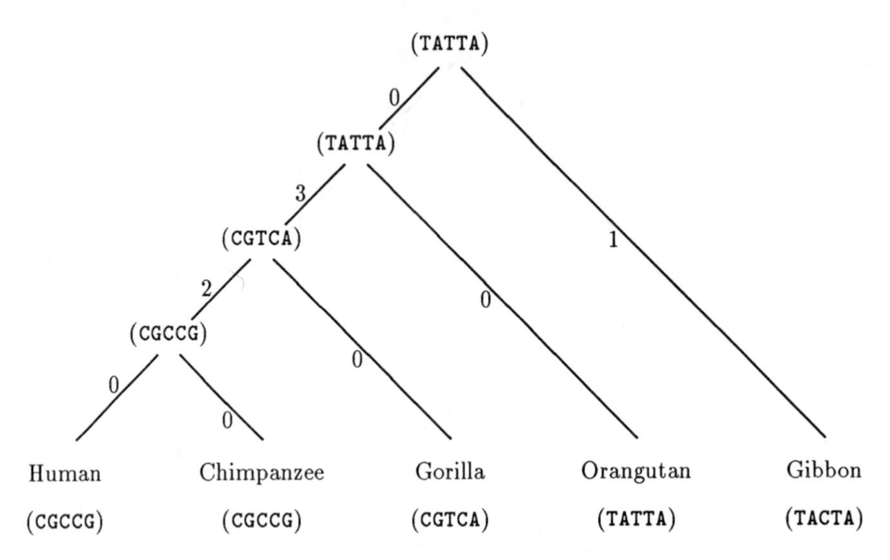

Figure 8.9 Maximum parsimony tree for mitochondrial sequence data of Figure 1.3. Figures are the numbers of changes from between nodes.

bases in the sequences being considered, with each different base occuring at least twice. A site that has a base in only one sequence is not informative since that unique base can be assumed to have arisen by a single change on the branch leading immediately to the sequence in which it is carried. Such a change is compatible with any topology.

For the mitochondrial sequences in Figure 1.3, there are five informative sites: 25, 39, 44, 47, and 54. A parsimonious tree based on these five positions is shown in Figure 8.9. It has six base changes, as do other possible trees. Although this has the same topology as the trees found using the matrix of distances, the very limited amount of data has resulted in some surprising effects. The branch between nodes E and G in Figure 8.6 was found to be shorter than that between nodes G and F, yet among the informative sites there have been three base changes on the former and none on the latter.

The mitochondrial data leading to Figure 8.9 are typical in that not all the informative sites support the same topology. Employing the notation of Nei (1987), site 25 can be designated as (Hu-Ch-Go) – (Or-Gi) since the human, chimpanzee, and gorilla sequences have the same base and so do the orangutan and gibbon sequences. This split supports a topology such as that in Figure 8.9 since it could result from a single base substitution on the path E–G. For this topology, site 25 is said to be compatible. Sites 39, 47, and 54 are also compatible with that topology. For a single change to be consistent with the observations at site 44, however, a different topology would be necessary. Although the parsimony method seeks to minimize the total number of changes necessary to produce the observed sequences, the COMPATIBILITY METHOD (e.g., Estabrook et al. 1975) maximizes the number of compatible sites.

Parsimony methods have been criticized by Felsenstein (e.g., Felsenstein 1983) on the grounds that they are not based on statistical principles. Felsenstein points out that parsimony methods, in trying to minimize the number of evolutionary events, implicitly assume that such events are improbable. If the amount of change over the evolutionary times being considered is small, the parsimony methods will be well-justified, but for cases in which there are large amounts of change, parsimony methods can give estimated trees that actually are more likely to be wrong as more data are used (Felsenstein 1978b).

LIKELIHOOD METHODS

Likelihood methods of constructing trees attempt to avoid the limitations of other methods, although they may require a prohibitive amount of computing. Unlike distance matrix methods, likelihood methods try to make explicit and efficient use of all the data, instead of reducing the data to a set of distances. They differ from parsimony methods by employing standard statistical methods for a probabilistic model of evolution (Felsenstein 1981).

As they are presently implemented, maximum likelihood methods of tree construction assume the form of the tree, and then choose the branch lengths to maximize the likelihood of the data given that tree. These likelihoods are then compared over different trees, and the tree with the greatest likelihood is taken as the best estimate. An immediate problem is that the number of possible trees increases very quickly as the number of OTU's increases. The number of unrooted bifurcating trees (two branches joining at each interior node) with n OTU's at the tips is $(2n-5)!/[(n-3)!2^{n-3}]$ (Edwards and Cavalli-Sforza 1964). The first few values of this number are 1, 3, 105, 10,395, and 2,027,025 for n=3, 4, 6, 8, and 10. The number of rooted trees with n tips is the same as the number of unrooted trees with $n+1$ tips (Felsenstein 1978a). In practice, only a subset of all trees is examined.

Likelihood Model for DNA Sequences

For DNA sequence data, the model on which the likelihood is based specifies the probability of one sequence changing by mutation to another sequence in a specified time. Although neighboring nucleotides in a DNA sequence are not independent, the models do assume independence of evolution at different sites so that the probability of a set of sequences for some tree is the product of the probabilities for each of the sites in the sequences. At any single site, the models work with probabilities $P_{ij}(T)$ that base i will have changed to base j after a time T. The subscripts i, j take the values 1, 2, 3, 4 for bases `A, C, G, T`.

The simplest base-substitution mutation model assumes a constant rate u of mutation. If a base mutates, it changes to type i with a constant probability π_i. This includes the case in which a base mutates to the same type, although this type of substitution is not observable. With u being the rate of base substitution per unit time (generation), the probability of no mutations at a site after T generations is $(1-u)^T$, so that the probability p of there being mutation is

$$p = 1 - (1-u)^T \approx 1 - e^{-uT}$$

The probabilities of changes from base i to base j after time T can therefore be written as (Felsenstein 1981)

$$\begin{aligned} P_{ii}(T) &= (1-p) + p\pi_i \\ P_{ij}(T) &= p\pi_j,\ j \neq i \end{aligned}$$

When the π_i's are all set equal to 1/4, this is just the Jukes-Cantor mutation model, but with a slightly different interpretation for the mutation rate. The rate u of this model is for all base substitutions and is equal to 4/3 times the rate μ for detectable substitutions in the Jukes-Cantor model (Chapter 7).

Note that the probabilities involve mutation rate and time only through their product. Neither can be estimated separately by the methods discussed here, so interest is confined to the products $v = uT$, which are the expected number of substitutions along the branches. If base substitutions occur at the same rate in all branches of the tree, then the branch lengths will indicate the relative times between each pair of nodes in the tree.

The likelihood method assumes the structure of the tree. The extant sequences form the tips of the tree, but none of the sequences at other nodes is known. The likelihood of the data for the tree must therefore consider all possibilities for the unknown sequences.

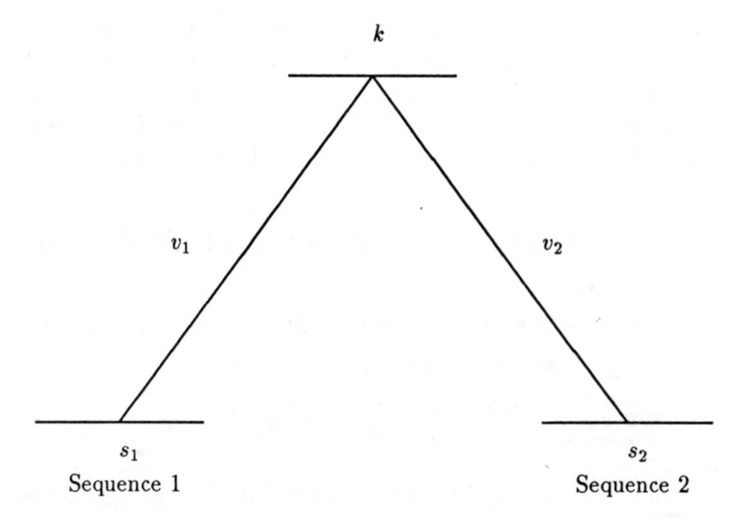

Figure 8.10 Rooted tree for two sequences. At site j, the two sequences have bases s_1, s_2 and the node has base k.

Under the one-parameter mutation model described here, each of the four base types is expected to become equally frequent, suggesting that π_i be set to 0.25 for $i = 1, 2, 3, 4$. An alternative would be to use the average base frequencies found in the set of sequences for which a tree is being constructed.

Two Sequences

There is one rooted tree for two sequences, shown in Figure 8.10. For nucleotide position j in these sequences, the observed bases are s_1, s_2 and the base in the unobserved root sequence is set to k. Adding over all possible values of k, the likelihood $L(j)$ is

$$L(j) = \sum_{k=1}^{4} \pi_k P_{ks_1}(v_1) P_{ks_2}(v_2)$$

and over all m sites the likelihood is

$$L = \prod_{j=1}^{m} L(j)$$

This likelihood is a function of two unknown branch lengths v_1, v_2.

As there is only one set of observed transitions, from sequence 1 to sequence 2, the interior node 0 cannot be located uniquely. A demonstration of this follows from Felsenstein's (1981) "pulley principle." Consider the likelihood for site j written out to show the sum over the four bases at the interior node in the case where sequence 1 has base A and sequence 2 has base C at that site

$$\begin{aligned} L(j) &= \pi_{\mathtt{A}} P_{\mathtt{AA}}(v_1) P_{\mathtt{AC}}(v_2) + \pi_{\mathtt{C}} P_{\mathtt{CA}}(v_1) P_{\mathtt{CC}}(v_2) + \pi_{\mathtt{G}} P_{\mathtt{GA}}(v_1) P_{\mathtt{GC}}(v_2) \\ &\quad + \pi_{\mathtt{T}} P_{\mathtt{TA}}(v_1) P_{\mathtt{TC}}(v_2) \\ &= \pi_{\mathtt{A}}[(1-p_1) + p_1 \pi_{\mathtt{A}}] p_2 \pi_{\mathtt{C}} + \pi_{\mathtt{C}} p_1 \pi_{\mathtt{A}}[(1-p_2) + p_2 \pi_{\mathtt{C}}] \\ &\quad + \pi_{\mathtt{G}} p_1 \pi_{\mathtt{A}} p_2 \pi_{\mathtt{C}} + \pi_{\mathtt{T}} p_1 \pi_{\mathtt{A}} p_2 \pi_{\mathtt{C}} \\ &= \pi_{\mathtt{A}}(p_1 + p_2 - p_1 p_2) \pi_{\mathtt{C}} \\ &= \pi_{\mathtt{A}} p_{12} \pi_{\mathtt{C}} \end{aligned}$$

In other words, the likelihood involving two paths (k to A and k to C) with mutation probabilities p_1 and p_2 is the same as that involving one path (A to C) with probability p_{12}. Note that

$$p_{12} = p_1 + p_2 - p_1 p_2 = 1 - e^{-(v_1+v_2)}$$

so that the likelihood for the tree in Figure 8.10 depends only on the total branch length $v_1 + v_2$ between the two species, 1 and 2, and not on the location of the node 0. It is not possible to estimate v_1 and v_2 separately, and the tree reduces to a single branch between the two sequences. In other words the estimable tree is unrooted.

When the four base probabilities are equal, $\pi_i = 1/4$, $i = 1,2,3,4$, and the likelihood for this one-branch tree reduces to

$$L = \left(\frac{4-3p}{16}\right)^s \left(\frac{p}{16}\right)^{m-s}$$

where p is the mutation probability for the branch and s of the m sites in the two sequences have the same base. This likelihood is maximized for

$$\hat{p} = \frac{4(m-s)}{3m}$$

The MLE for the branch length is therefore

$$\hat{v} = \ln\left(\frac{3}{4\tilde{q}-1}\right)$$

where

$$\tilde{q} = \frac{s}{m}$$

Recall that u corresponds to $4\mu/3$ in the Jukes-Cantor model, and that the time T between the two sequences was written as $2t$ in that previous model (twice the time from each sequence to the ancestral sequence). These relations provide that the branch length could also have been recovered from the Jukes-Cantor distance K between the two sequences:

$$\begin{aligned} v = uT &= \ln\left(\frac{3}{4q-1}\right) \\ K = 2\mu t &= \frac{3}{4}\ln\left(\frac{3}{4q-1}\right) \end{aligned}$$

Length v is the expected number of all substitutions, while length K is for detectable substitutions, and $v = 4K/3$. The estimate could have been found more directly from Bailey's method (Chapter 2) since there is one unknown (K) to estimate from one observed distance.

Three Sequences

There are three rooted trees for three sequences, one of which is shown in Figure 8.11. In addition to the three observed sequences, there are unobserved sequences at the nodes 0 and 4, and there are four branch lengths to be determined. Each of the three trees could be considered in turn, and the one giving the greatest maximum likelihood taken as the estimated tree. In fact, however, it is not necessary to do this as all three trees have the same likelihood.

For the arrangement shown in Figure 8.11, the likelihood for site j is expressed with base ℓ at node 4 and base k at node 0:

$$L(j) = \sum_k \sum_\ell \pi_k P_{k\ell}(v_4) P_{ks_3}(v_3) P_{\ell s_1}(v_1) P_{\ell s_2}(v_2)$$

Applying Felsenstein's pulley principle leaves the likelihood unchanged if node 0 is moved to any position between nodes 3 and 4. The likelihood depends only on the total distance $v_3 + v_4$. If nodes 0 and 4 are made coincident, the likelihood can be written as

$$L(j) = \sum_k \pi_k P_{ks_1}(v_1) P_{ks_2}(v_2) P_{ks_3}(v_3)$$

There is no way of locating the root 0 uniquely, and only the star phylogeny of Figure 8.12 needs to be considered for three sequences.

Under the assumption of equal base frequencies, since there are three unknown branch lengths, and three pairwise Jukes-Cantor distances available,

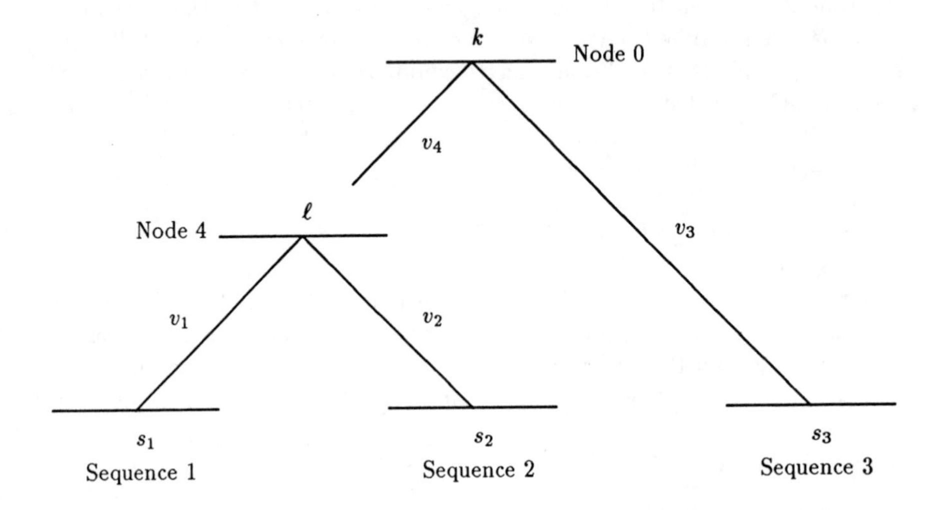

Figure 8.11 The three rooted trees for three sequences. At site j, the three sequences have bases s_1, s_2, s_3 and the nodes 0 and 4 have bases k and ℓ.

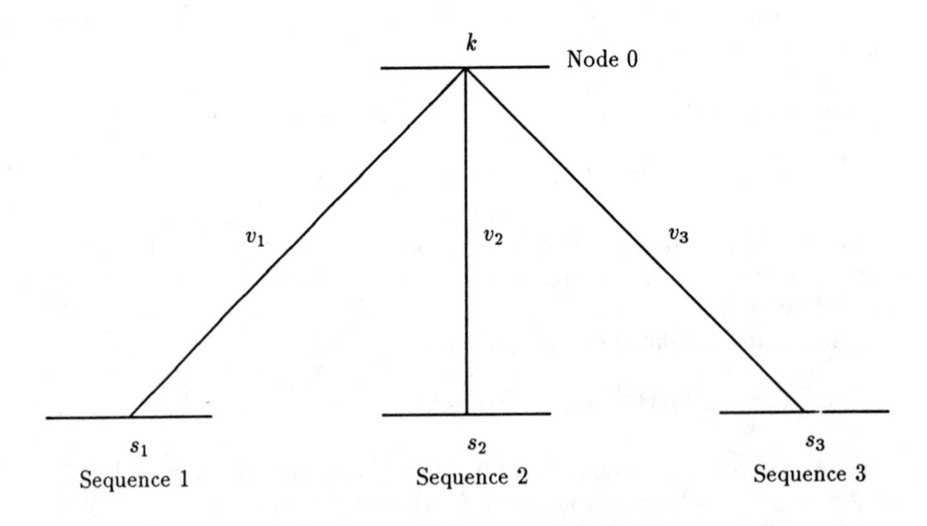

Figure 8.12 Star phylogeny for three sequences. Each of the three sequences 1, 2, 3 descends from the same ancestral sequence 0.

Bailey's method allows the maximum likelihood estimates to be found from solving

$$\begin{aligned}\hat{v}_1 + \hat{v}_2 &= K_{12}\\ \hat{v}_1 + \hat{v}_3 &= K_{13}\\ \hat{v}_2 + \hat{v}_3 &= K_{23}\end{aligned}$$

The estimates are

$$\begin{aligned}\hat{v}_1 &= \frac{1}{2}(K_{12} + K_{13} - K_{23})\\ \hat{v}_2 &= \frac{1}{2}(K_{12} + K_{23} - K_{13})\\ \hat{v}_3 &= \frac{1}{2}(K_{13} + K_{23} - K_{12})\end{aligned}$$

Actual sequences do not have equal base frequencies, and the Jukes-Cantor distances will not maximize the likelihood, but they do provide good initial values for an iterative solution. A direct method for finding maximum likelihood solutions numerically is provided by Newton-Raphson iteration, and this is phrased most simply in terms of looking for estimates of $p_i = 1 - e^{-v_i}$. An initial column vector of estimates $\mathbf{p} = (p_1, p_2, p_3)'$ is found from the Jukes-Cantor estimates, where $'$ denotes the transpose of a vector. The vector of scores is

$$\mathbf{S} = \begin{bmatrix} \dfrac{\partial \ln L}{\partial p_1} \\ \\ \dfrac{\partial \ln L}{\partial p_2} \\ \\ \dfrac{\partial \ln L}{\partial p_3} \end{bmatrix}$$

The necessary derivatives are

$$\begin{aligned}\frac{\partial L(j)}{\partial p_1} &= \sum_k \pi_k \frac{\partial P_{ks_1}(p_1)}{\partial p_1} P_{ks_2}(p_2) P_{ks_3}(p_3)\\ \frac{\partial L(j)}{\partial p_2} &= \sum_k \pi_k P_{ks_1}(p_1) \frac{\partial P_{ks_2}(p_2)}{\partial p_2} P_{ks_3}(p_3)\\ \frac{\partial L(j)}{\partial p_3} &= \sum_k \pi_k P_{ks_1}(p_1) P_{ks_2}(p_2) \frac{\partial P_{ks_3}(p_3)}{\partial p_3}\\ \frac{\partial^2 L(j)}{\partial p_i^2} &= 0, \;\; i = 1, 2, 3\end{aligned}$$

$$\frac{\partial^2 L(j)}{\partial p_1 \partial p_2} = \sum_k \pi_k \frac{\partial P_{ks_1}(p_1)}{\partial p_1} \frac{\partial P_{ks_2}(p_2)}{\partial p_2} P_{ks_3}$$

$$\frac{\partial^2 L(j)}{\partial p_1 \partial p_3} = \sum_k \pi_k \frac{\partial P_{ks_1}(p_1)}{\partial p_1} P_{ks_2} \frac{\partial P_{ks_3}(p_3)}{\partial p_3}$$

$$\frac{\partial^2 L(j)}{\partial p_2 \partial p_3} = \sum_k \pi_k P_{ks_1} \frac{\partial P_{ks_2}(p_1)}{\partial p_2} \frac{\partial P_{ks_3}(p_3)}{\partial p_3}$$

The vector of scores is

$$S_1 = \frac{\partial \ln L}{\partial p_1} = \sum_j \frac{1}{L(j)} \frac{\partial L(j)}{\partial p_1}$$

$$S_2 = \frac{\partial \ln L}{\partial p_2} = \sum_j \frac{1}{L(j)} \frac{\partial L(j)}{\partial p_2}$$

$$S_3 = \frac{\partial \ln L}{\partial p_3} = \sum_j \frac{1}{L(j)} \frac{\partial L(j)}{\partial p_3}$$

The probabilities $P_{ij}(v)$ are linear functions of the p_i's and so have very simple derivatives.

The information matrix consists of the derivatives of the scores with respect to each p_i

$$\mathbf{I} = \begin{bmatrix} -\frac{\partial S_1}{\partial p_1} & -\frac{\partial S_1}{\partial p_2} & -\frac{\partial S_1}{\partial p_3} \\ -\frac{\partial S_2}{\partial p_1} & -\frac{\partial S_2}{\partial p_2} & -\frac{\partial S_2}{\partial p_3} \\ -\frac{\partial S_3}{\partial p_1} & -\frac{\partial S_3}{\partial p_2} & -\frac{\partial S_3}{\partial p_3} \end{bmatrix}$$

and, for example,

$$\frac{\partial S_1}{\partial p_1} = -\sum_j \frac{1}{L(j)^2} \left[\frac{\partial L(j)}{\partial p_1}\right]^2$$

$$\frac{\partial S_1}{\partial p_2} = -\sum_j \frac{1}{L(j)^2} \left[\frac{\partial L(j)}{\partial p_1}\right] \left[\frac{\partial L(j)}{\partial p_2}\right] + \sum_j \frac{1}{L(j)} \frac{\partial^2 L(j)}{\partial p_1 \partial p_2}$$

The initial set of estimates is then modified to a new set $\mathbf{p}'$ by

$$\mathbf{p}' = \mathbf{p} + \mathbf{I}^{-1}\mathbf{S}$$

Table 8.4 Successive iterates for branch lengths in the constrained maximum likelihood tree for human, gorilla, and gibbon mitochondrial sequences from Figure 1.3.

Iterate	v_1	v_2	v_3	ln L
Initial	0.0423	0.0174	0.2215	
1	0.0420	0.0196	0.2230	−148.44
2	0.0420	0.0199	0.2299	−148.42
3	0.0420	0.0199	0.2299	−148.42
Std. Dev.	0.0297	0.0218	0.0600	

where all the terms on the right-hand side are evaluated with the initial estimates. The process is continued until successive values of the estimates are sufficiently close that convergence can be considered to have been achieved. As discussed in Chapter 2, the final value of the information matrix provides estimates of the variances and the covariance of the two estimates. An example of the convergence process is shown in Table 8.4 for the mitochondrial sequences for human (1), gorilla (2) and gibbon (3) in Figure 1.3. The overall average base frequencies among these three sequences were used as the π_i's in the probability terms of the model.

Relative Rate Test

Within the likelihood framework there is the opportunity for hypothesis testing from comparing likelihoods under different models. One such situation concerns the RELATIVE RATE TEST (Wu and Li 1985). The concept of testing for differences in evolutionary rates without referring to geological times was first discussed by Sarich and Wilson (1973) for amino acid sequences. The work of Wu and Li was for DNA sequences. The hypothesis to be tested is that the rates of mutation are the same in the branches leading to sequences 1 and 2 in Figure 8.11. As shown above, this is equivalent to testing that lengths v_1, v_2 are the same in Figure 8.12. Using the methodology described in the previous section, the unconstrained likelihood L_1 can be found. Under the constraint of the hypothesis the likelihood is L_0, and the quantity $-2\ln(L_0/L_1)$ may be used as a single degree-of-freedom chi-square test statistic, as discussed in Chapter 3.

Under the hypothesis, the probabilities $P_{ks_1}(v_1)$ and $P_{ks_2}(v_2)$ are both functions of the same parameter $v_1 = v_2$. Numerical methods are again needed to find the MLE's of branch lengths. The methodology of the previous section can be used with some minor changes.

The vector of scores now has two elements:

$$\mathbf{S} = \begin{bmatrix} \dfrac{\partial \ln L}{\partial p_1} \\ \\ \dfrac{\partial \ln L}{\partial p_3} \end{bmatrix}$$

Introducing the quantity X as

$$X = P_{ks_1}(p_1)\frac{\partial P_{ks_2}(p_1)}{\partial p_1} + P_{ks_2}(p_1)\frac{\partial P_{ks_1}(p_1)}{\partial p_1}$$

allows the necessary derivatives to be written as

$$\begin{aligned}
\frac{\partial L(j)}{\partial p_1} &= \sum_k \pi_k X P_{ks_3}(p_3) \\
\frac{\partial L(j)}{\partial p_3} &= \sum_k \pi_k P_{ks_1}(p_1) P_{ks_2}(p_1) \frac{\partial P_{ks_3}(p_3)}{\partial p_3} \\
\frac{\partial^2 L(j)}{\partial p_1^2} &= 2\sum_k \pi_k \left[\frac{\partial P_{ks_1}(p_1)}{\partial p_1}\right]\left[\frac{\partial P_{ks_2}(p_1)}{\partial p_1}\right] P_{ks_3}(p_3) \\
\frac{\partial^2 L(j)}{\partial p_1 \partial p_3} &= \sum_k \pi_k X \frac{\partial P_{ks_3}(p_3)}{\partial p_3} \\
\frac{\partial^2 L(j)}{\partial p_3^2} &= 0
\end{aligned}$$

The terms in the vector of scores are

$$\begin{aligned}
S_1 &= \frac{\partial \ln L}{\partial p_1} \\
S_2 &= \frac{\partial \ln L}{\partial p_3}
\end{aligned}$$

The information matrix is now 2×2

$$\mathbf{I} = \begin{bmatrix} -\dfrac{\partial S_1}{\partial p_1} & -\dfrac{\partial S_1}{\partial p_3} \\ \\ -\dfrac{\partial S_2}{\partial p_1} & -\dfrac{\partial S_2}{\partial p_3} \end{bmatrix}$$

Table 8.5 Successive iterates for two branch lengths in the constrained maximum likelihood tree for human, gorilla, and gibbon mitochondrial sequences from Figure 1.3.

Iterate	v_1	v_3	ln L
Initial	0.0299	0.2215	
1	0.0309	0.2300	−148.59
2	0.0309	0.2303	−148.58
3	0.0309	0.2303	−148.58
Std. Dev.	0.0175	0.0601	

with, for example,

$$\frac{\partial S_1}{\partial p_1} = -\sum_j \frac{1}{L(j)^2}\left[\frac{\partial L(j)}{\partial p_1}\right]^2 + \sum_j \frac{1}{L(j)}\frac{\partial^2 L(j)}{\partial p_1^2}$$

$$\frac{\partial S_1}{\partial p_3} = -\sum_j \frac{1}{L(j)^2}\left[\frac{\partial L(j)}{\partial p_1}\right]\left[\frac{\partial L(j)}{\partial p_3}\right] + \sum_j \frac{1}{L(j)}\frac{\partial^2 L(j)}{\partial p_1 \partial p_3}$$

The same Newton-Raphson iterative equation as for the unconstrained model is used, and the results for the human (1), gorilla (2) and gibbon (3) mitochondrial sequences, with equal rates assigned to the branches from the common ancestor to human and gorilla, are shown in Table 8.5. The initial value for the common value of v_1 and v_2 was set to the average of the initial values in the unconstrained case.

Twice the difference of the log-likelihoods in Tables 8.4 and 8.5 is only 0.31, indicating no significant differences in the rates of evolution leading to human and gorilla from their common ancestor on the basis of this very small set of data. Of course this result was expected, as the estimated rates in Table 8.4 for those two lineages were within one standard deviation of each other.

Many Sequences

Construction of trees with several sequences as tips follows the same general procedure as outlined above. A tree is specified, and then the branch lengths estimated by Newton-Raphson iteration of equations resulting from the likelihood of the tree. In theory, all possible trees should be studied to

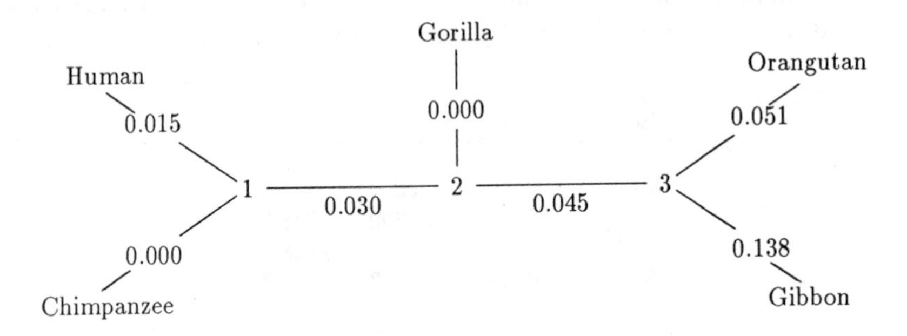

Figure 8.13 Maximum likelihood tree for mitochondrial sequence data of Figure 1.3. Constructed with PHYLIP package of J. Felsenstein.

find the one with the highest likelihood. Fukami and Tateno (1989) showed that there is at most one set of branch lengths that give a stationary value for L, and that this set provides the required MLE's.

In his computer programs, Felsenstein (1981) adopts a strategy that examines at least $2n^2 - 9n + 8$ trees for n sequences, but this can be a very much smaller number than the total number of trees possible. He proceeds by successively adding sequences to a tree already constructed, and starts with a two-sequence tree. The kth sequence can be added to any of $2k - 5$ segments on the tree found for the previous $k - 1$ sequences. For $k > 4$, local rearrangements of the tree are carried out to see if this improves the likelihood before the next sequence is added. Of course this strategy depends on the order in which sequences are added to the tree, and Felsenstein suggests that several different orderings be tried. He also uses an EM-algorithm approach by which one branch length is found at a time, instead of the Newton-Raphson method for all branches simultaneously. Application of his algorithm for the five mitochondrial sequences of Figure 1.3 results in the unrooted tree shown in Figure 8.13.

Other Mutation Models

It is unlikely that all types of base substitution occur at the same rate, and it would be desirable to include different rates in the likelihood model. Felsenstein allows for different rates of transitions (A $\leftrightarrow$ G, C $\leftrightarrow$ T) and transversions (A,G $\leftrightarrow$ C,T) in current versions of his computer programs (Felsenstein 1985).

Bootstrapping Phylogenies

Within any specified tree topology, maximum likelihood is known to provide consistent estimates of branch lengths, meaning that the estimates approach the true values as the amount of data increases. Moreover, the information matrix provides an indication of the sampling variances of these estimates. What about properties of the tree found to have the largest likelihood over all topologies? In what sense can it be regarded as estimating the true tree? Although this is a difficult theoretical question, empirical evidence can be provided in practice by numerical resampling.

Felsenstein (1985) suggested that bootstrapping be performed over the sites in the sequences being studied. When the sequences are of length m, a bootstrap sample consists of those bases in each sequence at each of m sites sampled with replacement from the original set of m sites. Each bootstrap sample is then subjected to the same likelihood estimation algorithm as was the original data set. Attention is paid to sets of monophyletic species over all the bootstrap samples. A group of species can be declared monophyletic at the 5% significance level if it is found together in 95% of the bootstrap trees. There is also the useful concept of a "majority rule" consensus tree (Margush and McMorris 1981) that consists of all groups of species that appear in a majority of the trees found from the bootstrap samples.

SUMMARY

The preceding discussion is a very brief look at the rich field of phylogeny construction. As with many areas of statistical analysis of genetic data, there are conflicts between computational expediency and biological realism. The magnitude of DNA sequence data makes it presently difficult to adopt approaches with the otherwise desirable properties of maximum likelihood. Fast algorithms have been developed to build trees from pairwise distances, and such methods have many advantages, even though they use only a summary of the data available. Parsimony methods can also be fast, but may be appropriate only for very slow rates of evolutionary change.

Several other methods are under active investigation, and the current literature (e.g., Saitou and Imanishi, 1989) contains comparisons of many techniques. The reader is directed to Nei (1987), Felsenstein (1988), and recent issues of *Molecular Biology and Evolution*.

EXERCISES

Exercise 8.1

The following data are for the shrub *Lisianthius skinneri* (Sytsma and Schaal 1985). Rogers' distances were calculated among 7 populations on the basis of frequencies at 12 isozyme loci.

Population	LS4	LS6	LS7	LJ1	LH1	LP1	LA1
LS4	–	0.083	0.250	0.458	0.509	0.563	0.399
LS6		–	0.167	0.392	0.425	0.563	0.337
LS7			–	0.392	0.342	0.479	0.307
LJ1				–	0.400	0.424	0.473
LH1					–	0.384	0.489
LP1						–	0.321

Apply the UPGMA and the Fitch-Margoliash method to construct a tree from this distance matrix. Slight differences may be found from the branch lengths reported by Sytsma and Schaal.

Exercise 8.2

The mitochondrial data described by Brown et al. (1982) have two other *tRNA* genes – *ser* and *leu*. These data are now shown:

```
Leu-tRNA
Hu ACTTTTAAAG GATAACAGCT ATCCATTGGT CTTAGGCCCC AAAAATTTTG GTGCAACTCC AAATAAAAGT A
Ch ACTTTTAAAG GATAACAGCC ATCCGTTGGT CTTAGGCCCC AAAAATTTTG GTGCAACTCC AAATAAAAGT A
Go ACTTTTAAAG GATAACAGCT ATCCATTGGT CTTAGGACCC AAAAATTTTG GTGCAACTCC AAATAAAAGT A
Or GCTTTTAAAG GATAACAGCT ATCCCTTGGT CTTAGGATCC AAAAATTTTG GTGCAACTCC AAATAAAAGT A
Gi ACTTTTAAAG GATAACAGCT ATCCATTGGT CTTAGGACCC AAAAATTTTG GTGCAACTCC AAATAAAAGT A

Ser-tRNA
Hu  GAGAAAGCTC ACAAGAACTG CTAACTCATG CCCCCATGTC TGACAACATG GCTTTCTCA
Ch  GAGAAAGCTT ATAAGAACTG CTAATTCATA TCCCCATGCC TGACAACATG GCTTTCTCA
Go  GAGAAAGCTC GTAAGAGCTG CTAACTCATA CCCCCGTGCT TGACAACATG GCTTTCTCA
Or  GAGAAAGCTC ACAAGAACTG CTAACTCTCA CT-CCATGTG TGACAACATG GCTTTCTCA
Gi  GAGAAAGCCC ACAAGAACTG CTAACTCACT ATCCCATGTA TGACAACATG GCTTTCTCA
```

The GenBank references for these sequences are: HUMMTTRPR (L00016 V00658), CHPMTTRCH (V00672), GORMTTGO (L00015 V00658), ORAMTTROR (V00675), GIBMTTGI (V00659). Note the gap in the orangutan sequence.

Use these data to construct distance matrix (UPGMA and Fitch-Margoliash) and maximum parsimony trees. Either use these data alone, or add them to the $his - tRNA$ sequences of Figure 1.3.

Appendix A

Statistical Tables

Normal Distribution

The probability that a normal random variable lies between x and ∞ is given by the area under the normal density curve $f(x)$. A standard normal variable has mean of 0 and a variance of 1, and a density function of

$$f(x) = \frac{1}{\sqrt{2\pi}} e^{-\frac{x^2}{2}}$$

This curve is plotted in Figure A.1, and the areas for positive x values shown in Table A.1 were calculated by an approximate polynomial expression given by Abramowitz and Stegun (1970 – formula 26.2.17).

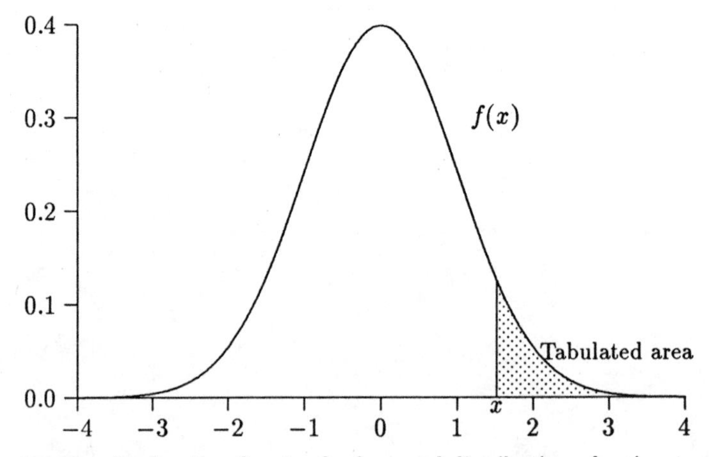

Figure A.1 Density function for standard normal distribution, showing areas tabulated in Table A.1.

Table A.1 Areas under the standard normal density curve, beyond value x.

x	0.00	0.01	0.02	0.03	0.04	0.05	0.06	0.07	0.08	0.09
0.0	0.5000	0.4960	0.4920	0.4880	0.4840	0.4801	0.4761	0.4721	0.4681	0.4641
0.1	0.4602	0.4562	0.4522	0.4483	0.4443	0.4404	0.4364	0.4325	0.4286	0.4247
0.2	0.4207	0.4168	0.4129	0.4090	0.4052	0.4013	0.3974	0.3936	0.3897	0.3859
0.3	0.3821	0.3783	0.3745	0.3707	0.3669	0.3632	0.3594	0.3557	0.3520	0.3483
0.4	0.3446	0.3409	0.3372	0.3336	0.3300	0.3264	0.3228	0.3192	0.3156	0.3121
0.5	0.3085	0.3050	0.3015	0.2981	0.2946	0.2912	0.2877	0.2843	0.2810	0.2776
0.6	0.2743	0.2709	0.2676	0.2643	0.2611	0.2578	0.2546	0.2514	0.2483	0.2451
0.7	0.2420	0.2389	0.2358	0.2327	0.2296	0.2266	0.2236	0.2206	0.2177	0.2148
0.8	0.2119	0.2090	0.2061	0.2033	0.2005	0.1977	0.1949	0.1922	0.1894	0.1867
0.9	0.1841	0.1814	0.1788	0.1762	0.1736	0.1711	0.1685	0.1660	0.1635	0.1611
1.0	0.1587	0.1562	0.1539	0.1515	0.1492	0.1469	0.1446	0.1423	0.1401	0.1379
1.1	0.1357	0.1335	0.1314	0.1292	0.1271	0.1251	0.1230	0.1210	0.1190	0.1170
1.2	0.1151	0.1131	0.1112	0.1093	0.1075	0.1056	0.1038	0.1020	0.1003	0.0985
1.3	0.0968	0.0951	0.0934	0.0918	0.0901	0.0885	0.0869	0.0853	0.0838	0.0823
1.4	0.0808	0.0793	0.0778	0.0764	0.0749	0.0735	0.0721	0.0708	0.0694	0.0681
1.5	0.0668	0.0655	0.0643	0.0630	0.0618	0.0606	0.0594	0.0582	0.0571	0.0559
1.6	0.0548	0.0537	0.0526	0.0516	0.0505	0.0495	0.0485	0.0475	0.0465	0.0455
1.7	0.0446	0.0436	0.0427	0.0418	0.0409	0.0401	0.0392	0.0384	0.0375	0.0367
1.8	0.0359	0.0351	0.0344	0.0336	0.0329	0.0322	0.0314	0.0307	0.0301	0.0294
1.9	0.0287	0.0281	0.0274	0.0268	0.0262	0.0256	0.0250	0.0244	0.0239	0.0233
2.0	0.0228	0.0222	0.0217	0.0212	0.0207	0.0202	0.0197	0.0192	0.0188	0.0183
2.1	0.0179	0.0174	0.0170	0.0166	0.0162	0.0158	0.0154	0.0150	0.0146	0.0143
2.2	0.0139	0.0136	0.0132	0.0129	0.0125	0.0122	0.0119	0.0116	0.0113	0.0110
2.3	0.0107	0.0104	0.0102	0.0099	0.0096	0.0094	0.0091	0.0089	0.0087	0.0084
2.4	0.0082	0.0080	0.0078	0.0075	0.0073	0.0071	0.0069	0.0068	0.0066	0.0064
2.5	0.0062	0.0060	0.0059	0.0057	0.0055	0.0054	0.0052	0.0051	0.0049	0.0048
2.6	0.0047	0.0045	0.0044	0.0043	0.0041	0.0040	0.0039	0.0038	0.0037	0.0036
2.7	0.0035	0.0034	0.0033	0.0032	0.0031	0.0030	0.0029	0.0028	0.0027	0.0026
2.8	0.0026	0.0025	0.0024	0.0023	0.0023	0.0022	0.0021	0.0021	0.0020	0.0019
2.9	0.0019	0.0018	0.0018	0.0017	0.0016	0.0016	0.0015	0.0015	0.0014	0.0014
3.0	0.0013	0.0013	0.0013	0.0012	0.0012	0.0011	0.0011	0.0011	0.0010	0.0010
3.1	0.0010	0.0009	0.0009	0.0009	0.0008	0.0008	0.0008	0.0008	0.0007	0.0007
3.2	0.0007	0.0007	0.0006	0.0006	0.0006	0.0006	0.0006	0.0005	0.0005	0.0005
3.3	0.0005	0.0005	0.0005	0.0004	0.0004	0.0004	0.0004	0.0004	0.0004	0.0003
3.4	0.0003	0.0003	0.0003	0.0003	0.0003	0.0003	0.0003	0.0003	0.0003	0.0002
3.5	0.0002	0.0002	0.0002	0.0002	0.0002	0.0002	0.0002	0.0002	0.0002	0.0002
4.0	0.0000	0.0000	0.0000	0.0000	0.0000	0.0000	0.0000	0.0000	0.0000	0.0000

Chi-square Distribution

The probability that a chi-square random variable lies between x and ∞ is given by the area under the chi-square density curve $f(x)$. For n degrees of freedom, this density function is

$$f(x) = \frac{1}{2^{n/2}\Gamma(n/2)} x^{n/2-1} e^{-x/2}$$

where the gamma function $\Gamma(n/2)$ is given by

$$\Gamma(n/2) = \begin{cases} \sqrt{\pi} & n = 1 \\ 1 & n = 2 \\ (m-1)! & n = 2m \\ (2m)!\sqrt{\pi}/2^{2m}m! & n = 2m+1 \end{cases}$$

and m is a positive integer. The density curve for $n = 10$ d.f. is shown in Figure A.2. The probabilities in Table A.2 were calculated with an algorithm given by Posten (1989).

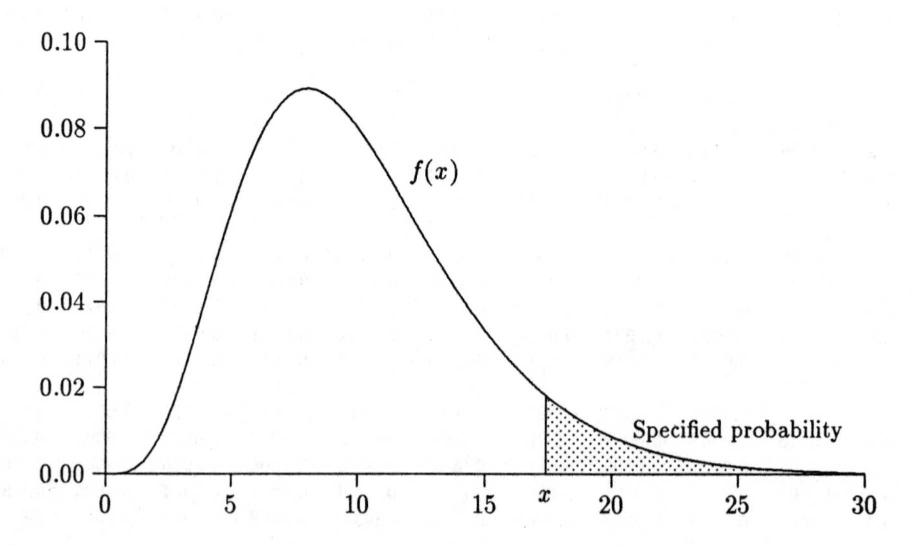

Figure A.2 Density function for chi-square distribution with 10 degrees of freedom, showing values exceeded with specified probabilities.

Table A.2 Chi-square values that are exceeded with specified probabilities.

	Probability									
d.f.	0.995	0.990	0.975	0.950	0.900	0.100	0.050	0.025	0.010	0.005
1	0.00	0.00	0.00	0.00	0.02	2.71	3.84	5.02	6.63	7.88
2	0.01	0.02	0.05	0.10	0.21	4.61	5.99	7.38	9.21	10.6
3	0.07	0.12	0.22	0.35	0.58	6.25	7.81	9.35	11.3	12.8
4	0.21	0.30	0.48	0.71	1.06	7.78	9.49	11.1	13.3	14.9
5	0.41	0.55	0.83	1.15	1.61	9.24	11.1	12.8	15.1	16.7
6	0.68	0.87	1.24	1.64	2.20	10.6	12.6	14.4	16.8	18.5
7	0.99	1.24	1.69	2.17	2.83	12.0	14.1	16.0	18.5	20.3
8	1.34	1.65	2.18	2.73	3.49	13.4	15.5	17.5	20.1	22.0
9	1.73	2.09	2.70	3.33	4.17	14.7	16.9	19.0	21.7	23.6
10	2.16	2.56	3.25	3.94	4.87	16.0	18.3	20.5	23.2	25.2
11	2.60	3.05	3.82	4.57	5.58	17.3	19.7	21.9	24.7	26.8
12	3.07	3.57	4.40	5.23	6.30	18.5	21.0	23.3	26.2	28.3
13	3.57	4.11	5.01	5.89	7.04	19.8	22.4	24.7	27.7	29.8
14	4.07	4.66	5.63	6.57	7.79	21.1	23.7	26.1	29.1	31.3
15	4.60	5.23	6.26	7.26	8.55	22.3	25.0	27.5	30.6	32.8
16	5.14	5.81	6.91	7.96	9.31	23.5	26.3	28.8	32.0	34.3
17	5.70	6.41	7.56	8.67	10.1	24.8	27.6	30.2	33.4	35.7
18	6.26	7.01	8.23	9.39	10.9	26.0	28.9	31.5	34.8	37.2
19	6.84	7.63	8.91	10.1	11.7	27.2	30.1	32.9	36.2	38.6
20	7.43	8.26	9.59	10.9	12.4	28.4	31.4	34.2	37.6	40.0
21	8.03	8.90	10.3	11.6	13.2	29.6	32.7	35.5	38.9	41.4
22	8.64	9.54	11.0	12.3	14.0	30.8	33.9	36.8	40.3	42.8
23	9.26	10.2	11.7	13.1	14.8	32.0	35.2	38.1	41.6	44.2
24	9.89	10.9	12.4	13.8	15.7	33.2	36.4	39.4	43.0	45.6
25	10.5	11.5	13.1	14.6	16.5	34.4	37.7	40.6	44.3	46.9
26	11.2	12.2	13.8	15.4	17.3	35.6	38.9	41.9	45.6	48.3
27	11.8	12.9	14.6	16.2	18.1	36.7	40.1	43.2	47.0	49.6
28	12.5	13.6	15.3	16.9	18.9	37.9	41.3	44.5	48.3	51.0
29	13.1	14.3	16.0	17.7	19.8	39.1	42.6	45.7	49.6	52.3
30	13.8	15.0	16.8	18.5	20.6	40.3	43.8	47.0	50.9	53.7
40	20.7	22.2	24.4	26.5	29.1	51.8	55.8	59.3	63.7	66.8
50	28.0	29.7	32.4	34.8	37.7	63.2	67.5	71.4	76.2	79.5
60	35.5	37.5	40.5	43.2	46.5	74.4	79.1	83.3	88.4	92.0
70	43.3	45.4	48.8	51.7	55.3	85.5	90.5	95.0	100.4	104.2
80	51.2	53.5	57.2	60.4	64.3	96.6	101.9	106.6	112.3	116.3
100	67.3	70.1	74.2	77.9	82.4	118.5	124.3	129.6	135.8	140.2

Noncentral Chi-square Distribution

The probability that a random variable with the noncentral chi-square distribution exceeds a value x is given by the sum of terms that are the probabilities that central (i.e., usual) chi-square variables exceed the same value. Writing these latter probabilities as $\Pr(\chi^2|n)$ for a central chi-square with n d.f., the required probability for the noncentral chi-square χ'^2 with n d.f. and noncentrality parameter ν is

$$\Pr(\chi'^2|n,\nu) = \sum_{j=0}^{\infty} e^{-\nu/2} \frac{(\nu/2)^j}{j!} \Pr(\chi^2|n+2j)$$

The noncentrality parameter values ν shown in Table A.3 were calculated with the algorithm of Posten (1989). These were chosen to give a probability β of the noncentral chi-square exceeding the $(1-\alpha)$th percentile of the central chi-square (i.e., that value of the central chi-square exceeded with probability α). When $\alpha = 0.05, \beta = 0.90$, for example, the probability is 0.90 that a noncentral chi-square exceeds the 95% percentile value of 3.84 if $\nu = 10.5$.

If a chi-square statistic is being calculated to perform a test, and large values of the statistic will cause rejection of a hypothesis, then α is the significance level of the test. It is the probability of rejection when the hypothesis is true. The probability β of rejecting the hypothesis when it is false is the power of the test. Table A.3 therefore relates significance level, power and noncentrality parameter.

Table A.3 Noncentrality parameters for power β and significance level α.

	$\alpha = 0.10$			$\alpha = 0.05$			$\alpha = 0.01$		
d.f.	β=0.50	β=0.90	β=0.99	β=0.50	β=0.90	β=0.99	β=0.50	β=0.90	β=0.99
1	2.70	8.56	15.8	3.84	10.5	18.4	6.64	14.9	24.0
2	3.56	10.5	18.6	4.96	12.7	21.4	8.19	17.4	27.4
3	4.18	11.8	20.5	5.76	14.2	23.5	9.31	19.2	29.8
4	4.69	12.9	22.1	6.42	15.4	25.2	10.2	20.7	31.8
5	5.14	13.8	23.4	6.99	16.5	26.7	11.0	22.0	33.5
6	5.53	14.7	24.6	7.50	17.4	28.0	11.8	23.2	35.0
7	5.90	15.4	25.7	7.97	18.3	29.2	12.4	24.2	36.4
8	6.24	16.1	26.8	8.41	19.1	30.4	13.0	25.2	37.7
9	6.55	16.8	27.7	8.81	19.8	31.4	13.6	26.1	38.9
10	6.85	17.4	28.6	9.19	20.5	32.4	14.1	27.0	40.0
11	7.13	18.0	29.4	9.56	21.2	33.3	14.6	27.8	41.1
12	7.40	18.5	30.2	9.90	21.8	34.2	15.1	28.6	42.1
13	7.66	19.1	31.0	10.2	22.4	35.0	15.6	29.3	43.1
14	7.91	19.6	31.7	10.6	23.0	35.8	16.0	30.0	44.0
15	8.15	20.1	32.4	10.9	23.6	36.6	16.5	30.7	44.9
16	8.38	20.5	33.1	11.2	24.1	37.3	16.9	31.4	45.8
17	8.60	21.0	33.8	11.4	24.7	38.1	17.3	32.0	46.6
18	8.82	21.4	34.4	11.7	25.2	38.8	17.7	32.7	47.5
19	9.03	21.9	35.0	12.0	25.7	39.4	18.1	33.3	48.3
20	9.24	22.3	35.6	12.3	26.2	40.1	18.5	33.9	49.0
21	9.44	22.7	36.2	12.5	26.6	40.7	18.8	34.4	49.8
22	9.64	23.1	36.8	12.8	27.1	41.4	19.2	35.0	50.5
23	9.83	23.5	37.3	13.0	27.5	42.0	19.5	35.5	51.2
24	10.0	23.9	37.9	13.3	27.9	42.6	19.9	36.1	51.9
25	10.2	24.3	38.4	13.5	28.4	43.2	20.2	36.6	52.6
26	10.4	24.6	38.9	13.7	28.8	43.7	20.5	37.1	53.3
27	10.6	25.0	39.5	14.0	29.2	44.3	20.8	37.6	53.9
28	10.7	25.3	40.0	14.2	29.6	44.9	21.1	38.1	54.6
29	10.9	25.7	40.4	14.4	30.0	45.4	21.5	38.6	55.2
30	11.1	26.0	40.9	14.6	30.4	45.9	21.8	39.1	55.9
40	12.6	29.1	45.4	16.6	33.9	50.8	24.5	43.5	61.6
50	14.0	31.9	49.3	18.3	37.1	55.1	27.0	47.3	66.6
60	15.2	34.3	52.8	19.9	39.9	59.0	29.2	50.8	71.1
70	16.3	36.6	56.0	21.3	42.5	62.5	31.2	54.0	75.3
80	17.4	38.7	59.0	22.7	44.9	65.8	33.1	57.0	79.1
100	19.3	42.6	64.5	25.1	49.3	71.8	36.6	62.4	86.2

Appendix B

Random Numbers

GENERATING RANDOM NUMBERS

Several procedures described in this book require the drawing of random numbers. It is not possible to specify a procedure for generating truly random numbers, but there are algorithms that give very long sequences of "pseudo-random" numbers that have the appearance of randomness. A complete discussion of the generation and testing of random numbers is given by Knuth (1981).

Random numbers uniformly distributed on the interval [0, 1] are often sought. As a first step, integers X are generated between 0 and m, where m is the largest integer that can be handled by the computer being used. For computers with a word-length of 32, this number is

$$m = 2^{31} - 1 = 2,147,483,647$$

The required numbers R are then

$$R = \frac{X}{m}$$

The most common algorithm uses the LINEAR CONGRUENTIAL METHOD to generate a sequence $\{R_n\}$ of random numbers. The random integers X_n are found from

$$X_{n+1} = (aX_n + c) \bmod m$$

with a specified starting value X_0. A particular case that has been found to have a long PERIOD, before the numbers start repeating themselves, was reported by Lewis et al. (1969) and has been used widely since then. This algorithm uses a "seed" a and is

$$\begin{aligned} X_0 &= a \\ X_{n+1} &= 7^5 X_n \bmod (2^{31} - 1) \\ R_n &= 2^{-31} X_n \end{aligned}$$

A convenient seed would be the date. With $a = 870309$, for example

$$\begin{aligned} X_0 &= 870,309 \\ X_1 &= 16,807 \times 870,309 (\text{mod} 2,147,483,647) \\ &= 1,741,371,481 \\ R_1 &= 0.810,889,286 \end{aligned}$$

To three decimal places the first random number is 0.811.

This procedure was modified by Schrage (1979) to allow FORTRAN programs to be more readily portable to a range of computers. It is necessary that the computer be able to represent all integers in the range $[0, 2^{31} - 1]$, and then the generator is full cycle – i.e., every integer from 1 to $2^{31} - 1$ is generated exactly once in a cycle. Schrage's implementation is used in the programs listed in this Appendix, and requires an inital integer as seed.

BOOTSTRAPPING

Drawing a bootstrap sample from a data set of size n is equivalent to drawing n integers from the set $1, 2, \ldots, n$ with replacement. This can be achieved by taking the product of n times each of a set of numbers drawn at random on the unit interval and rounding up to the nearest integer. If twelve random numbers 0.240, 0.827, 0.950, 0.240, 0.066, 0.855, 0.239, 0.416, 0.207, 0.239, 0.417, 0.450 are drawn, then a bootstrap sample of size 12 would consist of items 3, 10, 12, 3, 1, 11, 3, 5, 3, 3, 6, 6.

RANDOM SEQUENCES

Testing hypotheses about DNA sequences often makes use of random sequences. Two possibilities are those of REPRESENTATIVE and SHUFFLED sequences (Fitch 1983). A representative sequence is one drawn at random from a population that has the same characteristics as the sequence at hand, while a shuffled sequence is just a permutation of the elements in that sequence. There may be very little difference in results based on the two approaches for long sequences.

To draw a representative sequence of length m that has the same expected base frequencies as those in the actual sequence, an infinite population of bases is imagined. This has the same base frequencies p_i, $i = 1, 2, 3, 4$ for bases `A, C, G, T` as does the actual sequence. These enable the unit interval to be partitioned into four segments, with boundaries

$$0,\ p_1,\ p_1 + p_2,\ p_1 + p_2 + p_3,\ 1$$

These segments therefore have lengths p_1, p_2, p_3, p_4 respectively. For the mth site in the random sequence a random number is drawn from the unit interval. If this number lies in the ith segment, then the site is filled with base i.

The following algorithm for shuffling a sequence was given by Knuth (1981) and depends on the exchange of random pairs of elements in the sequence. The algorithm for a sequence of length m is as follows. Each sequence element is considered in turn and exchanged with another element chosen at random.

1. Set $j = m$
2. Choose a random number u from the interval [0,1].
   Set $k = \lfloor ju \rfloor + 1$, i.e. the integral part of $j \times u$ plus 1.
3. Exchange elements k and j.
   Set $j = j - 1$.
4. Return to step 2 if $j > 1$

To shuffle a sequence of twelve elements, suppose the twelve random numbers 0.240, 0.827, 0.950, 0.240, 0.066, 0.855, 0.239, 0.416, 0.207, 0.239, 0.417, 0.450 are drawn. The numbers k for step 2 of the algorithm are 3, 10, 12, 3, 1, 11, 3, 5, 3, 3, 6, 6. The twelve steps in the shuffling produce the following set of sequences, with the final one being the desired complete shuffle.

1	2	12	4	5	6	7	8	9	10	11	3
1	2	12	4	5	6	7	8	9	11	10	3
1	2	12	4	5	6	7	8	9	3	10	11
1	2	9	4	5	6	7	8	12	3	10	11
8	2	9	4	5	6	7	1	12	3	10	11
8	2	9	4	5	6	10	1	12	3	7	11
8	2	6	4	5	9	10	1	12	3	7	11
8	2	6	4	5	9	10	1	12	3	7	11
8	2	4	6	5	9	10	1	12	3	7	11
8	2	4	6	5	9	10	1	12	3	7	11
8	9	4	6	5	2	10	1	12	3	7	11
2	9	4	6	5	8	10	1	12	3	7	11

Constrained Random Sequences

When sequence statistics, such as the length of the longest direct repeat, are studied by calculating them on a set of random sequences, it may be appropriate to ensure that dinucleotide frequencies are preserved. Such constraints may be built into the algorithms for representative or shuffled sequences (Fitch 1983).

For a representative sequence to preserve dinucleotide frequencies, it is necessary to find these frequencies, p_{ij} for bases ij, in the original sequence in addition to the individual base frequencies p_i. The random sequence is begun by drawing a random base as described in the previous section. If this is of type i, then the next base is selected to be of type j with probability p_{ij}/p_i. In other words the unit interval is partitioned into four segments in a way that depends on frequencies of dinucleotides beginning with base i. At the third site, the base drawn is of type k with probability p_{jk}/p_j, and so on.

Shufflings with constraints are a little more difficult to construct, and Fitch (1983) describes a method that first counts dinucleotides, including that formed by the last base followed by the first. The initial base i is chosen, without replacement, from the set of m bases in the original sequence. The next base j is chosen, without replacement, from the set of of dinucleotides that start with i.

Appendix C

Computer Programs

LINKAGE DISEQUILIBRIUM

Gametic Disequilibrium

The program GAMETE.FOR computes the gametic linkage disequilibria at all sets of 2, 3 and 4 loci from data on multilocus gamete (haplotype) *counts.* Variances are estimated and chi-square statistics for testing that the coefficients are zero are calculated.

The input for the program consists of one line for each gamete/haplotye. The alleles (coded as 1 or 2) are read in order across the line, and there is an identifier at the beginning of the line. The output consists of disequilibria, estimated standard deviations, chi-square statistics and the number of gametes or haplotypes that had that set of loci.

As an example, the following set of data is the first five loci of the set reported by Macpherson et al. (1990). The data are in the file GAMETE.DAT in the following format:

```
␣1␣1␣1␣2␣1␣1
␣2␣1␣1␣2␣1␣1
␣3␣1␣1␣2␣1␣2
␣4␣2␣2␣1␣2␣2
␣5␣1␣1␣2␣1␣2
␣6␣2␣2␣1␣2␣2
␣7␣1␣1␣2␣1␣1
␣8␣2␣2␣1␣2␣2
␣9␣1␣1␣2␣1␣2
10␣1␣1␣2␣1␣1
11␣1␣1␣2␣1␣1
12␣1␣1␣1␣1␣2
13␣1␣1␣2␣1␣1
14␣1␣1␣2␣1␣1
15␣2␣2␣1␣2␣2
16␣2␣1␣1␣1␣2
```

```
17␣1␣1␣2␣1␣2
18␣1␣1␣2␣1␣1
19␣2␣2␣1␣2␣2
20␣9␣9␣1␣1␣2
21␣1␣1␣2␣1␣2
22␣1␣1␣2␣1␣2
23␣9␣9␣2␣9␣1
24␣9␣9␣2␣9␣2
25␣9␣9␣2␣1␣2
26␣2␣1␣1␣1␣2
27␣9␣9␣1␣1␣2
28␣1␣1␣1␣1␣2
29␣2␣1␣1␣1␣2
30␣1␣1␣2␣1␣1
31␣1␣1␣2␣1␣1
32␣1␣1␣2␣1␣1
33␣9␣1␣1␣1␣2
34␣1␣9␣2␣1␣2
35␣1␣1␣2␣1␣2
36␣2␣2␣1␣2␣2
37␣9␣1␣2␣1␣2
38␣9␣1␣2␣1␣2
39␣1␣1␣1␣1␣2
40␣1␣1␣2␣1␣1
41␣1␣1␣1␣1␣2
42␣9␣1␣2␣1␣1
43␣1␣1␣2␣1␣2
44␣9␣1␣2␣1␣1
```

The output file, `GAMETE.OUT` is

```
Two-locus analyses
       1  2    0.1322     0.0390      19.56    33.
       1  3   -0.1635     0.0331      19.77    34.
       1  4    0.1298     0.0387      20.24    34.
       1  5    0.0934     0.0259       6.68    34.
       2  3   -0.0997     0.0320      12.21    38.
       2  4    0.1330     0.0405      38.00    38.
       2  5    0.0582     0.0212       4.16    38.
       3  4   -0.0884     0.0293      11.37    42.
       3  5   -0.1240     0.0242      13.00    44.
       4  5    0.0476     0.0180       3.50    42.

Three-locus analyses
    1  2  3    0.0501     0.0151       3.27    33.
    1  2  4   -0.0841     0.0086       6.93    33.
    1  2  5   -0.0301     0.0106       2.87    33.
    1  3  4    0.0513     0.0147       3.50    34.
    1  3  5    0.0220     0.0149       1.58    34.
    1  4  5   -0.0293     0.0101       2.93    34.
    2  3  4    0.0682     0.0114       7.21    38.
    2  3  5    0.0153     0.0095       1.30    38.
    2  4  5   -0.0398     0.0095       4.46    38.
    3  4  5    0.0113     0.0076       1.12    42.
```

```
Four-locus analyses
 1  2  3  4   -0.0027    0.0199     0.02   33.
 1  2  3  5    0.0155    0.0142     0.94   33.
 1  2  4  5    0.0016    0.0121     0.02   33.
 1  3  4  5    0.0135    0.0134     0.82   34.
 2  3  4  5    0.0011    0.0097     0.01   38.
```

The program listing follows:

```
  1 c     GAMETE.FOR
  2 c
  3 c     Gametic disequilibrium at two, three and four loci
  4 c     Written by B.S. Weir
  5 c     Use in conjunction with Weir and Golding, 1990
  6 c
  7       implicit real*8(a-h,o-z)
  8       dimension ic4(7,7,7,7,16),ic3(7,7,7,8),ic2(7,7,4),id(7)
  9 c
 10 c
 11 c     Number of loci in data set is nloci
 12 c     Each locus has two alleles - scored as 1 or 2
 13 c         missing data indicated by blank or zero
 14 c         any other allele scored as 9
 15 c
 16 c
 17       nloci=7
 18       nloci1=nloci-1
 19       nloci2=nloci-2
 20       nloci3=nloci-3
 21 c
 22 c
 23 c     Initialize counts
 24 c
 25 c
 26       do 3 i1=1,nloci
 27          do 3 i2=1,nloci
 28             do 1 j=1,4
 29     1          ic2(i1,i2,j)=0
 30                do 2 i3=1,nloci
 31                do 2 k=1,8
 32     2             ic3(i1,i2,i3,k)=0
 33                   do 3 i4=1,nloci
 34                   do 3 l=1,16
 35                      ic4(i1,i2,i3,i4,l)=0
 36     3 continue
 37 c
 38 c
 39 c     Read data
 40 c
 41 c
 42       open(1,file='GAMETE.DAT',status='old')
 43       open(3,file='GAMETE.OUT',status='unknown')
 44 c
 45     4 read(1,100,end=1000)(id(j),j=1,nloci)
 46   100 format(2x,7i2)
 47 c
 48 c
```

```
 49 c     Increment two-locus counts
 50 c
 51 c
 52       do 10 i1=1,nloci1
 53          i11=i1+1
 54          do 10 i2=i11,nloci
 55             if(id(i1).eq.0.or.id(i1).eq.9)go to 10
 56             if(id(i2).eq.0.or.id(i2).eq.9)go to 10
 57             if(id(i1).eq.1.and.id(i2).eq.1)ic2(i1,i2,1)=ic2(i1,i2,1)+1
 58             if(id(i1).eq.1.and.id(i2).eq.2)ic2(i1,i2,2)=ic2(i1,i2,2)+1
 59             if(id(i1).eq.2.and.id(i2).eq.1)ic2(i1,i2,3)=ic2(i1,i2,3)+1
 60             if(id(i1).eq.2.and.id(i2).eq.2)ic2(i1,i2,4)=ic2(i1,i2,4)+1
 61    10 continue
 62 c
 63 c
 64 c     Increment three-locus counts
 65 c
 66 c
 67       do 11 i1=1,nloci2
 68          i11=i1+1
 69          do 11 i2=i11,nloci1
 70             i21=i2+1
 71             do 11 i3=i21,nloci
 72                if(id(i1).eq.0.or.id(i1).eq.9)go to 11
 73                if(id(i2).eq.0.or.id(i2).eq.9)go to 11
 74                if(id(i3).eq.0.or.id(i3).eq.9)go to 11
 75                if(id(i1).eq.1.and.id(i2).eq.1.and.id(i3).eq.1)
 76      *                   ic3(i1,i2,i3,1)=ic3(i1,i2,i3,1)+1
 77                if(id(i1).eq.1.and.id(i2).eq.1.and.id(i3).eq.2)
 78      *                   ic3(i1,i2,i3,2)=ic3(i1,i2,i3,2)+1
 79                if(id(i1).eq.1.and.id(i2).eq.2.and.id(i3).eq.1)
 80      *                   ic3(i1,i2,i3,3)=ic3(i1,i2,i3,3)+1
 81                if(id(i1).eq.1.and.id(i2).eq.2.and.id(i3).eq.2)
 82      *                   ic3(i1,i2,i3,4)=ic3(i1,i2,i3,4)+1
 83                if(id(i1).eq.2.and.id(i2).eq.1.and.id(i3).eq.1)
 84      *                   ic3(i1,i2,i3,5)=ic3(i1,i2,i3,5)+1
 85                if(id(i1).eq.2.and.id(i2).eq.1.and.id(i3).eq.2)
 86      *                   ic3(i1,i2,i3,6)=ic3(i1,i2,i3,6)+1
 87                if(id(i1).eq.2.and.id(i2).eq.2.and.id(i3).eq.1)
 88      *                   ic3(i1,i2,i3,7)=ic3(i1,i2,i3,7)+1
 89                if(id(i1).eq.2.and.id(i2).eq.2.and.id(i3).eq.2)
 90      *                   ic3(i1,i2,i3,8)=ic3(i1,i2,i3,8)+1
 91    11 continue
 92 c
 93 c
 94 c     Increment four-locus counts
 95 c
 96 c
 97       do 12 i1=1,nloci3
 98          i11=i1+1
 99          do 12 i2=i11,nloci2
100             i21=i2+1
101             do 12 i3=i21,nloci1
102                i31=i3+1
103                do 12 i4=i31,nloci
104                if(id(i1).eq.0.or.id(i1).eq.9)go to 12
105                if(id(i2).eq.0.or.id(i2).eq.9)go to 12
```

```
106               if(id(i3).eq.0.or.id(i3).eq.9)go to 12
107               if(id(i4).eq.0.or.id(i4).eq.9)go to 12
108          if(id(i1).eq.1.and.id(i2).eq.1.and.id(i3).eq.1.and.id(i4).eq.1)
109      *           ic4(i1,i2,i3,i4,1)=ic4(i1,i2,i3,i4,1)+1
110          if(id(i1).eq.1.and.id(i2).eq.1.and.id(i3).eq.1.and.id(i4).eq.2)
111      *           ic4(i1,i2,i3,i4,2)=ic4(i1,i2,i3,i4,2)+1
112          if(id(i1).eq.1.and.id(i2).eq.1.and.id(i3).eq.2.and.id(i4).eq.1)
113      *           ic4(i1,i2,i3,i4,3)=ic4(i1,i2,i3,i4,3)+1
114          if(id(i1).eq.1.and.id(i2).eq.1.and.id(i3).eq.2.and.id(i4).eq.2)
115      *           ic4(i1,i2,i3,i4,4)=ic4(i1,i2,i3,i4,4)+1
116          if(id(i1).eq.1.and.id(i2).eq.2.and.id(i3).eq.1.and.id(i4).eq.1)
117      *           ic4(i1,i2,i3,i4,5)=ic4(i1,i2,i3,i4,5)+1
118          if(id(i1).eq.1.and.id(i2).eq.2.and.id(i3).eq.1.and.id(i4).eq.2)
119      *          ic4(i1,i2,i3,i4,6)=ic4(i1,i2,i3,i4,6)+1
120          if(id(i1).eq.1.and.id(i2).eq.2.and.id(i3).eq.2.and.id(i4).eq.1)
121      *           ic4(i1,i2,i3,i4,7)=ic4(i1,i2,i3,i4,7)+1
122          if(id(i1).eq.1.and.id(i2).eq.2.and.id(i3).eq.2.and.id(i4).eq.2)
123      *           ic4(i1,i2,i3,i4,8)=ic4(i1,i2,i3,i4,8)+1
124          if(id(i1).eq.2.and.id(i2).eq.1.and.id(i3).eq.1.and.id(i4).eq.1)
125      *           ic4(i1,i2,i3,i4,9)=ic4(i1,i2,i3,i4,9)+1
126          if(id(i1).eq.2.and.id(i2).eq.1.and.id(i3).eq.1.and.id(i4).eq.2)
127      *           ic4(i1,i2,i3,i4,10)=ic4(i1,i2,i3,i4,10)+1
128          if(id(i1).eq.2.and.id(i2).eq.1.and.id(i3).eq.2.and.id(i4).eq.1)
129      *           ic4(i1,i2,i3,i4,11)=ic4(i1,i2,i3,i4,11)+1
130          if(id(i1).eq.2.and.id(i2).eq.1.and.id(i3).eq.2.and.id(i4).eq.2)
131      *           ic4(i1,i2,i3,i4,12)=ic4(i1,i2,i3,i4,12)+1
132          if(id(i1).eq.2.and.id(i2).eq.2.and.id(i3).eq.1.and.id(i4).eq.1)
133      *           ic4(i1,i2,i3,i4,13)=ic4(i1,i2,i3,i4,13)+1
134          if(id(i1).eq.2.and.id(i2).eq.2.and.id(i3).eq.1.and.id(i4).eq.2)
135      *           ic4(i1,i2,i3,i4,14)=ic4(i1,i2,i3,i4,14)+1
136          if(id(i1).eq.2.and.id(i2).eq.2.and.id(i3).eq.2.and.id(i4).eq.1)
137      *           ic4(i1,i2,i3,i4,15)=ic4(i1,i2,i3,i4,15)+1
138          if(id(i1).eq.2.and.id(i2).eq.2.and.id(i3).eq.2.and.id(i4).eq.2)
139      *           ic4(i1,i2,i3,i4,16)=ic4(i1,i2,i3,i4,16)+1
140    12 continue
141 c
142       go to 4
143  1000 continue
144 c
145 c
146 c     Calculate two-locus disequilibria
147 c
148 c
149       write(3,200)
150   200 format(' Two-locus analyses')
151       do 23 i1=1,nloci1
152          i11=i1+1
153          do 23 i2=i11,nloci
154            an=0d0
155            do 20 j=1,4
156    20         an=an+ic2(i1,i2,j)
157            if(an.eq.0d0)go to 23
158            p1=ic2(i1,i2,1)+ic2(i1,i2,2)
159            p2=ic2(i1,i2,1)+ic2(i1,i2,3)
160            p12=ic2(i1,i2,1)
161            p1=p1/an
162            p2=p2/an
```

```
163            p12=p12/an
164            d12=p12-p1*p2
165 c
166 c
167 c     Calculate variance of two-locus disequilibria
168 c     Test statistics are squared estimates over variances
169 c         (A test statistic of 9.99 indicates that variance)
170 c          was negative and no test is possible)
171 c
172 c
173            pi1=p1*(1d0-p1)
174            pi2=p2*(1d0-p2)
175            tau1=1d0-2d0*p1
176            tau2=1d0-2d0*p2
177            v12=pi1*pi2/an
178            v12d=v12+d12*(tau1*tau2-d12)/an
179            if(v12d.ge.1d-8)go to 21
180            v12d=9.9999
181            chis=9.99
182            go to 22
183     21     v12d=dsqrt(v12d)
184            chis=d12**2/v12
185     22     continue
186 c
187            write(3,210)i1,i2,d12,v12d,chis,an
188    210     format(6x,2i3,2f10.4,f10.2,f6.0)
189     23 continue
190 c
191 c
192 c     Calculate three-locus disequilibria
193 c
194 c
195        write(3,300)
196    300 format(' Three-locus analyses')
197        do 33 i1=1,nloci2
198           i11=i1+1
199           do 33 i2=i11,nloci1
200              i21=i2+1
201              do 33 i3=i21,nloci
202                 an=0d0
203                 do 30 k=1,8
204     30             an=an+ic3(i1,i2,i3,k)
205                 if(an.eq.0d0)go to 33
206                 p1=ic3(i1,i2,i3,1)+ic3(i1,i2,i3,2)
207      *             +ic3(i1,i2,i3,3)+ic3(i1,i2,i3,4)
208                 p2=ic3(i1,i2,i3,1)+ic3(i1,i2,i3,2)
209      *             +ic3(i1,i2,i3,5)+ic3(i1,i2,i3,6)
210                 p3=ic3(i1,i2,i3,1)+ic3(i1,i2,i3,3)
211      *             +ic3(i1,i2,i3,5)+ic3(i1,i2,i3,7)
212                 p12=ic3(i1,i2,i3,1)+ic3(i1,i2,i3,2)
213                 p13=ic3(i1,i2,i3,1)+ic3(i1,i2,i3,3)
214                 p23=ic3(i1,i2,i3,1)+ic3(i1,i2,i3,5)
215                 p123=ic3(i1,i2,i3,1)
216                 p1=p1/an
217                 p2=p2/an
218                 p3=p3/an
219                 p12=p12/an
```

```
220                 p13=p13/an
221                 p23=p23/an
222                 p123=p123/an
223                 d12=p12-p1*p2
224                 d13=p13-p1*p3
225                 d23=p23-p2*p3
226                 d123=p123-p1*d23-p2*d13-p3*d12-p1*p2*p3
227 c
228 c
229 c     Calculate variance of three-locus disequilibria
230 c     Test statistic is squared estimate over variance
231 c         (A test statistic of 9.99 indicates that variance)
232 c          was negative and no test is possible)
233 c
234 c
235                  pi1=p1*(1d0-p1)
236                  pi2=p2*(1d0-p2)
237                  pi3=p3*(1d0-p3)
238                  tau1=1d0-2d0*p1
239                  tau2=1d0-2d0*p2
240                  tau3=1d0-2d0*p3
241                  v123=pi1*pi2*pi3+6d0*d12*d13*d23
242      *                 +pi1*(tau2*tau3*d23-d23**2)
243      *                 +pi2*(tau1*tau3*d13-d13**2)
244      *                 +pi3*(tau1*tau2*d12-d12**2)
245                  v123=v123/an
246                  v123d=v123+d123*(tau1*tau2*tau3
247      *                     -2d0*(tau1*d23+tau2*d13+tau3*d12)-d123)/an
248                  if(v123d.ge.1d-8)go to 31
249                  v123d=9.9999
250                  chis=9.99
251                  go to 32
252    31            v123d=dsqrt(v123d)
253                  chis=d123**2/v123
254    32            continue
255 c
256                  write(3,310)i1,i2,i3,d123,v123d,chis,an
257   310            format(3x,3i3,2f10.4,f10.2,f6.0)
258    33 continue
259 c
260 c
261 c     Calculate four-locus disequilibria
262 c
263 c
264       write(3,400)
265   400 format(' Four-locus analyses')
266       do 43 i1=1,nloci3
267          i11=i1+1
268          do 43 i2=i11,nloci2
269             i21=i2+1
270             do 43 i3=i21,nloci1
271                i31=i3+1
272                do 43 i4=i31,nloci
273                an=0d0
274                do 40 k=1,16
275    40             an=an+ic4(i1,i2,i3,i4,k)
276                if(an.eq.0d0)go to 43
```

```
277              p1=ic4(i1,i2,i3,i4,1) +ic4(i1,i2,i3,i4,2)
278       *        +ic4(i1,i2,i3,i4,3) +ic4(i1,i2,i3,i4,4)
279       *        +ic4(i1,i2,i3,i4,5) +ic4(i1,i2,i3,i4,6)
280       *        +ic4(i1,i2,i3,i4,7) +ic4(i1,i2,i3,i4,8)
281              p2=ic4(i1,i2,i3,i4,1) +ic4(i1,i2,i3,i4,2)
282       *        +ic4(i1,i2,i3,i4,3) +ic4(i1,i2,i3,i4,4)
283       *        +ic4(i1,i2,i3,i4,9) +ic4(i1,i2,i3,i4,10)
284       *        +ic4(i1,i2,i3,i4,11)+ic4(i1,i2,i3,i4,12)
285              p3=ic4(i1,i2,i3,i4,1) +ic4(i1,i2,i3,i4,2)
286       *        +ic4(i1,i2,i3,i4,5) +ic4(i1,i2,i3,i4,6)
287       *        +ic4(i1,i2,i3,i4,9) +ic4(i1,i2,i3,i4,10)
288       *        +ic4(i1,i2,i3,i4,13)+ic4(i1,i2,i3,i4,14)
289              p4=ic4(i1,i2,i3,i4,1) +ic4(i1,i2,i3,i4,3)
290       *        +ic4(i1,i2,i3,i4,5) +ic4(i1,i2,i3,i4,7)
291       *        +ic4(i1,i2,i3,i4,9) +ic4(i1,i2,i3,i4,11)
292       *        +ic4(i1,i2,i3,i4,13)+ic4(i1,i2,i3,i4,15)
293              p12=ic4(i1,i2,i3,i4,1)+ic4(i1,i2,i3,i4,2)
294       *         +ic4(i1,i2,i3,i4,3)+ic4(i1,i2,i3,i4,4)
295              p13=ic4(i1,i2,i3,i4,1)+ic4(i1,i2,i3,i4,2)
296       *         +ic4(i1,i2,i3,i4,5)+ic4(i1,i2,i3,i4,6)
297              p14=ic4(i1,i2,i3,i4,1)+ic4(i1,i2,i3,i4,3)
298       *         +ic4(i1,i2,i3,i4,5)+ic4(i1,i2,i3,i4,7)
299              p23=ic4(i1,i2,i3,i4,1)+ic4(i1,i2,i3,i4,2)
300       *         +ic4(i1,i2,i3,i4,9)+ic4(i1,i2,i3,i4,10)
301              p24=ic4(i1,i2,i3,i4,1)+ic4(i1,i2,i3,i4,3)
302       *         +ic4(i1,i2,i3,i4,9)+ic4(i1,i2,i3,i4,11)
303              p34=ic4(i1,i2,i3,i4,1)+ic4(i1,i2,i3,i4,5)
304       *         +ic4(i1,i2,i3,i4,9)+ic4(i1,i2,i3,i4,13)
305              p123=ic4(i1,i2,i3,i4,1)+ic4(i1,i2,i3,i4,2)
306              p124=ic4(i1,i2,i3,i4,1)+ic4(i1,i2,i3,i4,3)
307              p134=ic4(i1,i2,i3,i4,1)+ic4(i1,i2,i3,i4,5)
308              p234=ic4(i1,i2,i3,i4,1)+ic4(i1,i2,i3,i4,9)
309              p1234=ic4(i1,i2,i3,i4,1)
310              p1=p1/an
311              p2=p2/an
312              p3=p3/an
313              p4=p4/an
314              p12=p12/an
315              p13=p13/an
316              p14=p14/an
317              p23=p23/an
318              p24=p24/an
319              p34=p34/an
320              p123=p123/an
321              p124=p124/an
322              p134=p134/an
323              p234=p234/an
324              p1234=p1234/an
325              d12=p12-p1*p2
326              d13=p13-p1*p3
327              d14=p14-p1*p4
328              d23=p23-p2*p3
329              d24=p24-p2*p4
330              d34=p34-p3*p4
331              d123=p123-p1*d23-p2*d13-p3*d12-p1*p2*p3
332              d124=p124-p1*d24-p2*d14-p4*d12-p1*p2*p4
333              d134=p134-p1*d34-p3*d14-p4*d13-p1*p3*p4
```

```
334             d234=p234-p2*d34-p3*d24-p4*d23-p2*p3*p4
335             d1234=p1234-p1*d234-p2*d134-p3*d124-p4*d123
336      *           -p1*p2*d34-p1*p3*d24-p1*p4*d23
337      *           -p2*p3*d14-p2*p4*d13-p3*p4*d12
338      *           -d12*d34-d13*d24-d14*d23-p1*p2*p3*p4
339 c
340 c
341 c    Calculate variance of four-locus disequilibria
342 c    Test statistic is squared estimate over variance
343 c       (A test statistic of 9.99 indicates that variance)
344 c        was negative and no test is possible)
345 c
346 c
347      pi1=p1*(1d0-p1)
348      pi2=p2*(1d0-p2)
349      pi3=p3*(1d0-p3)
350      pi4=p4*(1d0-p4)
351      tau1=1d0-2d0*p1
352      tau2=1d0-2d0*p2
353      tau3=1d0-2d0*p3
354      tau4=1d0-2d0*p4
355 c
356        v1234 = pi1*pi2*pi3*pi4
357      *          + 2d0*d12*d13*d24*d34
358      *          + 2d0*d12*d14*d23*d34
359      *          + 2d0*d13*d14*d23*d24
360      *      + 6d0*pi1*d23*d24*d34 + 6d0*pi2*d13*d14*d34
361      *      + 6d0*pi3*d12*d14*d24 + 6d0*pi4*d12*d13*d23
362      *          - 2d0*tau1*tau2*d13*d34*d24
363      *          - 2d0*tau1*tau3*d12*d24*d34
364      *          - 2d0*tau1*tau4*d12*d23*d34
365      *          - 2d0*tau2*tau3*d12*d14*d34
366       v1234=v1234
367      *  - 2d0*tau2*tau4*d12*d13*d34 - 2d0*tau3*tau4*d12*d13*d24
368      *  - 2d0*tau1*tau2*d14*d23*d34 - 2d0*tau1*tau3*d14*d23*d24
369      *  - 2d0*tau1*tau4*d13*d23*d24 - 2d0*tau2*tau3*d13*d14*d24
370      *  - 2d0*tau2*tau4*d13*d14*d23 - 2d0*tau3*tau4*d12*d14*d23
371      *         + tau1*tau2*tau3*tau4*d12*d34
372      *         + tau1*tau2*tau3*tau4*d13*d24
373      *         + tau1*tau2*tau3*tau4*d14*d23
374      *  + d12*tau1*tau2*pi3*pi4 + d13*tau1*tau3*pi2*pi4
375      *  + d14*tau1*tau4*pi2*pi3 + d23*tau2*tau3*pi1*pi4
376      *  + d24*tau2*tau4*pi1*pi3 + d34*tau3*tau4*pi1*pi2
377      *       + (d34**2 - tau3*tau4*d34 - pi3*pi4)*d12**2
378      *       + (d24**2 - tau2*tau4*d24 - pi2*pi4)*d13**2
379      *       + (d23**2 - tau2*tau3*d23 - pi2*pi3)*d14**2
380      *       + (d14**2 - tau1*tau4*d14 - pi1*pi4)*d23**2
381      *       + (d13**2 - tau1*tau3*d13 - pi1*pi3)*d24**2
382      *       + (d12**2 - tau1*tau2*d12 - pi1*pi2)*d34**2
383       v1234 = v1234
384      *           + 2d0*d123*tau3*d14*d24
385      *           + 2d0*d123*tau2*d14*d34
386      *           + 2d0*d123*tau1*d24*d34
387      *           + 2d0*d124*tau4*d13*d23
388      *           + 2d0*d124*tau2*d13*d34
389      *           + 2d0*d124*tau1*d23*d34
390      *           + 2d0*d134*tau4*d12*d23
```

```
391      *          + 2d0*d134*tau1*d23*d24
392      *          + 2d0*d134*tau3*d12*d24
393      *          + 2d0*d234*tau3*d12*d14
394      *          + 2d0*d234*tau4*d12*d13
395      *          + 2d0*d234*tau2*d13*d14
396       v1234=v1234
397      *            - 2d0*d123*d12*pi4*tau3
398      *            - 2d0*d123*d23*pi4*tau1
399      *            - 2d0*d123*d13*pi4*tau2
400      *            - 2d0*d124*d12*pi3*tau4
401      *            - 2d0*d124*d24*pi3*tau1
402      *            - 2d0*d124*d14*pi3*tau2
403      *            - 2d0*d134*d13*pi2*tau4
404      *            - 2d0*d134*d34*pi2*tau1
405      *            - 2d0*d134*d14*pi2*tau3
406      *            - 2d0*d234*d24*pi1*tau3
407      *            - 2d0*d234*d23*pi1*tau4
408      *            - 2d0*d234*d34*pi1*tau2
409      *             + 6d0*d14*d123*d234
410      *             + 6d0*d34*d123*d124
411      *             + 6d0*d24*d123*d134
412      *             + 6d0*d23*d134*d124
413      *             + 6d0*d12*d234*d134
414      *             + 6d0*d13*d124*d234
415      *         + d12c*pi4*tau1*tau2*tau3
416      *         + d12d*pi3*tau1*tau2*tau4
417      *         + d13d*pi2*tau1*tau3*tau4
418      *         + d23d*pi1*tau2*tau3*tau4
419      *  - pi4*d123**2 - pi3*d124**2 - pi2*d134**2 - pi1*d234**2
420 c
421       v1234=v1234/an
422       v1234d = v1234 + d1234*( tau1*tau2*tau3*tau4
423      *            - 2d0*d12*tau3*tau4 - 2d0*d13*tau2*tau4
424      *            - 2d0*d14*tau2*tau3 - 2d0*d23*tau1*tau4
425      *            - 2d0*d24*tau1*tau3 - 2d0*d34*tau1*tau2
426      *        + 4d0*d12*d34 + 4d0*d13*d24 + 4d0*d14*d23
427      *            - 2d0*d123*tau4 - 2d0*d124*tau3
428      *            - 2d0*d134*tau2 - 2d0*d234*tau1 -d1234)/an
429 c
430          if(v1234d.ge.1d-8)go to 41
431          v1234d=9.9999
432          chis=9.99
433          go to 42
434    41    v1234d=dsqrt(v1234d)
435          chis=d1234**2/v1234
436    42    continue
437          write(3,410)i1,i2,i3,i4,d1234,v1234d,chis,an
438   410    format(4i3,2f10.4,f10.2,f6.0)
439    43 continue
440 c
441       stop
442       end
```

Assuming HWE

The program LD79.FOR calculates gametic linkage disequilibrium for two loci with two alleles when Hardy-Weinberg equilibrium is assumed. This allows the estimation of gametic frequencies from genotypic frequencies without needing to distinguish between the two types of double heterozygote. The cubic resulting from Equation 2.23 is solved, the likelihoods at all valid roots are examined, and the maximum likelihood solution given. The results are contrasted with those using the composite disequilibrium coefficient approach.

As an example, the counts on page 107 are entered in the file LD79.DAT in the following format:

```
␣␣1␣␣2␣␣91␣147␣␣85␣␣32␣␣78␣␣75␣␣␣5␣␣17␣␣␣7
```

The results are placed in file LD79.OUT:

```
Estimation of linkage disequilibrium
Loci     1  2
Counts  91. 147.  85.  32.  78.  75.   5.  17.   7.
Maximum likelihood approach - assumes HWE
          Pa       Pb        D         S.D.     Chis      Chisg       r         z
       0.7737   0.4637   0.0235    0.0087   8.0116    6.7840    0.1124    2.6083
Composite coefficient approach - does not assume HWE
                             D         S.D.     Chis                  r         z
                          0.0274    0.0094   8.2888               0.1242    2.8859
```

In the maximum likelihood results, D is the MLE of gametic linkage disequilibrium and S.D. is its standard deviation estimated from the information matrix. The Chis test statistic is a chi-square calculated from the likelihood ratio, while Chisg is a goodness-of-fit statistic calculated as nr^2 where n is the sample size and r is the correlation of gene frequencies calculated as $r = D/\sqrt{p_A(1-p_A)p_B(1-p_B)}$. The value of r is also shown, along with Fisher's z-transformed value of a correlation coefficient (see page 112).

In the composite coefficient results, D is the composite coefficient Δ and S.D is its standard deviation based on multinomial sampling theory. The Chis test statistic is calculated as nr^2, where r is now calculated as $r = \Delta/\sqrt{[p_A(1-p_A)+D_A][p_B(1-p_B)+D_B]}$ taking into account the sample departures from HWE (page 103). The quantity z is the z-transformation of r.

The program listing follows:

```
 1 c     LD79.FOR
 2 c
 3 c     Use in conjunction with B.S. Weir and C.C. Cockerham
 4 c     Heredity 42:105-111, 1979
 5 c
 6 c     Estimation of linkage disequilibrium from genotypic data
 7 c
 8 c
 9 c     Genotypic data is read in the following order
10 c        g(1)=AABB     g(2)=AABb     g(3)=AAbb
11 c        g(4)=AaBB     g(5)=AaBb     g(6)=Aabb
12 c        g(7)=aaBB     g(8)=aaBb     g(9)=aabb
13 c
14 c
15       implicit real*8(a-h,o-z)
16       common prod,chsg,chis,rlk,daa,dbb,sdh,f11,ann,an,q,p,d,c,z,ix22,
17      *ix21,ix12,ix11,ng5,ian,ixz
18       real*8 g(9),s(13)
19 c
20 c
21 c     Write labels
22 c
23 c
24       open(1,file='LD79.DAT',status='old')
25       open(3,file='LD79.out',status='unknown')
26       write(3,300)
27   300    format(' Estimation of linkage disequilibrium')
28 c
29 c
30 c     Read and write data
31 c
32 c     l1, l2 is to identify pair of loci
33 c
34 c
35     1 read(1,600,end=1000)l1,l2,(g(i),i=1,9)
36   600    format(2i3,9f4.0)
37       write(3,601)l1,l2
38   601    format(' Loci  ',2i3)
39       write(3,602)(g(i),i=1,9)
40   602    format(' Counts',9f5.0)
41 c
42       an=0d0
43       do 2 i=1,9
44     2    an=an+g(i)
45       ann=2d0*an
46       ian=an
47 c
48 c
49 c     Form allelic frequencies and summary statistics
50 c
51 c
52       p=(2d0*(g(1)+g(2)+g(3))+(g(4)+g(5)+g(6)))/ann
53       q=(2d0*(g(1)+g(4)+g(7))+(g(2)+g(5)+g(8)))/ann
54       s(1)=p
55       s(2)=q
56       prod=p*(1d0-p)*q*(1d0-q)
57       if(prod.lt.1d-8)go to 50
```

```
 58          ix11=2d0*g(1)+g(2)+g(4)
 59          ix12=2d0*g(3)+g(6)+g(2)
 60          ix21=2d0*g(7)+g(4)+g(8)
 61          ix22=2d0*g(9)+g(6)+g(8)
 62          ng5=g(5)
 63          daa=(g(1)+g(2)+g(3))/an-p**2
 64          dbb=(g(1)+g(4)+g(7))/an-q**2
 65 c
 66 c
 67 c        Maximum likelihood procedure
 68 c
 69 c
 70          call mleest
 71             d=f11-p*q
 72             s(3)=d
 73          call mlevar
 74             s(4)=sdh
 75          call mletst
 76             s(5)=chis
 77          call mlecor
 78             s(6)=chsg
 79             s(7)=c
 80             s(8)=z
 81 c
 82 c
 83 c        Composite procedure
 84 c
 85 c
 86          call comsit
 87             s(9)=d
 88             s(10)=sdh
 89             s(11)=chis
 90             s(12)=c
 91             s(13)=z
 92 c
 93 c
 94 c        Write results
 95 c
 96 c
 97          if(ixz.eq.1)go to 50
 98          write(3,400)
 99     400     format(' Maximum likelihood approach - assumes HWE')
100          write(3,500)
101     500     format(9x,'Pa',6x,'Pb',7x,'D',8x,'S.D.',4x,
102        *    'Chis',5x,'Chisg',5x,'r',8x,'z')
103          write(3,800)(s(i),i=1,8)
104     800     format(5x,2f8.4,11f9.4)
105          write(3,401)
106     401     format(' Composite coefficient approach - does not assume HWE')
107          write(3,501)
108     501     format(26x,'D',8x,'S.D.',4x,'Chis',15x,'r',8x,'z')
109          write(3,802)(s(i),i=9,13)
110     802 format(21x,3f9.4,9x,2f9.4)
111          go to 1
112      50 continue
113          write(3,900)(s(i),i=1,2)
114     900     format(' allele freqeuncies',2f8.4)
```

```
115         go to 1
116  1000 continue
117         stop
118         end
119 c
120 c
121         subroutine mleest
122 c
123 c
124         implicit real*8(a-h,o-z)
125         common prod,chsg,chis,rlk,daa,dbb,sdh,f11,ann,an,q,p,d,c,z,ix22,
126        *ix21,ix12,ix11,ng5,ian,ixz
127         real*8 r(3),rl(3),xcof(4),zz(3)
128 c
129         x11=ix11
130         x12=ix12
131         x21=ix21
132         x22=ix22
133         g5=ng5
134         sz=p*q/2d0
135         w=(1d0-2d0*p-2d0*q)/2d0
136         y=g5/(4d0*an)
137         if(p-q)11,11,10
138     10 pqmax=q
139         go to 12
140     11 pqmax=p
141     12 pq=p+q-1d0
142         if(pq)14,14,13
143     13 pqmin=pq
144         go to 15
145     14 pqmin=0d0
146     15 continue
147         ixz=0
148         inx1=0
149         inx2=0
150 c
151 c
152 c       g5 = 0
153 c
154 c
155         if(ng5.ne.0)go to 100
156         f11=x11/ann
157         rlk=0d0
158         iprod=ix11*ix12*ix21*ix22
159         if(iprod.ne.0)go to 16
160         ixz=1
161     16 continue
162         if(ix11.eq.0)go to 17
163         rlk=x11*dlog(f11)
164     17 if(ix12.eq.0)go to 20
165         rlk=rlk+x12*dlog(p-f11)
166     20 if(ix21.eq.0)go to 30
167         rlk=rlk+x21*dlog(q-f11)
168     30 if(ix22.eq.0)go to 40
169         rlk=rlk+x22*dlog(1d0-p-q+f11)
170     40 return
171 c
```

```
172 c
173 c      x11 = 0
174 c
175 c
176   100 if(ix11.ne.0)go to 200
177 c
178 c       x11 = x22 = 0
179 c
180        if(ix22.ne.0)go to 110
181        f11=0d0
182        rlk=(x12+g5)*dlog(p)+(x21+g5)*dlog(q)
183        ixz=1
184        return
185 c
186 c       x11 = x12 = 0
187 c
188   110 if(ix12.ne.0) go to 120
189        if(ix21.eq.0)go to 201
190        r(1)=0d0
191        r(2)=p
192        r(3)=q-(1d0-p)/2d0
193        rl(1)=g5*dlog(p)+(g5+x21)*dlog(q)+x22*dlog(1d0-p-q)
194        rl(2)=x21*dlog(q-p)+(x22+g5)*dlog(1d0-q)+g5*dlog(p)
195        rl(3)=(x21+x22+g5)*dlog((1d0-p)/2d0)+g5*dlog(p)
196        inx1=1
197        inx2=1
198        go to 700
199 c
200 c      x11 = x21 = 0
201 c
202   120 if(ix21.ne.0)go to 130
203        r(1)=0d0
204        r(2)=q
205        r(3)=p-(1d0-q)/2d0
206        rl(1)=(x12+g5)*dlog(p)+x22*dlog(1d0-p-q)+g5*dlog(q)
207        rl(2)=(x22+g5)*dlog(1d0-p)+g5*dlog(q)+x12*dlog(p-q)
208        rl(3)=(x12+x22+g5)*dlog((1d0-q)/2d0)+g5*dlog(q)
209        inx1=1
210        inx2=1
211        go to 700
212 c
213 c      only x11 = 0
214 c
215   130 m=2
216        xcof(1)=1d0
217        xcof(2)=w-y
218        xcof(3)=sz+y*(p+q-1d0)
219        call quad(xcof,zz,ier)
220        if(ier.eq.0)go to 134
221        write(3,1000)
222  1000 format(' only x11=0, no real roots for quadratic')
223        return
224   134 continue
225        r(1)=0d0
226        r(2)=zz(1)
227        r(3)=zz(2)
228        do 132 i=2,3
```

```
229           if(r(i).lt.0d0)go to 133
230           if(r(i).gt.pqmax)go to 133
231           rl(i)=x12*dlog(p-r(i))+x21*dlog(q-r(i))+x22*dlog(1d0-p-q+r(i))+
232      *    g5*dlog(r(i)*(1d0-p-q+r(i))+(p-r(i))*(q-r(i)))
233           go to 132
234   133     rl(i)=-1d10
235   132 continue
236       rl(1)=(x12+g5)*dlog(p)+(x21+g5)*dlog(q)+x22*dlog(1d0-p-q)
237       inx1=1
238       go to 700
239 c
240 c
241 c     x12 = 0
242 c
243 c
244   200 if(ix12.ne.0)go to 300
245 c
246 c      x12 = x21 = 0
247 c
248       if(ix21.ne.0) go to 210
249   201 f11=p
250       rlk=(x11+g5)*dlog(p)+(x22+g5)*dlog(1d0-q)
251       ixz=1
252       return
253 c
254 c     x12 = x22 =0
255 c
256   210 if(ix22.ne.0)go to 220
257       r(1)=p
258       r(2)=p+q-1d0
259       r(3)=q/2d0
260       rl(1)=(x11+g5)*dlog(p)+x21*dlog(q-p)+g5*dlog(1d0-q)
261       rl(2)=x11*dlog(p+q-1d0)+(x21+g5)*dlog(1d0-p)+g5*dlog(1d0-q)
262       rl(3)=(x11+x21+g5)*dlog(q/2d0)+g5*dlog(1d0-q)
263       inx1=1
264       inx2=1
265       go to 700
266 c
267 c     only x12 = 0
268 c
269   220 m=2
270       xcof(1)=1d0
271       xcof(2)=w+y
272       xcof(3)=sz-y*q
273       call quad(xcof,zz,ier)
274       if(ier.eq.0)go to 224
275       write(3,1001)
276  1001 format(' only x12=0, no real roots for quadratic')
277       return
278   224 continue
279       r(1)=p
280       r(2)=zz(1)
281       r(3)=zz(2)
282       do 222 i=2,3
283       if(r(i).gt.p)go to 223
284       if(r(i).lt.pqmin)go to 223
285       rl(i)=x11*dlog(r(i))+x21*dlog(q-r(i))+x22*dlog(1d0-p-q+r(i))+
```

```
286      * g5*dlog(r(i)*(1d0-p-q+r(i))+(p-r(i))*(q-r(i)))
287        go to 222
288    223 rl(i)=-1d10
289    222 continue
290        rl(1)=(x11+g5)*dlog(p)+(x22+g5)*dlog(1d0-q)+x21*dlog(q-p)
291        inx1=1
292        go to 700
293 c
294 c
295 c      x21 = 0
296 c
297 c
298    300 if(ix21.ne.0)go to 400
299 c
300 c      x21 = x22 = 0
301 c
302        if(ix22.ne.0)go to 310
303        r(1)=q
304        r(2)=p+q-1d0
305        r(3)=p/2d0
306        rl(1)=(x11+g5)*dlog(q)+x12*dlog(p-q)+g5*dlog(1d0-p)
307        rl(2)=x11*dlog(p+q-1d0)+(x12+g5)*dlog(1d0-q)+g5*dlog(1d0-p)
308        rl(3)=(x11+x12+g5)*dlog(p/2d0)+g5*dlog(1d0-p)
309        inx1=1
310        inx2=1
311        go to 700
312 c
313 c      only x21 = 0
314 c
315    310 m=2
316        xcof(1)=1d0
317        xcof(2)=w+y
318        xcof(3)=sz-y*p
319        call quad(xcof,zz,ier)
320        if(ier.eq.0) go to 314
321        write(3,1002)
322   1002 format(' x12=x22=0, no real roots for quadratic')
323        ixz=1
324        return
325    314 continue
326        r(1)=q
327        r(2)=zz(1)
328        r(3)=zz(2)
329        do 312 i=2,3
330        if(r(i).gt.q)go to 313
331        if(r(i).lt.pqmin)go to 313
332        rl(i)=x11*dlog(r(i))+x12*dlog(p-r(i))+x22*dlog(1d0-p-q+r(i))+
333      * g5*dlog(r(i)*(1d0-p-q+r(i))+(p-r(i))*(q-r(i)))
334        go to 312
335    313 rl(i)=-1d10
336    312 continue
337        rl(1)=(x11+g5)*dlog(q)+x12*dlog(p-q)+(x22+g5)*dlog(1d0-p)
338        inx1=1
339        go to 700
340 c
341 c
342 c      x22 = 0
```

```
343 c
344 c
345   400 if(ix22.ne.0)go to 500
346       m=2
347       xcof(1)=1d0
348       xcof(2)=w-y
349       xcof(3)=sz
350       call quad(xcof,zz,ier)
351       if(ier.eq.0)go to 404
352       write(3,1003)
353  1003 format(' x22=0, no real roots for quadratic')
354       ixz=1
355       return
356   404 continue
357       r(1)=p+q-1d0
358       r(2)=zz(1)
359       r(3)=zz(2)
360       do 402 i=2,3
361       if(r(i).gt.pqmax)go to 403
362       if(r(i).lt.pqmin)go to 403
363       rl(i)=x11*dlog(r(i))+x12*dlog(p-r(i))+x21*dlog(q-r(i))+
364      * g5*dlog(r(i)*(1d0-p-q+r(i))+(p-r(i))*(q-r(i)))
365       go to 402
366   403 rl(i)=-1d10
367   402 continue
368       rl(1)=x11*dlog(p+q-1d0)+(x12+g5)*dlog(1d0-q)+(x21+g5)*dlog(1d0-p)
369       inx1=1
370       go to 700
371 c
372 c
373 c     no zero x's
374 c
375 c
376   500 m=3
377       xcof(1)=4d0*an
378       xcof(2)=ann*(1d0-2d0*p-2d0*q)-2d0*x11-g5
379       xcof(3)=ann*p*q-x11*(1d0-2d0*p-2d0*q)-g5*(1d0-p-q)
380       xcof(4)=-x11*p*q
381       call cubic(xcof,zz,ier)
382       if(ier.eq.0)go to 504
383       write(3,1004)
384  1004 format(' no zero x s, no real roots for cubic')
385       ixz=1
386       return
387   504 continue
388       do 502 i=1,3
389          r(i)=zz(i)
390          if(r(i).gt.pqmax)go to 503
391          if(r(i).lt.pqmin)go to 503
392          rl(i)=x11*dlog(r(i))+x12*dlog(p-r(i))+x21*dlog(q-r(i))
393      *   +x22*dlog(1d0-p-q+r(i))+g5*dlog(r(i)*(1d0-p-q+r(i))
394      *   +(p-r(i))*(q-r(i)))
395          go to 502
396   503    rl(i)=-1d10
397   502 continue
398 c
399 c
```

```
400 c     find maximum likelihood root
401 c
402 c
403   700 if(rl(1).gt.rl(2))go to 701
404       if(rl(2).gt.rl(3))go to 702
405       f11=r(3)
406       rlk=rl(3)
407       return
408   701 if(rl(1).gt.rl(3)) go to 703
409       f11=r(3)
410       rlk=rl(3)
411       return
412   702 f11=r(2)
413       rlk=rl(2)
414       if(inx2.eq.1)go to 704
415       return
416   703 f11=r(1)
417       rlk=rl(1)
418       if(inx1.eq.1)go to 704
419       return
420   704 ixz=1
421       return
422       end
423 c
424 c
425       subroutine mletst
426 c
427 c
428       implicit real*8(a-h,o-z)
429       common prod,chsg,chis,rlk,daa,dbb,sdh,f11,ann,an,q,p,d,c,z,ix22,
430      *ix21,ix12,ix11,ng5,ian,ixz
431       g5=ng5
432       ala=ann*(p*dlog(p)+(1d0-p)*dlog(1d0-p))
433       alb=ann*(q*dlog(q)+(1d0-q)*dlog(1d0-q))
434       chis=2d0*(rlk-ala-alb-g5*dlog(2d0))
435       chsg=an*d**2/prod
436       return
437       end
438 c
439 c
440       subroutine mlevar
441 c
442 c
443       implicit real*8(a-h,o-z)
444       common prod,chsg,chis,rlk,daa,dbb,sdh,f11,ann,an,q,p,d,c,z,ix22,
445      *ix21,ix12,ix11,ng5,ian,ixz
446       real*8 y(3,3,3),ay(3,3),am(3,3)
447 c
448 c
449 c     One gamete frequency is zero
450 c
451 c
452       if(ixz.ne.1)go to 101
453    50 continue
454       prudb=(1d0-2d0*p)*(1d0-2d0*q)*d
455       sdh=ann**2*(ann-1d0)*prod+ann*(ann-1d0)**2*prudb
456      * -2d0*ann*(ann-1d0)*(an-1d0)*d**2
```

```
457       if(sdh.gt.0d0)go to 51
458       sdh=0d0
459       go to 52
460    51 sdh=dsqrt(sdh/ann**4)
461    52 continue
462       return
463 c
464 c
465 c     no zero gametic frequencies
466 c
467 c
468   101 f12=p-f11
469       f21=q-f11
470       f22=1d0-f11-f12-f21
471       y(1,1,1)=q*f11
472       y(1,1,2)=q*f12+(1d0-q)*f11
473       y(1,1,3)=(1d0-q)*f12
474       y(1,2,1)=q*(f21-f11)
475       y(1,2,2)=q*(f22-f12)+(1d0-q)*(f21-f11)
476       y(1,2,3)=(1d0-q)*(f22-f12)
477       y(1,3,1)=-q*f21
478       y(1,3,2)=-q*f22-(1d0-q)*f21
479       y(1,3,3)=-(1d0-q)*f22
480       y(2,1,1)=p*f11
481       y(2,1,2)=p*(f12-f11)
482       y(2,1,3)=-p*f12
483       y(2,2,1)=p*f21+(1d0-p)*f11
484       y(2,2,2)=p*(f22-f21)+(1d0-p)*(f12-f11)
485       y(2,2,3)=-p*f22-(1d0-p)*f12
486       y(2,3,1)=(1d0-p)*f21
487       y(2,3,2)=(1d0-p)*(f22-f21)
488       y(2,3,3)=-(1d0-p)*f22
489       y(3,1,1)=f11
490       y(3,1,2)=f12-f11
491       y(3,1,3)=-f12
492       y(3,2,1)=f21-f11
493       y(3,2,2)=f11-f12-f21+f22
494       y(3,2,3)=f12-f22
495       y(3,3,1)=-f21
496       y(3,3,2)=f21-f22
497       y(3,3,3)=f22
498       do 110 i=1,3
499       do 110 j=1,3
500       do 110 k=1,3
501   110 y(i,j,k)=y(i,j,k)*2d0
502       ay(1,1)=f11**2
503       ay(1,2)=2d0*f11*f12
504       ay(1,3)=f12**2
505       ay(2,1)=2d0*f11*f21
506       ay(2,2)=2d0*(f11*f22+f12*f21)
507       ay(2,3)=2d0*f12*f22
508       ay(3,1)=f21**2
509       ay(3,2)=2d0*f21*f22
510       ay(3,3)=f22**2
511 c
512 c
513 c     Define information matrix
```

```
514 c
515 c
516       do 120 k=1,3
517          do 120 l=1,3
518             am(k,l)=0d0
519             do 120 i=1,3
520                do 120 j=1,3
521                   if(ay(i,j).lt.1d-8)go to 50
522                   am(k,l)=am(k,l)+an*y(k,i,j)*y(l,i,j)/ay(i,j)
523   120 continue
524 c
525 c
526 c     Inverse of information matrix
527 c
528 c
529       deta=am(1,1)*am(2,2)*am(3,3)-am(1,1)*am(3,2)*am(2,3)
530      *    -am(2,2)*am(1,3)*am(3,1)-am(3,3)*am(1,2)*am(2,1)
531      *    +am(1,2)*am(2,3)*am(3,1)+am(2,1)*am(1,3)*am(3,2)
532       if(deta.lt.1d-8) go to 50
533       ay(1,1)=am(2,2)*am(3,3)-am(2,3)*am(3,2)
534       ay(2,2)=am(3,3)*am(1,1)-am(3,1)*am(1,3)
535       ay(3,3)=am(1,1)*am(2,2)-am(1,2)*am(2,1)
536       ay(1,2)=am(3,2)*am(1,3)-am(1,2)*am(3,3)
537       ay(2,1)=am(2,3)*am(3,1)-am(2,1)*am(3,3)
538       ay(1,3)=am(1,2)*am(2,3)-am(1,3)*am(2,2)
539       ay(3,1)=am(2,1)*am(3,2)-am(3,1)*am(2,2)
540       ay(2,3)=am(1,3)*am(2,1)-am(2,3)*am(1,1)
541       ay(3,2)=am(3,1)*am(1,2)-am(3,2)*am(1,1)
542       do 130 i=1,3
543          do 130 j=1,3
544            ay(i,j)=ay(i,j)/deta
545   130 continue
546       sdh=dsqrt(ay(3,3))
547       return
548       end
549 c
550 c
551       subroutine mlecor
552 c
553 c
554       implicit real*8(a-h,o-z)
555       common prod,chsg,chis,rlk,daa,dbb,sdh,f11,ann,an,q,p,d,c,z,ix22,
556      *ix21,ix12,ix11,ng5,ian,ixz
557       c=d/dsqrt(prod)
558       if(c.ne.1d0)go to 100
559       z=1d2
560       return
561   100 ct=dabs(1d0-c)
562       if(ct.gt.1d-8)go to 200
563       z=1d2
564       return
565   200 ct=dabs(1d0+c)
566       if(ct.gt.1d-8)go to 300
567       z=-1d2
568       return
569   300 z=dlog((1d0+c)/(1d0-c))*dsqrt(an-3d0)/2d0
570       return
```

```
571       end
572 c
573 c
574       subroutine comsit
575 c
576 c
577       implicit real*8(a-h,o-z)
578       common prod,chsg,chis,rlk,daa,dbb,sdh,f11,ann,an,q,p,d,c,z,ix22,
579      *ix21,ix12,ix11,ng5,ian,ixz
580       x11=ix11
581       g5=ng5
582       d=(2d0*x11+g5)/(ann)-2d0*p*q
583       prodb=(p*(1d0-p)+daa)*(q*(1d0-q)+dbb)
584       prudb=(1d0-2d0*p)*(1d0-2d0*q)*d
585       chis=an*d**2/prodb
586       c=d/dsqrt(prodb)
587       d=an*d/(an-1d0)
588       if(c.ne.1d0)go to 100
589       z=1d2
590       return
591   100 ct=dabs(c-1d0)
592       if(ct.gt.1d-8)go to 200
593       z=1d2
594       return
595   200 ct=dabs(1d0+c)
596       if(ct.gt.1d-8)go to 300
597       z=-1d2
598       return
599   300 z=dlog((1d0+c)/(1d0-c))*dsqrt(an-3d0)/2d0
600       sdh=prodb/(an-1d0)+prudb/ann+d**2/(an*(an-1d0))
601       sdh=dsqrt(sdh)
602       return
603       end
604 c
605 c
606       subroutine quad(u,v,i)
607 c
608 c     Solves quadratic equation
609 c
610 c
611       implicit real*8(a-h,o-z)
612       dimension u(3),v(2)
613 c
614       i=0
615       v(1)=-1d0
616       v(2)=-1d0
617 c
618       a=u(1)
619       b=u(2)
620       c=u(3)
621       d=b**2-4d0*a*c
622       if(d.ge.0d0)go to 10
623       i=1
624       return
625 c
626    10 v(1)=(-b-dsqrt(d))/(2d0*a)
627       v(2)=(-b+dsqrt(d))/(2d0*a)
```

```
628         return
629 c
630         end
631 c
632 c
633         subroutine cubic(u,v,i)
634 c
635 c       Solves cubic equation
636 c
637 c
638         implicit real*8(a-h,o-z)
639         dimension u(4),v(3)
640 c
641         i=1
642         v(1)=-1d0
643         v(2)=-1d0
644         v(3)=-1d0
645 c
646         pi=3.141592653
647         pi2=2d0*pi/3d0
648         pi4=4d0*pi/3d0
649 c
650         do 10 i=2,4
651            u(i)=u(i)/u(1)
652      10 continue
653         u(1)=1d0
654 c
655 c       remove quadratic term
656 c
657         cp=u(3)-u(2)**2/3d0
658         cq=2d0*u(2)**3/27d0 - u(2)*u(3)/3d0 + u(4)
659         cd=4d0*cp**3 + 27d0*cq**2
660         if(cd.gt.0d0)go to 30
661 c
662 c       three real roots
663 c
664         ak=dsqrt(-4d0*cp/3d0)
665         c3=-4d0*cq/ak**3
666         theta=dacos(c3)/3d0
667         v(1)=ak*dcos(theta)-u(2)/3d0
668         v(2)=ak*dcos(theta+pi2)-u(2)/3d0
669         v(3)=ak*dcos(theta+pi4)-u(2)/3d0
670         i=0
671         return
672 c
673      30 continue
674         cd=cq**2+4d0*cp**3/27d0
675         y1=(-cq+dsqrt(cd))/2d0
676         y1=y1**(0.3333)
677         y2=-cp/(3d0*y1)
678         v(1)=y1+y2-u(2)/3d0
679         i=0
680         return
681 c
682         end
```

Composite Disequilibria

The program LD86.FOR computes the complete set of two-locus disequilibrium coefficients at two loci when the double heterozygotes cannot be distinguished. Two alleles are assumed per locus – if there are more than two, the data need to be collapsed into two allelic classes.

The input for the program consists of one line for each pair of loci. Identifiers for the two loci are followed by the nine genotypic *counts* in the order shown on page 102. The output consists of the six disequilibrium coefficients, standard errors calculated according to formulas given in the text, and chi-square statistics calculated as the squared estimates divided by the sampling variances (calculated under the hypothesis that the true value of the disequilibrium is zero). A value of 99.99 indicates that the chi-square could not be calculated because the variance was too small.

As an example, the data on page 107 are put in file LD86.DAT with format

```
␣␣1␣␣2␣␣91␣147␣␣85␣␣32␣␣78␣␣75␣␣␣5␣␣17␣␣␣7
```

and result (Table 3.7) in the output file LD86.OUT with format

```
LOCI    1  2
COUNTS     91.  147.   85.   32.   78.   75.    5.   17.    7.
ESTIMATES      0.0028   0.0234   0.0273   0.0063   0.0022  -0.0034
STD. DEVS.     0.0077   0.0107   0.0090   0.0026   0.0032   0.0020
CHI-SQUARES    0.14     4.74     8.21     5.00     0.46     3.00
```

for $D_A, D_B, \Delta_{AB}, D_{AAB}, D_{ABB}$ and Δ_{AABB}, respectively.

The program listing follows:

```
 1 c
 2 c      LD86.FOR
 3 c      Complete characterization of disequilibrium at two loci
 4 c      using nine genotypes
 5 c
 6 c      Written by B.S. Weir
 7 c      Use in conjunction with
 8 c      "Complete characterization of disequilibrium at two loci"
 9 c      B.S. Weir and C.C. Cockerham
10 c      Pp 86-110
11 c      Mathematical Evolutionary Biology
12 c      Edited by M.W. Feldman
13 c      Princeton University Press, 1989
14 c
15        implicit real*8(a-h,o-z)
16        common ang(9),x(6),y(6),z(6),d(6),ix(6),tch,p,q,an,ann,ig
17        real*8 bng(10)
18 c
19 c
20 c      Read data
```

```
21 c      data are in file LD86.DAT
22 c      results are in file LD86.OUT
23 c      data are read in as locus identifiers
24 c          in format 2i3
25 c      followed by nine genotypic counts
26 c          in format 9f4.0
27 c          in the order
28 c             aabb aabB aaBB aAbb aAbB aABB AAbb AAbB AABB
29 c
30 c
31        open(1,file='LD86.DAT',status='old')
32        open(3,file='LD86.OUT',status='unknown')
33        tch=384d-2
34      1 read(1,100,end=1000)l1,l2,(ang(i),i=1,9)
35    100    format(2i3,9f4.0)
36        write(3,200)l1,l2
37    200    format(' loci ',2i3)
38        write(3,300)(ang(i),i=1,9)
39    300    format(' counts ',9f6.0)
40        an=0d0
41        do 2 i=1,9
42      2    an=an+ang(i)
43        ann=2d0*an
44 c
45 c
46 c      Sample gene frequencies
47 c
48 c
49        p=2d0*(ang(1)+ang(2)+ang(3))+(ang(4)+ang(5)+ang(6))
50        q=2d0*(ang(1)+ang(4)+ang(7))+(ang(2)+ang(5)+ang(8))
51        p=p/ann
52        q=q/ann
53 c
54 c
55 c      If both loci have a zero gene freq, no calculations
56 c
57 c
58        cab=p*(1d0-p)+q*(1d0-q)
59        if(cab.lt.1d-8)go to 90
60 c
61 c
62 c      Six disequilibrium estimates and test statistics
63 c
64 c
65        call oneloc
66        if(ig.eq.2)go to 90
67        call forgen
68        call thrgen
69        call lindis
70 c
71 c
72 c      Write results
73 c
74 c
75        write(3,400)(d(i),i=1,6)
76    400    format(' estimates',2x,6f9.4)
77        write(3,500)(z(i),i=1,6)
```

```
 78   500    format(' std. devs.',1x,6f9.4)
 79         write(3,600)(x(i),i=1,6)
 80   600    format(' chi-squares',f7.2,5f9.2)
 81 c
 82     90 continue
 83        go to 1
 84   1000 continue
 85        stop
 86        end
 87 c
 88 c
 89        subroutine oneloc
 90 c
 91 c
 92        implicit real*8(a-h,o-z)
 93        common ang(9),x(6),y(6),z(6),d(6),ix(6),tch,p,q,an,ann,ig
 94 c
 95        ix(1)=1
 96        ix(2)=1
 97        ig=6
 98        a=ang(1)+ang(2)+ang(3)
 99 c
100        d(1)=a/an-p**2
101 c
102        vd1=p**2*(1d0-p)**2/an
103        y(1)=vd1
104        z(1)=y(1)+((1d0-2d0*p)**2*d(1)-d(1)**2)/an
105        if(z(1).lt.0d0)z(1)=9998.0001
106        z(1)=dsqrt(z(1))
107        if(vd1.lt.1d-8)go to 10
108        x(1)=d(1)**2/vd1
109        go to 11
110     10 x(1)=99.99
111     11 continue
112 c
113        b=ang(1)+ang(4)+ang(7)
114 c
115        d(2)=b/an-q**2
116 c
117        vd2=q**2*(1d0-q)**2/an
118        y(2)=vd2
119        z(2)=y(2)+((1d0-2d0*q)**2*d(2)-d(2)**2)/an
120        if(z(2).lt.0d0)z(2)=9998.0001
121        z(2)=dsqrt(z(2))
122        if(vd2.lt.1d-8)go to 20
123        x(2)=d(2)**2/vd2
124        go to 21
125     20 x(2)=99.99
126     21 continue
127 c
128 c
129 c      If either locus has a zero gene freq, do no more
130 c
131 c
132        cab=p*(1d0-p)*q*(1d0-q)
133        if(cab.gt.1d-8)go to 30
134        ig=2
```

```
135     30 continue
136 c
137        return
138        end
139 c
140 c
141        subroutine lindis
142 c
143 c
144        implicit real*8(a-h,o-z)
145        common ang(9),x(6),y(6),z(6),d(6),ix(6),tch,p,q,an,ann,ig
146 c
147        ix(3)=1
148        pia=p*(1d0-p)
149        pib=q*(1d0-q)
150        ta=1d0-2d0*p
151        tb=1d0-2d0*q
152 c
153        d(3)=(2d0*ang(1)+ang(2)+ang(4)+ang(5)/2d0-2d0*an*p*q)/an
154 c
155        aab=2d0*ang(1)+ang(2)
156        d(4)=aab/(2d0*an)-p*d(3)-q*d(1)-p**2*q
157 c
158        abb=2d0*ang(1)+ang(4)
159        d(5)=abb/(2d0*an)-p*d(2)-q*d(3)-p*q**2
160 c
161        d(6)=ang(1)/an-2d0*p*d(5)-2d0*q*d(4)-2d0*p*q*d(3)
162       *             -p**2*d(2)-q**2*d(1)-d(1)*d(2)-d(3)**2-p**2*q**2
163 c
164        vd3=((pia+d(1))*(pib+d(2)))/an
165        y(3)=vd3
166        z(3)=ta*tb*d(3)/2d0+ta*d(5)+tb*d(4)+d(6)
167        z(3)=vd3+z(3)/an
168        if(z(3).lt.0d0)z(3)=9998.0001
169        z(3)=dsqrt(z(3))
170        vd3=0d0
171        if(ix(4).eq.1)go to 12
172        vd3=vd3+tb*d(4)
173     12 if(ix(5).eq.1)go to 13
174        vd3=vd3+ta*d(5)
175     13 if(ix(6).eq.1)go to 14
176        vd3=vd3+d(6)
177     14 continue
178        vd3=y(3)+vd3/an
179        if(vd3.lt.1d-8)go to 10
180        x(3)=d(3)**2/vd3
181        go to 11
182     10 x(3)=99.99
183     11 continue
184        return
185        end
186 c
187 c
188        subroutine thrgen
189 c
190 c
191        implicit real*8(a-h,o-z)
```

```
192         common ang(9),x(6),y(6),z(6),d(6),ix(6),tch,p,q,an,ann,ig
193 c
194         tch=384d-2
195         ix(4)=1
196         ix(5)=1
197         pia=p*(1d0-p)
198         pib=q*(1d0-q)
199         ta=1d0-2d0*p
200         tb=1d0-2d0*q
201 c
202         aab=2d0*ang(1)+ang(2)
203         d(4)=aab/(2d0*an)-p*d(3)-q*d(1)-p**2*q
204 c
205         abb=2d0*ang(1)+ang(4)
206         d(5)=abb/(2d0*an)-p*d(2)-q*d(3)-p*q**2
207 c
208         d(6)=ang(1)/an-2d0*p*d(5)-2d0*q*d(4)-2d0*p*q*d(3)
209        *            -p**2*d(2)-q**2*d(1)-d(1)*d(2)-d(3)**2-p**2*q**2
210 c
211         vd4=((pia**2+ta**2*d(1)-d(1)**2)*(pib+d(2))+pia*ta*tb*d(3)
212        *    + (1d0-5d0*pia+d(1))*d(3)**2)/(2d0*an)
213         y(4)=vd4
214         z(4)=2d0*pia*ta*d(5)+(ta**2*tb-2d0*d(1)*tb-4d0*ta*d(3))*d(4)
215        *     -2d0*d(4)**2+(ta**2-2d0*d(1))*d(6)
216         z(4)=vd4+z(4)/(2d0*an)
217         if(z(4).lt.0d0)z(4)=9998.0001
218         z(4)=dsqrt(z(4))
219         vd4=2d0*pia*ta*d(5)
220      12 if(ix(6).eq.1)go to 13
221         vd4=vd4+(ta**2-2d0*d(1))*d(6)
222      13 continue
223         vd4=y(4)+vd4/(2d0*an)
224         if(vd4.lt.1d-8)go to 10
225         x(4)=d(4)**2/vd4
226         if(x(4).gt.tch)ix(4)=2
227         go to 11
228      10 x(4)=99.99
229      11 continue
230 c
231 c
232         vd5=((pib**2+tb**2*d(2)-d(2)**2)*(pia+d(1))+pib*tb*ta*d(3)
233        *    + (1d0-5d0*pib+d(2))*d(3)**2)/(2d0*an)
234         y(5)=vd5
235         z(5)=2d0*pib*tb*d(4)+(ta*tb**2-2d0*d(2)*ta-4d0*tb*d(3))*d(5)
236        *     -2d0*d(5)**2+(tb**2-2d0*d(2))*d(6)
237         z(5)=vd5+z(5)/(2d0*an)
238         if(z(5).lt.0d0)z(5)=9998.0001
239         z(5)=dsqrt(z(5))
240         vd5=0d0
241         if(ix(4).eq.1)go to 22
242         vd5=vd5+2d0*pib*tb*d(4)
243      22 if(ix(6).eq.1)go to 23
244         vd5=vd5+(tb**2-2d0*d(2))*d(6)
245      23 continue
246         vd5=y(5)+vd5/(2d0*an)
247         if(vd5.lt.1d-8)go to 20
248         x(5)=d(5)**2/vd5
```

```
249       if(x(5).gt.tch)ix(5)=2
250       go to 21
251    20 x(5)=99.99
252    21 continue
253 c
254       if(ix(5).eq.2)return
255 c
256       vd4=0d0
257       if(ix(6).eq.1)go to 33
258       vd4=vd4+(ta**2-2d0*d(1))*d(6)
259    33 continue
260       vd4=y(4)+vd4/(2d0*an)
261       if(vd4.lt.1d-8)go to 30
262       x(4)=d(4)**2/vd4
263       if(x(4).gt.tch)ix(4)=2
264       go to 31
265    30 x(4)=99.99
266    31 continue
267 c
268       return
269       end
270 c
271 c
272       subroutine forgen
273 c
274 c
275       implicit real*8(a-h,o-z)
276       common ang(9),x(6),y(6),z(6),d(6),ix(6),tch,p,q,an,ann,ig
277 c
278       tch=384d-2
279       ix(6)=1
280 c
281       pia=p*(1d0-p)
282       pib=q*(1d0-q)
283       ta=1d0-2d0*p
284       tb=1d0-2d0*q
285 c
286       d(3)=(2d0*ang(1)+ang(2)+ang(4)+ang(5)/2d0-2d0*an*p*q)/an
287 c
288       aab=2d0*ang(1)+ang(2)
289       d(4)=aab/(2d0*an)-p*d(3)-q*d(1)-p**2*q
290 c
291       abb=2d0*ang(1)+ang(4)
292       d(5)=abb/(2d0*an)-p*d(2)-q*d(3)-p*q**2
293 c
294       d(6)=ang(1)/an-2d0*p*d(5)-2d0*q*d(4)-2d0*p*q*d(3)
295      *           -p**2*d(2)-q**2*d(1)-d(1)*d(2)-d(3)**2-p**2*q**2
296 c
297       d1=d(1)
298       d2=d(2)
299       d3=d(3)
300       d4=d(4)
301       d5=d(5)
302       d6=d(6)
303       vd6=(pia**2+ta**2*d1-d1**2)*(pib**2+tb**2*d2-d2**2)
304      *    +d3*(2d0*pia*ta*pib*tb-8d0*ta*tb*d1*d2)
305      *    +d3**2*(ta**2*tb**2-4d0*tb**2*d1-4d0*ta**2*d2+4d0*d1*d2
```

```
306      *            +2d0*d1+2d0*d2)-6d0*ta*tb*d3**3+3d0*d3**4
307      *     +d4*(2d0*ta**2*tb*pib-4d0*tb*pib*d1+8d0*ta*d2*d3
308      *          -8d0*pib*ta*d3+4d0*tb*d3**2)
309      *     +d5*(2d0*ta*pia*tb**2-4d0*ta*pia*d2+8d0*tb*d1*d3
310      *          -8d0*pia*tb*d3+4d0*ta*d3**2)
311      *     +d4**2*(6d0*d2-2d0*pib)+d5**2*(6d0*d1-2d0*pia)
312      *     +20d0*d3*d4*d5
313         vd6=vd6/an
314         y(6)=vd6
315         z(6)=d6*(ta**2*tb**2-2d0*ta**2*d2-2d0*tb**2*d1+4d0*d1*d2
316      *               -8d0*ta*tb*d3+6d0*d3**2-4d0*ta*d5-4d0*tb*d4)
317      *     -d6**2
318         z(6)=y(6)+z(6)/an
319         if(z(6).lt.0d0)z(6)=9998.0001
320         z(6)=dsqrt(z(6))
321         if(vd6.lt.1d-8)go to 10
322         x(6)=d(6)**2/vd6
323         if(x(6).gt.tch)ix(6)=2
324         go to 11
325      10 x(6)=99.99
326      11 continue
327 c
328         return
329         end
```

F-STATISTICS

Haploid data

The program HAPLOID.FOR estimates the parameter $\theta = F_{ST}$ from gametic data. The estimate on page 147 is used. Properties of this estimate are determined by jackknifing over loci or populations, and by bootstrapping over loci.

The data are put in file HAPLOID.DAT, one line per gamete, with a gamete identifier followed by the alleles at each locus. As an example, the data

```
␣␣6␣␣5␣␣4
␣␣1␣␣4␣␣3␣␣3␣␣3␣␣4
␣␣1␣␣4␣␣4␣␣3␣␣3␣␣4
␣␣1␣␣4␣␣4␣␣3␣␣3␣␣4
␣␣1␣␣4␣␣4␣␣␣␣␣3␣␣4
␣␣1␣␣4␣␣4␣␣2␣␣4␣␣4
␣␣1␣␣4␣␣4␣␣4␣␣4␣␣4
␣␣1␣␣4␣␣4␣␣3␣␣4␣␣4
␣␣1␣␣4␣␣4␣␣3␣␣4␣␣4
␣␣1␣␣4␣␣4␣␣␣␣␣3␣␣4
␣␣1␣␣4␣␣4␣␣3␣␣3␣␣4
␣␣2␣␣4␣␣4␣␣2␣␣2␣␣4
␣␣2␣␣4␣␣3␣␣4␣␣3␣␣4
␣␣2␣␣4␣␣4␣␣3␣␣3␣␣4
␣␣2␣␣4␣␣4␣␣4␣␣4␣␣4
␣␣2␣␣4␣␣3␣␣4␣␣4␣␣4
␣␣2␣␣4␣␣4␣␣4␣␣2␣␣4
```

```
␣␣2␣␣4␣␣4␣␣3␣␣3␣␣4
␣␣2␣␣4␣␣4␣␣4␣␣4␣␣4
␣␣3␣␣4␣␣4␣␣4␣␣3␣␣4
␣␣3␣␣4␣␣4␣␣4␣␣3␣␣4
␣␣3␣␣4␣␣3␣␣4␣␣4␣␣4
␣␣3␣␣4␣␣4␣␣4␣␣4␣␣4
␣␣3␣␣4␣␣4␣␣3␣␣1␣␣4
␣␣3␣␣4␣␣4␣␣3␣␣3␣␣4
␣␣4␣␣4␣␣4␣␣3␣␣4␣␣4
␣␣4␣␣4␣␣4␣␣3␣␣3␣␣4
␣␣4␣␣4␣␣4␣␣3␣␣3␣␣4
␣␣4␣␣4␣␣4␣␣3␣␣4␣␣4
␣␣4␣␣4␣␣4␣␣4␣␣3␣␣4
␣␣4␣␣4␣␣4␣␣4␣␣4␣␣4
␣␣5␣␣4␣␣4␣␣4␣␣1␣␣4
␣␣5␣␣4␣␣4␣␣4␣␣3␣␣4
␣␣5␣␣4␣␣4␣␣2␣␣3␣␣4
␣␣5␣␣4␣␣4␣␣3␣␣3␣␣4
␣␣5␣␣4␣␣4␣␣3␣␣3␣␣4
␣␣5␣␣4␣␣4␣␣4␣␣4␣␣4
␣␣5␣␣4␣␣4␣␣3␣␣3␣␣4
␣␣5␣␣4␣␣4␣␣4␣␣␣␣␣4
␣␣6␣␣4␣␣4␣␣4␣␣3␣␣4
␣␣6␣␣4␣␣4␣␣3␣␣3␣␣4
␣␣6␣␣4␣␣4␣␣2␣␣2␣␣4
␣␣6␣␣4␣␣4␣␣3␣␣1␣␣4
␣␣6␣␣4␣␣4␣␣4␣␣4␣␣4
␣␣6␣␣4␣␣4␣␣4␣␣4␣␣4
␣␣6␣␣4␣␣4␣␣4␣␣1␣␣4
```

results in the output file **HAPLOID.OUT**:

```
For locus   2
allele   theta
   1     0.0000
   2     0.0000
   3    -0.0011
   4    -0.0011
 All    -0.0011

For locus   3
allele   theta
   1     0.0000
   2    -0.1069
   3     0.0465
   4     0.0364
 All     0.0202

For locus   4
allele   theta
   1     0.0345
   2     0.0537
   3    -0.0503
   4    -0.0901
 All    -0.0424
```

```
Over all loci
theta   -0.0107


Jackknifing over populations
  locus  theta
   2   -0.0027 Mean
        0.0689 Std. Dev.
  locus  theta
   3    0.0269 Mean
        0.0976 Std. Dev.
  locus  theta
   4   -0.0393 Mean
        0.0329 Std. Dev.


Jackknifing over loci
         theta
 All   -0.0122 Mean
        0.0313 Std. Dev.


Bootstrapping over loci

 theta 95% CI:    -0.0424     0.0202
```

Note that an estimate of θ is shown for every allele, but that these are combined over alleles (`All`) and then over loci. When there only two alleles, each allele gives the same estimate. Jackknifing over populations allows estimates at different loci to be compared. Resampling over loci uses all the data to give properties of the overall estimate.

The program listing follows:

```
 1 c      HAPLOID.FOR
 2 c
 3 c      Estimation of theta for haploid data
 4 c
 5 c      Use in conjunction with B.S. Weir
 6 c      "Intraspecic Differentiation"
 7 c      in "Molecular Systematics"
 8 c      Edited by D. Hillis and C. Moritz
 9 c      Sinauer, Sunderland, MA. 1990
10 c
11 c
12        implicit real*8(a-h,o-z)
13        common mb,x(1000)
14        real ran,x
15        dimension an(6,5),p(6,5,4)
16        dimension ia(5)
17        dimension anbar(5),annbar(5),anc(5)
18        dimension term1(5),term2(5)
19        dimension pbar(5,4),ppbar(5,4),varp(5,4)
20        dimension tht(5,4)
21        dimension thtl(5),thtb(1000)
```

```
22 c
23 c     np populations, indexed by ip.
24 c        Replace 6 by at least np in dimension statements.
25 c     nl loci, indexed by il.
26 c        Replace 5 by at least nl in dimension statements.
27 c     up to nu alleles per locus, indexed by iu
28 c        Repalce 4 by at least nu in dimension statements.
29 c
30 c
31 c     Data is read in from file HAPLOID.DAT
32 c     First line contains three parameters np, nl and nu
33 c        (in format 3i3)
34 c     Each succeeding line contains population identifier
35 c        (in format i3)
36 c     and then the nl alleles, one per locus
37 c        (in format i3 for each locus)
38 c        alleles are called ia(il) for locus il
39 c        use a zero, or blank, for locus that is not scored
40 c
41        open(1,file='haploid.dat',status='old')
42        open(3,file='haploid.out',status='unknown')
43        read(1,100)np,nl,nu
44    100 format(3i3)
45        anp=np
46        anl=nl
47        nb=1000
48 c
49 c
50 c     Set counts to zero
51 c     for locus il in population ip:
52 c        an(ip,il) is sample size
53 c        p(ip,il,iu) is number af alleles of type iu
54 c
55 c
56        do 10 ip=1,np
57           do 10 il=1,nl
58              an(ip,il)=0d0
59              do 10 iu=1,nu
60                 p(ip,il,iu)=0d0
61     10 continue
62 c
63     20 read(1,200,end=22)ip,(ia(il),il=1,nl)
64    200 format(i3,15i3)
65 c
66 c
67 c     Increase counts of individuals and alleles
68 c
69 c
70        do 21 il=1,nl
71           if(ia(il).eq.0)go to 21
72           an(ip,il)=an(ip,il)+1d0
73           iu=ia(il)
74           p(ip,il,iu)=p(ip,il,iu)+1d0
75     21 continue
76        go to 20
77 c
78     22 continue
```

```
 79 c
 80 c
 81 c     Change counts to frequencies
 82 c     Form average sample sizes for each locus
 83 c     Form averages and variances of allelic frequencies
 84 c
 85 c
 86       do 30 ip=1,np
 87          do 30 il=1,nl
 88             do 30 iu=1,nu
 89                p(ip,il,iu)=p(ip,il,iu)/an(ip,il)
 90    30 continue
 91 c
 92       do 35 il=1,nl
 93          anbar(il)=0d0
 94          annbar(il)=0d0
 95          do 31 ip=1,np
 96             anbar(il)=anbar(il)+an(ip,il)
 97             annbar(il)=annbar(il)+an(ip,il)**2
 98    31    continue
 99          if(anp.le.1d0)go to 99
100          anbar(il)=anbar(il)/anp
101          anc(il)=(anp*anbar(il)-annbar(il)/(anp*anbar(il)))/(anp-1d0)
102 c
103          do 33 iu=1,nu
104             pbar(il,iu)=0d0
105             ppbar(il,iu)=0d0
106             do 32 ip=1,np
107                pbar(il,iu)=pbar(il,iu)+an(ip,il)*p(ip,il,iu)
108                ppbar(il,iu)=ppbar(il,iu)+an(ip,il)*p(ip,il,iu)**2
109    32       continue
110             pbar(il,iu)=pbar(il,iu)/(anp*anbar(il))
111             varp(il,iu)=ppbar(il,iu)-anp*anbar(il)*pbar(il,iu)**2
112             varp(il,iu)=varp(il,iu)/((anp-1d0)*anbar(il))
113    33    continue
114    35 continue
115 c
116 c
117 c     Form variance components for each allele at each locus
118 c     Estimate theta for each allele
119 c     Then combine over alleles
120 c     Check for fixed loci
121 c     Write estimates
122 c
123 c
124        tterm1=0d0
125        tterm2=0d0
126        do 42 il=1,nl
127           term1(il)=0d0
128           term2(il)=0d0
129           do 40 iu=1,nu
130              if(pbar(il,iu).le.1d-4.or.pbar(il,iu).gt.9999d-4)go to 40
131              if(ftest.eq.2)go to 40
132              d=anp*anbar(il)*pbar(il,iu)*(1d0-pbar(il,iu))
133                 d=d-(anp-1d0)*anbar(il)*varp(il,iu)
134                 d=d/(anp*(anbar(il)-1d0))
135              a=anbar(il)*varp(il,iu)
```

```
136                 a=(a-d)/anc(il)
137              tht(il,iu)=a/(a+d)
138              term1(il)=term1(il)+a
139              term2(il)=term2(il)+a+d
140     40    continue
141           if(term2(il).eq.0d0)go to 42
142           write(3,300)
143    300    format('    ')
144           write(3,400)il
145    400    format(' For locus ',i3)
146           write(3,410)
147    410    format(' allele',3x,'theta')
148           thtl(il)=term1(il)/term2(il)
149 c
150           do 41 iu=1,nu
151              write(3,420)iu,tht(il,iu)
152    420       format(i5,f10.4)
153     41    continue
154           write(3,430)thtl(il)
155    430    format('  All ',f9.4)
156           tterm1=tterm1+term1(il)
157           tterm2=tterm2+term2(il)
158     42 continue
159        thtt=tterm1/tterm2
160        write(3,300)
161        write(3,300)
162        write(3,440)
163    440 format(' Over all loci')
164        write(3,450)thtt
165    450 format(' theta',f10.4)
166 c
167 c
168 c      Jackknife over populations
169 c      Leave out each population in turn
170 c
171 c
172        write(3,300)
173        write(3,300)
174        write(3,600)
175    600 format(' Jackknifing over populations')
176        do 65 il=1,nl
177           anbar(il)=anbar(il)*anp
178           anp1=anp-1d0
179           anp2=anp1-1d0
180           thtj=0d0
181           thtjj=0d0
182           do 61 iu=1,nu
183              pbar(il,iu)=pbar(il,iu)*anbar(il)
184     61    continue
185           do 63 ip=1,np
186              anbarj=(anbar(il)-an(ip,il))/anp1
187              annbaj=annbar(il)-an(ip,il)**2
188              ancj=(anp1*anbarj-annbaj/(anp1*anbarj))/(anp1-1d0)
189 c
190              term1j=0d0
191              term2j=0d0
192              do 62 iu=1,nu
```

```
193                 pbarj=pbar(il,iu)-an(ip,il)*p(ip,il,iu)
194                 ppbarj=ppbar(il,iu)-an(ip,il)*p(ip,il,iu)**2
195 c
196                 pbarj=pbarj/(anp1*anbarj)
197                 varpj=(ppbarj-anp1*anbarj*pbarj**2)/(anp2*anbarj)
198                 if(pbarj.le.1d-4.or.pbarj.ge.9999d-4)go to 62
199 c
200                 d=anp1*anbarj*pbarj*(1d0-pbarj)
201                    d=d-(anp1-1d0)*anbarj*varpj
202                    d=d/(anp1*(anbarj-1d0))
203                 a=anbarj*varpj
204                    a=(a-d)/ancj
205 c
206                 term1j=term1j+a
207                 term2j=term2j+a+d
208    62        continue
209              if(term2j.eq.0)go to 63
210              thtj=thtj+term1j/term2j
211              thtjj=thtjj+(term1j/term2j)**2
212    63     continue
213           if(term2j.eq.0d0)go to 65
214           write(3,610)
215   610     format(2x,' locus',' theta')
216 C
217           thtjj=dsqrt(anp1*(thtjj-thtj**2/anp)/anp)
218           thtj=anp*thtl(il)-anp1*thtj/anp
219 c
220           write(3,620)il,thtj
221   620     format(2x,i3,f10.4,' Mean')
222           write(3,630)thtjj
223   630     format(5x,f10.4,' Std. Dev.')
224    65 continue
225 c
226 c
227 c      Jackknife over loci
228 c      Leave out each locus in turn
229 c
230 c
231        write(3,300)
232        write(3,300)
233        write(3,700)
234   700 format(' Jackknifing over loci')
235        anl1=anl-1d0
236 c
237        thtj=0d0
238        thtjj=0d0
239 c
240        do 75 il=1,nl
241           term1j=tterm1-term1(il)
242           term2j=tterm2-term2(il)
243 c
244           if(term2j.eq.0)go to 75
245           thtj=thtj+term1j/term2j
246           thtjj=thtjj+(term1j/term2j)**2
247 C
248    75     continue
249        write(3,710)
```

```
250    710 format(10x,'theta')
251        thtjj=dsqrt(anl1*(thtjj-thtj**2/anl)/anl)
252        thtj=anl*thtt-anl1*thtj/anl
253 c
254        write(3,720)thtj
255    720 format('  All ',f9.4,' Mean')
256        write(3,730)thtjj
257    730 format(5x,f10.4,' Std. Dev.')
258 c
259 c
260     99 continue
261 c
262 c
263 c      Bootstrap over loci
264 c      nb=bootstrap samples, indexed by ib
265 c      ixx is seed for random number generator
266 c
267 c
268        write(3,300)
269        write(3,300)
270        write(3,800)
271    800    format(' Bootstrapping over loci')
272        write(3,300)
273 c
274        ixx=900307
275        il1=ran(ixx)
276 c
277        mb=0
278        do 85 ib=1,nb
279           term1b=0d0
280           term2b=0d0
281           do 80 il=1,nl
282              il1=int(anl*ran(il))+1
283              term1b=term1b+term1(il1)
284              term2b=term2b+term2(il1)
285     80    continue
286 c
287              if(term2b.eq.0)go to 85
288              mb=mb+1
289              thtb(mb)=term1b/term2b
290 c
291     85 continue
292        mbl=int(float(mb)/40d0)
293        mbu=int(39d0*float(mb)/40d0)+1
294 c
295        do 86 ib=1,mb
296           x(ib)=thtb(ib)
297     86 continue
298        call sort
299        thtbl=x(mbl)
300        thtbu=x(mbu)
301 c
302        write(3,810)thtbl,thtbu
303    810   format('  theta 95% CI:',2f10.4)
304 c
305        stop
306        end
```

```
307 c
308 c
309 c     Sorting subroutine
310 c     orders a set of numbers x(i), i=1,...,n from
311 c        smallest to largest
312 c
313 c
314       subroutine sort
315 c
316 c
317       common mb,x(1000)
318       dimension y(1000)
319 c
320       do 15 i=1,mb
321          y(i)=x(1)
322          index=1
323          m=mb-i+1
324          do 10 j=2,m
325             if(x(j).ge.y(i))go to 10
326             y(i)=x(j)
327             index=j
328    10    continue
329          k=1
330          do 15 j=1,m
331             if(j.eq.index)go to 15
332             x(k)=x(j)
333             k=k+1
334    15 continue
335 c
336       do 20 i=1,mb
337          x(i)=y(i)
338    20 continue
339 c
340       return
341       end
342 c
343 c
344 c     Uniform pseudorandom number generator
345 c     Provided by J. Monahan, Statistics Dept, N.C. State University
346 c       From Schrage, ACM Trans. Math. Software 5:132-138 (1979)
347 c     first call sets seed to ixx, later ixx ignored
348 c
349 c
350       real function ran(ixx)
351 c
352 c
353       integer a,p,ix,b15,b16,xhi,xalo,leftlo,fhi,k
354       data a/16807/,b15/32768/,b16/65536/,p/2147483647/
355       data ix/0/
356 c
357       if(ix.eq.0)ix=ixx
358       xhi=ix/b16
359       xalo=(ix-xhi*b16)*a
360       leftlo=xalo/b16
361       fhi=xhi*a+leftlo
362       k=fhi/b15
363       ix=(((xalo-leftlo*b16)-p)+(fhi-k*b15)*b16)+k
```

```
364        if(ix.lt.0)ix=ix+p
365        ran=float(ix)*4.656612875e-10
366        return
367        end
```

Diploid data

The program DIPLOID.DAT estimates the three F-statistics, $F = F_{IT}, \theta = F_{ST}, f = F_{IS}$ from diploid data collected on individuals within populations. The estimation procedure on page 154 is used. Numerical resampling over loci or populations provides information about properties of the estimates.

As an example, the input data file DIPLOID.DAT

```
␣6␣5␣4
␣␣1␣44␣43␣43␣33␣44
␣␣1␣44␣44␣43␣33␣44
␣␣1␣44␣44␣43␣43␣44
␣␣1␣44␣44␣␣␣␣33␣44
␣␣1␣44␣44␣24␣34␣44
␣␣1␣44␣44␣␣␣␣43␣44
␣␣1␣44␣44␣43␣43␣44
␣␣1␣44␣44␣␣␣␣43␣44
␣␣2␣44␣44␣33␣32␣44
␣␣2␣44␣33␣44␣43␣44
␣␣2␣44␣43␣44␣43␣44
␣␣2␣44␣44␣33␣33␣44
␣␣2␣44␣43␣44␣44␣44
␣␣2␣44␣44␣44␣22␣44
␣␣2␣44␣44␣43␣43␣44
␣␣2␣44␣44␣44␣44␣44
␣␣3␣44␣44␣44␣43␣44
␣␣3␣44␣44␣44␣44␣44
␣␣3␣44␣44␣43␣21␣44
␣␣3␣44␣44␣33␣43␣44
␣␣3␣44␣44␣43␣21␣44
␣␣4␣44␣44␣43␣44␣44
␣␣4␣44␣44␣43␣43␣44
␣␣4␣44␣44␣43␣43␣44
␣␣4␣44␣44␣43␣44␣44
␣␣4␣44␣44␣43␣44␣44
␣␣4␣44␣44␣44␣33␣44
␣␣4␣44␣44␣44␣44␣44
␣␣5␣44␣44␣44␣21␣44
␣␣5␣44␣44␣44␣33␣44
␣␣5␣44␣44␣43␣43␣44
␣␣5␣44␣44␣43␣43␣44
␣␣5␣44␣44␣44␣44␣44
␣␣5␣44␣44␣44␣43␣44
␣␣5␣44␣44␣43␣43␣44
␣␣5␣44␣44␣44␣␣␣␣44
␣␣5␣44␣43␣44␣43␣44
␣␣6␣44␣44␣44␣43␣44
␣␣6␣44␣44␣43␣33␣44
␣␣6␣44␣44␣44␣32␣44
```

```
␣␣6␣44␣44␣43␣41␣44
␣␣6␣44␣44␣44␣44␣44
␣␣6␣44␣44␣44␣42␣44
␣␣6␣44␣44␣44␣43␣44
```

leads to the output in file DIPLOID.OUT:

```
For locus   2
allele    capf      theta     smallf
   1      0.0000    0.0000    0.0000
   2      0.0000    0.0000    0.0000
   3      0.3025    0.0694    0.2505
   4      0.3025    0.0694    0.2505
All       0.3025    0.0694    0.2505

For locus   3
allele    capf      theta     smallf
   1      0.0000    0.0000    0.0000
   2      0.0057    0.0369   -0.0324
   3     -0.0165   -0.0102   -0.0062
   4     -0.0453    0.0171   -0.0635
All      -0.0301    0.0046   -0.0348

For locus   4
allele    capf      theta     smallf
   1     -0.0298    0.0465   -0.0800
   2      0.1858    0.0111    0.1767
   3      0.0072    0.0478   -0.0426
   4      0.0268    0.0018    0.0250
All       0.0367    0.0243    0.0128

Over all loci
          capf      theta     smallf
          0.0420    0.0222    0.0202


Jackknifing over populations

For locus   2
          capf      theta     smallf
Total     0.6688    0.4164    0.3567 Means
          0.2256    0.0000    0.2256 Std. Devs.

For locus   3
          capf      theta     smallf
Total     0.0032    0.0276   -0.0253 Means
          0.2636    0.0000    0.2636 Std. Devs.

For locus   4
          capf      theta     smallf
Total     0.1580    0.1458    0.0141 Means
          0.1061    0.0000    0.1061 Std. Devs.


Jackknifing over loci
```

```
        capf      theta     smallf
Total   0.0306    0.0209    0.0096 Means
        0.0462    0.0105    0.0377 Std. Devs.


Bootstrapping over loci

  cap-f 95% CI:   -0.0301    0.3025
  theta 95% CI:    0.0046    0.0694
small-f 95% CI:   -0.0348    0.2505
```

The program listing follows:

```
  1 c      DIPLOID.FOR
  2 c
  3 c      Estimation of f-statistics for one-level hierarchy
  4 c
  5 c      Use in conjunction with B.S. Weir
  6 c      "Intraspecfic Differentiation"
  7 c      in
  8 c      "Molecular Systematics"
  9 c      Edited by D. Hillis and C. Moritz
 10 c      Sinauer, Sunderland, MA 1990
 11 c
 12 c
 13 c      np populations, indexed by ip
 14 c         Replace 6 by at least np in dimension statements
 15 c      nl loci, indexed by il
 16 c         Replace 5 by at least nl in dimension statements
 17 c      up to nu alleles per locus, indexed by iu
 18 c         Replace 4 by at least nu in dimension statements
 19 c
 20 c
 21        implicit real*8(a-h,o-z)
 22        common mb,x(1000)
 23        real ran,x
 24        dimension an(6,5),h(6,5,4),p(6,5,4)
 25        dimension ia1(5),ia2(5)
 26        dimension anbar(5),annbar(5),anc(5)
 27        dimension term1(5),term2(5),term3(5),term4(5),term5(5)
 28        dimension pbar(5,4),ppbar(5,4),varp(5,4),hbar(5,4)
 29        dimension capf(5,4),theta(5,4),smallf(5,4)
 30        dimension capfl(5),thetal(5),smalfl(5)
 31        dimension capfb(1000),thetb(1000),smlfb(1000)
 32 c
 33        open(1,file='diploid.dat',status='old')
 34        open(3,file='diploid.out',status='unknown')
 35 c
 36 c
 37 c      Data is read in from file DIPLOID.DAT
 38 c      First line contains three parameters: np, nl and nu
 39 c          (in format 3i3)
 40 c      Each suceeding line contains population identifier
 41 c          (in format i3)
 42 c      and then the nl genortypes
 43 c          (in format 2i1, with a blank between each locus)
 44 c      Alleles are called ia1(il) and ia2(il) for locus il
```

```
 45 c     Use a zero or a blank for a locus that is not scored
 46 c
 47 c
 48       read(1,100)np,nl,nu
 49   100    format(3i2)
 50       anp=np
 51       anl=nl
 52       nb=1000
 53 c
 54 c
 55 c     Set counts to zero
 56 c     For locus il in population ip:
 57 c          an(ip,il) is sample size
 58 c          h(ip,il,iu) is number of heterozygotes for allele iu
 59 c          p(ip,il,iu) is number of alleles of type iu
 60 c
 61 c
 62       do 10 ip=1,np
 63          do 10 il=1,nl
 64             an(ip,il)=0d0
 65             do 10 iu=1,nu
 66                h(ip,il,iu)=0d0
 67                p(ip,il,iu)=0d0
 68    10 continue
 69 c
 70    20 read(1,200,end=25)ip,(ia1(il),ia2(il),il=1,nl)
 71   200 format(i3,6(1X,2i1))
 72 c
 73 c     The "6" in the last line needs to be set to np
 74 c
 75 c
 76 c     Increase counts of individuals, heterozygotes and alleles
 77 c
 78 c
 79       do 21 il=1,nl
 80          if(ia1(il).eq.0.or.ia2(il).eq.0)go to 21
 81          an(ip,il)=an(ip,il)+1d0
 82          iu=ia1(il)
 83          iv=ia2(il)
 84          p(ip,il,iu)=p(ip,il,iu)+1d0
 85          p(ip,il,iv)=p(ip,il,iv)+1d0
 86          if(iu.eq.iv)go to 21
 87          h(ip,il,iu)=h(ip,il,iu)+1d0
 88          h(ip,il,iv)=h(ip,il,iv)+1d0
 89    21 continue
 90       go to 20
 91    25 continue
 92 c
 93 c
 94 c     Change counts to frequencies
 95 c     Form average sample sizes for each locus
 96 c     Form average and variance of allelic frequencies
 97 c     Form average heterozygosities
 98 c
 99 c
100       do 30 ip=1,np
101          do 30 il=1,nl
```

```
102           do 30 iu=1,nu
103              p(ip,il,iu)=p(ip,il,iu)/(2d0*an(ip,il))
104              h(ip,il,iu)=h(ip,il,iu)/an(ip,il)
105    30 continue
106 c
107       do 35 il=1,nl
108          anbar(il)=0d0
109          annbar(il)=0d0
110          do 31 ip=1,np
111             anbar(il)=anbar(il)+an(ip,il)
112             annbar(il)=annbar(il)+an(ip,il)**2
113    31    continue
114          if(anp.le.1d0)go to 99
115          anbar(il)=anbar(il)/anp
116          anc(il)=(anp*anbar(il)-annbar(il)/(anp*anbar(il)))/(anp-1d0)
117 c
118          do 33 iu=1,nu
119             pbar(il,iu)=0d0
120             ppbar(il,iu)=0d0
121             hbar(il,iu)=0d0
122             do 32 ip=1,np
123                pbar(il,iu)=pbar(il,iu)+an(ip,il)*p(ip,il,iu)
124                ppbar(il,iu)=ppbar(il,iu)+an(ip,il)*p(ip,il,iu)**2
125                hbar(il,iu)=hbar(il,iu)+an(ip,il)*h(ip,il,iu)
126    32       continue
127             pbar(il,iu)=pbar(il,iu)/(anp*anbar(il))
128             varp(il,iu)=ppbar(il,iu)-anp*anbar(il)*pbar(il,iu)**2
129             varp(il,iu)=varp(il,iu)/((anp-1d0)*anbar(il))
130             hbar(il,iu)=hbar(il,iu)/(anp*anbar(il))
131    33    continue
132    35 continue
133 c
134 c
135 c     Form variance components for each allele at each locus
136 c     Estimate f-statistics for each allele
137 c     Then combine over alleles
138 c     Test for fixed loci
139 c     Write estimates
140 c
141 c
142       tterm1=0d0
143       tterm2=0d0
144       tterm3=0d0
145       tterm4=0d0
146       tterm5=0d0
147       do 42 il=1,nl
148          term1(il)=0d0
149          term2(il)=0d0
150          term3(il)=0d0
151          term4(il)=0d0
152          term5(il)=0d0
153          do 40 iu=1,nu
154             if(pbar(il,iu).le.1d-4.or.pbar(il,iu).gt.9999d-4)go to 40
155             a=pbar(il,iu)*(1d0-pbar(il,iu))-(anp-1d0)*varp(il,iu)/anp
156             b=a
157             a=a-hbar(il,iu)/4d0
158             a=anbar(il)*(varp(il,iu)-a/(anbar(il)-1d0))/anc(il)
```

```
159             b=b-(2d0*anbar(il)-1d0)*hbar(il,iu)/(4d0*anbar(il))
160             b=anbar(il)*b/(anbar(il)-1d0)
161             c=hbar(il,iu)/2d0
162             capf(il,iu)=(a+b)/(a+b+c)
163             theta(il,iu)=a/(a+b+c)
164             smallf(il,iu)=b/(b+c)
165             term1(il)=term1(il)+a+b
166             term2(il)=term2(il)+a+b+c
167             term3(il)=term3(il)+a
168             term4(il)=term4(il)+b
169             term5(il)=term5(il)+b+c
170    40     continue
171           if(term2(il).eq.0d0)go to 42
172           if(term5(il).eq.0d0)go to 42
173           write(3,300)
174   300     format('   ')
175           write(3,305)il
176   305     format(' For locus ',i3)
177           write(3,310)
178   310     format(' allele',3x,'capf',6x,'theta',4x,'smallf')
179           capfl(il)=term1(il)/term2(il)
180           thetal(il)=term3(il)/term2(il)
181           smalfl(il)=term4(il)/term5(il)
182 c
183
184           do 41 iu=1,nu
185             write(3,330)iu,capf(il,iu),theta(il,iu),smallf(il,iu)
186   330       format(i5,3f10.4)
187    41     continue
188           write(3,340)capfl(il),thetal(il),smalfl(il)
189   340     format(' All    ',f9.4,2f10.4)
190           tterm1=tterm1+term1(il)
191           tterm2=tterm2+term2(il)
192           tterm3=tterm3+term3(il)
193           tterm4=tterm4+term4(il)
194           tterm5=tterm5+term5(il)
195    42 continue
196       tcapf=tterm1/tterm2
197       ttheta=tterm3/tterm2
198       tsmalf=tterm4/tterm5
199       write(3,300)
200       write(3,350)
201   350 format(' Over all loci')
202       write(3,360)
203   360 format(9x,' capf',6x,'theta',4x,'smallf')
204       write(3,370)tcapf,ttheta,tsmalf
205   370 format(5x,3f10.4)
206 c
207 c
208 c     Jackknife over populations
209 c     Leave out each population in turn
210 c
211 c
212       write(3,300)
213       write(3,300)
214       write(3,400)
215   400 format(' Jackknifing over populations')
```

```
216        do 65 il=1,nl
217           anbar(il)=anbar(il)*anp
218           anp1=anp-1d0
219           anp2=anp1-1d0
220           capflj=0d0
221           capfjj=0d0
222           thetlj=0d0
223           thtljj=0d0
224           smlflj=0d0
225           smfljj=0d0
226           do 61 iu=1,nu
227              pbar(il,iu)=pbar(il,iu)*anbar(il)
228              hbar(il,iu)=hbar(il,iu)*anbar(il)
229     61    continue
230           do 63 ip=1,np
231              anbarj=(anbar(il)-an(ip,il))/anp1
232              annbaj=annbar(il)-an(ip,il)**2
233              ancj=(anp1*anbarj-annbaj/(r1*anbarj))/(anp1-1d0)
234 c
235              term1j=0d0
236              term2j=0d0
237              term3j=0d0
238              term4j=0d0
239              term5j=0d0
240              do 62 iu=1,nu
241                 pbarj=pbar(il,iu)-an(ip,il)*p(ip,il,iu)
242                 ppbarj=ppbar(il,iu)-an(ip,il)*p(ip,il,iu)**2
243                 hbarj=hbar(il,iu)-an(ip,il)*h(ip,il,iu)
244 c
245                 pbarj=pbarj/(anp1*anbarj)
246                 varpj=(ppbarj-anp1*anbarj*pbarj**2)/(anp2*anbarj)
247                 hbarj=hbarj/(anp1*anbarj)
248                 if(pbarj.le.1d-4.or.pbarj.ge.9999d-4)go to 62
249 c
250                 a=pbarj*(1d0-pbarj)-(anp1-1d0)*varpj/anp1
251                 b=a
252                 a=anbarj*(varpj-(a-hbarj/4d0)/(anbarj-1d0))/ancj
253                 b=b-(2d0*anbarj-1d0)*hbarj/(4d0*anbarj)
254                 b=anbarj*b/(anbarj-1d0)
255                 c=hbarj/2d0
256 c
257                 term1j=term1j+a+b
258                 term2j=term2j+a+b+c
259                 term3j=term3j+a
260                 term4j=term4j+b
261                 term5j=term5j+b+c
262     62       continue
263              if(term2j.eq.0)go to 63
264              if(term5j.eq.0)go to 63
265              capflj=capflj+term1j/term2j
266              capfjj=capfjj+(term1j/term2j)**2
267              thetlj=thetlj+term3j/term2j
268              thtljj=thtljj+(term3j/term2j)**2
269              smlflj=smlflj+term4j/term5j
270              smfljj=smfljj+(term4j/term5j)**2
271     63    continue
272           if(term2j.eq.0d0)go to 65
```

```
273           if(term5j.eq.0d0)go to 65
274           write(3,300)
275           write(3,305)il
276           write(3,410)
277   410     format(10x,'capf',6x,'theta',4x,'smallf')
278 c
279           capfjj=dsqrt(anp1*(capfjj-capflj**2/anp)/anp)
280           capflj=anp*capfl(il)-anp1*capflj/anp
281           thtljj=dsqrt(anp1*(thtljj-thetlj**2/anp)/anp)
282           thetlj=anp*thetal(il)-anp1*thetlj/anp
283           smfljj=dsqrt(anp1*(smfljj-smlflj**2/anp)/anp)
284           smlflj=anp*smalfl(il)-anp1*smlflj/anp
285 c
286           write(3,420)capflj,thetlj,smlflj
287   420     format(' Total',f9.4,2f10.4,' Means')
288           write(3,430)capfjj,thtljj,smfljj
289   430     format(5x,3f10.4,' Std. Devs.')
290    65 continue
291 c
292 c
293 c     Jackknife over loci
294 c     Leave out each locus in turn
295 c
296 c
297       write(3,300)
298       write(3,300)
299       write(3,500)
300   500 format(' Jackknifing over loci')
301       anl1=anl-1d0
302 c
303       capflj=0d0
304       capfjj=0d0
305       thetlj=0d0
306       thetjj=0d0
307       smlflj=0d0
308       smlfjj=0d0
309 c
310       do 75 il=1,nl
311          term1j=tterm1-term1(il)
312          term2j=tterm2-term2(il)
313          term3j=tterm3-term3(il)
314          term4j=tterm4-term4(il)
315          term5j=tterm5-term5(il)
316 c
317          if(term2j.eq.0)go to 75
318          if(term5j.eq.0)go to 75
319          capflj=capflj+term1j/term2j
320          capfjj=capfjj+(term1j/term2j)**2
321          thetlj=thetlj+term3j/term2j
322          thetjj=thetjj+(term3j/term2j)**2
323          smlflj=smlflj+term4j/term5j
324          smlfjj=smlfjj+(term4j/term5j)**2
325 c
326     75   continue
327 c
328       write(3,300)
329       write(3,510)
```

```
330    510 format(10x,'capf',6x,'theta',4x,'smallf')
331        capfjj=dsqrt(anl1*(capfjj-capflj**2/anl)/anl)
332        capflj=anl*tcapf-anl1*capflj/anl
333        thetjj=dsqrt(anl1*(thetjj-thetlj**2/anl)/anl)
334        thetlj=anl*ttheta-anl1*thetlj/anl
335        smlfjj=dsqrt(anl1*(smlfjj-smlflj**2/anl)/anl)
336        smlflj=anl*tsmalf-anl1*smlflj/anl
337 c
338        write(3,520)capflj,thetlj,smlflj
339    520 format(' Total',f9.4,2f10.4,' Means')
340        write(3,530)capfjj,thetjj,smlfjj
341    530 format(5x,3f10.4,' Std. Devs.')
342 c
343 c
344     99 continue
345 c
346 c
347 c      Bootstrap over loci
348 c      nb=bootstrap samples, indexed by ib
349 c      ixx is seed for random number generator
350 c
351 c
352        write(3,300)
353        write(3,300)
354        write(3,600)
355   600     format(' Bootstrapping over loci')
356        write(3,300)
357 c
358        ixx=890613
359        il1=ran(ixx)
360 c
361        mb=0
362        do 85 ib=1,nb
363           term1b=0d0
364           term2b=0d0
365           term3b=0d0
366           term4b=0d0
367           term5b=0d0
368           do 80 il=1,nl
369              il1=int(anl*ran(il))+1
370              term1b=term1b+term1(il1)
371              term2b=term2b+term2(il1)
372              term3b=term3b+term3(il1)
373              term4b=term4b+term4(il1)
374              term5b=term5b+term5(il1)
375    80     continue
376 c
377              if(term2b.eq.0)go to 85
378              if(term5b.eq.0)go to 85
379              mb=mb+1
380              capfb(mb)=term1b/term2b
381              thetb(mb)=term3b/term2b
382              smlfb(mb)=term4b/term5b
383 c
384     85 continue
385       mbl=int(float(mb)/40d0)
386       mbu=int(39d0*float(mb)/40d0)+1
```

```
387 c
388       do 86 ib=1,mb
389          x(ib)=capfb(ib)
390    86 continue
391       call sort
392       cafl=x(mbl)
393       cafu=x(mbu)
394 c
395       do 87 ib=1,mb
396          x(ib)=thetb(ib)
397    87 continue
398       call sort
399       thetl=x(mbl)
400       thetu=x(mbu)
401 c
402       do 88 ib=1,mb
403          x(ib)=smlfb(ib)
404    88 continue
405       call sort
406       smlfl=x(mbl)
407       smlfu=x(mbu)
408 c
409       write(3,610)cafl,cafu
410   610   format('   cap-f 95% CI:',2f10.4)
411       write(3,620)thetl,thetu
412   620   format('   theta 95% CI:',2f10.4)
413       write(3,630)smlfl,smlfu
414   630   format(' small-f 95% CI:',2f10.4)
415 c
416       stop
417       end
418 c
419 c
420 c     Sorting subroutine
421 c     orders a set of numbers x(i), i=1,...,n from
422 c        smallest to largest
423 c
424 c
425       subroutine sort
426 c
427 c
428       common mb,x(1000)
429       dimension y(1000)
430 c
431       do 15 i=1,mb
432          y(i)=x(1)
433          index=1
434          m=mb-i+1
435          do 10 j=2,m
436             if(x(j).ge.y(i))go to 10
437             y(i)=x(j)
438             index=j
439    10    continue
440          k=1
441          do 15 j=1,m
442             if(j.eq.index)go to 15
443             x(k)=x(j)
```

```
444              k=k+1
445      15 continue
446 c
447         do 20 i=1,mb
448            x(i)=y(i)
449      20 continue
450 c
451         return
452         end
453 c
454 c
455 c       Uniform pseudorandom number generator
456 c       Provided by J. Monahan, Statistics Dept, N.C. State University
457 c         From Schrage, ACM Trans. Math. Software 5:132-138 (1979)
458 c
459 c       first call sets seed to ixx, later ixx ignored
460 c
461 c
462         real function ran(ixx)
463 c
464 c
465         integer a,p,ix,b15,b16,xhi,xalo,leftlo,fhi,k
466         data a/16807/,b15/32768/,b16/65536/,p/2147483647/
467         data ix/0/
468 c
469         if(ix.eq.0)ix=ixx
470         xhi=ix/b16
471         xalo=(ix-xhi*b16)*a
472         leftlo=xalo/b16
473         fhi=xhi*a+leftlo
474         k=fhi/b15
475         ix=(((xalo-leftlo*b16)-p)+(fhi-k*b15)*b16)+k
476         if(ix.lt.0)ix=ix+p
477         ran=float(ix)*4.656612875e-10
478         return
479         end
```

SEQUENCE MANIPULATIONS

Shuffling Sequences

The program SHUFFLE.FOR shuffles a sequence of length n by a series of $n-1$ transpositions of the bases at pairs of positions.

As an example, the sequence in Figure 7.2 is placed in file SHUFFLE.DAT in the following format:

```
␣␣438␣␣␣␣1
␣GTGCACTGGA␣CTGCTGAGGA␣GAAGCAGCTC␣ATCACCGGCC␣TCTGGGGCAA␣GGTCAATGTG
␣GCCGAATGTG␣GGGCCGAAGC␣CCTGGCCAGG␣CTGCTGATCG␣TCTACCCCTG␣GACCCAGAGG
␣TTCTTTGCGT␣CCTTTGGGAA␣CCTCTCCAGC␣CCCACTGCCA␣TCCTTGGCAA␣CCCCATGGTC
␣CGCGCCCACG␣GCAAGAAAGT␣GCTCACCTCC␣TTTGGGGATG␣CTGTGAAGAA␣CCTGGACAAC
␣ATCAAGAACA␣CCTTCTCCCA␣ACTGTCCGAA␣CTGCATTGTG␣ACAAGCTGCA␣TGTGGACCCC
␣GAGAACTTCA␣GGCTCCTGGG␣TGACATCCTC␣ATCATTGTCC␣TGGCCGCCCA␣CTTCAGCAAG
␣GACTTCACTC␣CTGAATGCCA␣GGCTGCCTGG␣CAGAAGCTGG␣TCCGCGTGGT␣GGCCCATGCC
```

```
␣CTGGCTCGCA␣AGTACCAC
```

and the shuffled sequence is written to file SHUFFLE.OUT:

```
GGGGACTTTC CGGCGCGCTT CTTGGCGTAT CTTTCACTCC ATCGAAAGGG GGTCCAACAG
CCTGTCGCTG CGACGCCGTT TGGGAGACGT GTGATATGGG CCGCATGCTG CTTGCACGCG
CGCCCGTGCC GATCTTAGCC CGTTTGCGCA CAGAATCGAA CCTTAGTCCA GAGCTGCGAA
TCTCGCGGGT ATATGAGCAA AGTAGCACCC CGTAGTACCC CGACCGGGTC TACGACGACT
CACGGCACTC CACGGAGCCC CAGATACCCC GCGCTACACG CCGTACGTAA ACTCGCACCC
TGCTGGGTTG CGCCGCTGGC TTTGGCAGAG TATCGCAGCC ACGCATAAGA AGCCCGTGCA
CCAACACGGA TCGACTCCAC ATGACGGTAG CGCAATCACC GGCCGCTTGT ATTTCCTAAG
TCCTTTAGAC TACTTGCC
```

The program listing follows:

```
 1 c      SHUFFLE.FOR
 2 c
 3 c      Shuffling a sequence of n bases
 4 c
 5 c
 6        implicit real*8(a-h,o-z)
 7        character s(1000),mhold
 8 c
 9 c      Read sequence length and number of shuffled sequences required
10 c      Then read sequence
11 c
12        open(1,file='SHUFFLE.DAT',status='old')
13        open(3,file='SHUFFLE.OUT',status='unknown')
14        read(1,110)n,ishuff
15   110     format(2i5)
16        read(1,120)(s(i),i=1,n)
17   120     format(6(1x,10a1))
18 c
19 c
20 c      Shuffle ishuff times
21 c
22 c      ixx is seed for random number generator
23 c
24 c
25        ixx=900307
26        iy=ran(ixx)
27        do 30 iss=1,ishuff
28 C
29 C      Draw n-1 transpositions
30 C
31
32        n1=n-1
33        do 10 i=1,n1
34           j=n-i+1
35           aj=j
36           iy=int(aj*ran(iy))+1
37           mhold=s(iy)
38           s(iy)=s(j)
39           s(j)=mhold
40    10     continue
41           write(3,120)(s(k),k=1,n)
```

```
42 c
43     30 continue
44 c
45        stop
46        end
47 c
48 c
49 c
50 c     Uniform pseudorandom number generator
51 c
52 c     Provided by J. Monahan, Statistics Dept, N.C. State University
53 c       From Schrage, ACM Trans. Math. Software 5:132-138 (1979)
54 c
55 c
56        real function ran(ixx)
57 c
58 c
59        integer a,p,ix,b15,b16,xhi,xalo,leftlo,fhi,k
60        data a/16807/,b15/32768/,b16/65536/,p/2147483647/
61        data ix/0/
62 c
63        if(ix.eq.0)ix=ixx
64        xhi=ix/b16
65        xalo=(ix-xhi*b16)*a
66        leftlo=xalo/b16
67        fhi=xhi*a+leftlo
68        k=fhi/b15
69        ix=(((xalo-leftlo*b16)-p)+(fhi-k*b15)*b16)+k
70        if(ix.lt.0)ix=ix+p
71        ran=float(ix)*4.656612875e-10
72        return
73        end
```

Karlin's Algorithm

The program KARLIN.FOR finds all direct repeats, of length greater than a specified value k, along a DNA sequence. The algorithm described on pages 240–243 is used.

As an example, the chicken β-globin sequence of Figure 7.2 is put in the input file KARLIN.DAT in the format:

```
ACGT
␣␣438␣␣␣␣8
␣GTGCACTGGA␣CTGCTGAGGA␣GAAGCAGCTC␣ATCACCGGCC␣TCTGGGGCAA␣GGTCAATGTG
␣GCCGAATGTG␣GGGCCGAAGC␣CCTGGCCAGG␣CTGCTGATCG␣TCTACCCCTG␣GACCCAGAGG
␣TTCTTTGCGT␣CCTTTGGGAA␣CCTCTCCAGC␣CCCACTGCCA␣TCCTTGGCAA␣CCCCATGGTC
␣CGCGCCCACG␣GCAAGAAAGT␣GCTCACCTCC␣TTTGGGGATG␣CTGTGAAGAA␣CCTGGACAAC
␣ATCAAGAACA␣CCTTCTCCCA␣ACTGTCCGAA␣CTGCATTGTG␣ACAAGCTGCA␣TGTGGACCCC
␣GAGAACTTCA␣GGCTCCTGGG␣TGACATCCTC␣ATCATTGTCC␣TGGCCGCCCA␣CTTCAGCAAG
␣GACTTCACTC␣CTGAATGCCA␣GGCTGCCTGG␣CAGAAGCTGG␣TCCGCGTGGT␣GGCCCATGCC
␣CTGGCTCGCA␣AGTACCAC
```

and results in the output file KARLIN.OUT:

```
k is  8
In positions   79  418
         match is GCCCTGGC
In positions   85  377
         match is GCCAGGCT
In positions   86  378
         match is CCAGGCTG
In positions   87  379
         match is CAGGCTGC
In positions  130  208
         match is TCCTTTGG
In positions  131  209
         match is CCTTTGGG
In positions  176  398
         match is TGGTCCGC
In positions  177  399
         match is GGTCCGCG
k is  9
In positions   85  377
         match is GCCAGGCTG
In positions   86  378
         match is CCAGGCTGC
In positions  130  208
         match is TCCTTTGGG
In positions  176  398
         match is TGGTCCGCG
k is 10
In positions   85  377
         match is GCCAGGCTGC
k is 11
longest length is 10
```

The program listing follows:

```
  1 c      KARLIN.FOR
  2 c
  3 c      Karlin's method for searching for direct repeats
  4 c
  5 c      Use in conjunction with S. Karlin et al.
  6 c      Proc. Natl. Acad. Sci. USA 80:5660-5664, 1983
  7 c
  8 c
  9        implicit real*8(a-h,o-z)
 10        character s(1000)
 11        character b(4)
 12        dimension ipos(1000,5),ipos2(1000,5)
 13        dimension ialpha(1000),ia(4),ic(1000),iw(1000),iws(1000),inw(1000)
 14        dimension nw(1000),iwn(1000),ic2(1000)
 15 c
 16        open(1,file='KARLIN.DAT',status='old')
 17        open(3,file='KARLIN.OUT',status='unknown')
 18 c
 19 c
 20 c      Read data
 21 c
 22 c      First line lists the four base types
 23 c      Second line lists sequence length and initial word length
```

```
24 c      Subsequent lines list sequence: 6 blocks of 10 bases per
25 c      line, with a blank before each block.
26 c
27 c
28        read(1,100)(b(i),i=1,4)
29   100     format(4a1)
30        ia(1)=0
31        ia(2)=1
32        ia(3)=2
33        ia(4)=3
34      1 read(1,110,end=1000)n,k
35   110     format(2i5)
36        read(1,130)(s(i),i=1,n)
37   130     format(6(1x,10a1))
38 c
39 c
40 c      convert sequence of letters to numbers
41 c
42 c
43        do 5 i=1,n
44           do 5 j=1,4
45              if(s(i).eq.b(j))ialpha(i)=ia(j)
46      5 continue
47 c
48 c
49 c      list all nwk distinct k-words and their positions
50 c
51 c
52        write(3,200)k
53   200     format(' k is',i3)
54        nk1=n-k+1
55        do 10 i=1,nk1
56           ic(i)=0
57    10 continue
58 c
59           do 15 iwpos=1,nk1
60              iword=1
61              do 11 i=1,k
62                 isite=iwpos+i-1
63                 iword=iword+ialpha(isite)*4**(k-i)
64    11        continue
65              iw(iwpos)=iword
66              if(iwpos.eq.1)then
67                 nwk=1
68                 nw(1)=1
69                 iwn(1)=iword
70                 go to 14
71              endif
72              iwpos1=iwpos-1
73              do 12 j=1,iwpos1
74                 if(iword.eq.iw(j))go to 13
75    12        continue
76              nwk=nwk+1
77              nw(iwpos)=nwk
78              iwn(nwk)=iword
79              go to 14
80    13        continue
```

```
 81                   nw(iwpos)=nw(j)
 82     14            continue
 83                   iws(iwpos)=iword
 84                   nwipos=nw(iwpos)
 85                   ic(nwipos)=ic(nwipos)+1
 86                     icw=ic(nwipos)
 87                   ipos(nwipos,icw)=iwpos
 88     15         continue
 89                do 16 i=1,nwk
 90                   icw=ic(i)
 91                   if(icw.eq.1)go to 16
 92                   is1=ipos(i,1)
 93                   is2=ipos(i,1)+k-1
 94                   write(3,210)(ipos(i,j),j=1,icw)
 95    210               format(' In positions', 10i5)
 96                   write(3,220)(s(j),j=is1,is2)
 97    220               format(10x,'match is ',20a1)
 98     16         continue
 99 c
100 c
101 c     increment k
102 c
103 c
104     19 continue
105        k=k+1
106        write(3,200)k
107        nk1=n-k+1
108        do 20 i=1,nk1
109           ic2(i)=0
110     20 continue
111        jpi=0
112        do 26 i=1,nwk
113           icw=ic(i)
114           if(icw.lt.2)go to 26
115           do 25 j=1,icw
116           iwpos=ipos(i,j)
117           if(iwpos.gt.nk1)go to 25
118           jpi=jpi+1
119           iword=1
120           do 21 ij=1,k
121              isite=iwpos+ij-1
122              iword=iword+ialpha(isite)*4**(k-ij)
123     21    continue
124              iw(jpi)=iword
125              if(jpi.eq.1)then
126                 nwk2=1
127                 nw(1)=1
128                 iwn(1)=iword
129                 go to 24
130              endif
131              jpi1=jpi-1
132              do 22 ij=1,jpi1
133                 if(iword.eq.iw(ij))go to 23
134     22       continue
135              nwk2=nwk2+1
136              nw(jpi)=nwk2
137              iwn(nwk2)=iword
```

```
138              go to 24
139     23       continue
140              nw(jpi)=nw(ij)
141     24       continue
142              iws(jpi)=iword
143              nwipos=nw(jpi)
144              ic2(nwipos)=ic2(nwipos)+1
145                icw=ic2(nwipos)
146              ipos2(nwipos,icw)=iwpos
147     25    continue
148     26    continue
149           do 27 i=1,nwk2
150              icw=ic2(i)
151              if(icw.eq.1)go to 27
152              is1=ipos2(i,1)
153              is2=ipos2(i,1)+k-1
154              write(3,210)(ipos2(i,j),j=1,icw)
155              write(3,220)(s(j),j=is1,is2)
156     27    continue
157 c
158           do 30 i=1,nwk2
159              if(ic2(i).gt.1)go to 31
160     30    continue
161           l=k-1
162           write(3,300)l
163    300       format(' longest length is',i3)
164           go to 1
165     31    do 32 i=1,nwk2
166              ic(i)=ic2(i)
167              icw=ic(i)
168              do 32 j=1,icw
169                 ipos(i,j)=ipos2(i,j)
170     32    continue
171           nwk=nwk2
172           go to 19
173 c
174   1000 continue
175        stop
176        end
```

Bibliography

Abramowitz, M. and I.A. Stegun (Editors). 1970. *Handbook of Mathematical Functions with Formulas, Graphs, and Mathematical Tables.* National Bureau of Standards Applied Math. Series 55. U.S. Government Printing Office, Washington, DC.

Agard, D.A., R.A. Steinberg and R.M. Stroud. 1981. Quantitative analysis of electrophoretograms: A mathematical approach to super-resolution. Anal. Biochem. 111:257–268.

Allard, R.W., A.L. Kahler and B.S. Weir. 1972. The effect of selection on esterase allozymes in a barley population. Genetics 72:489–503.

Arnold, J., A.J. Cuticchia, D.A. Newsome, W.W. Jennings III and R. Ivarie. 1988. Mono- through hexanucleotide composition of the sense strand of yeast DNA: A Markov chain analysis. Nucleic Acids Res. 16:7145–7158.

Arratia, R. and M.S. Waterman. 1985. An Erdös-Rényi law with shifts. Adv. Math. 55:13–23.

Bailey, N.T.J. 1951. Testing the solubility of maximum likelihood equations in the routine application of scoring methods. Biometrics 7:268–274.

Bailey, N.T.J. 1961. *Introduction to the Mathematical Theory of Genetic Linkage.* Oxford Univ. Press, Oxford.

Balakrishnan, V. and L.D. Sanghvi. 1968. Distance between populations on the basis of attribute data. Biometrics 24:859–865.

Balasz, I., M. Baird, M. Clyne and E. Meade. 1989. Human population genetic studies of five hypervariable DNA loci. Am. J. Human Genet. 44:182–190.

Barker, J.S.F., P.D. East and B.S. Weir. 1986. Temporal and microgeographic variation in allozyme frequencies in a natural population of *Drosophila buzzatii.* Genetics 112:577–611.

Bennett, J.H. 1954. On the theory of random mating. Ann. Eugen. 18:311–317.

Bernstein, F. 1925. Zusammenfassende Betrachutungen über die erblichen Blutenstructuren des Menschen. Z. Abstamm. Vererbgsl. 37:237–270.

Bishop, D.T., J.A. Williamson and M.H. Skolnick. 1983. A model for restriction fragment length distribution. Am. J. Human Genet. 35:795–815.

Brooks, L.D., B.S. Weir and H.E. Schaffer. 1988. The probabilities of similarities in DNA sequence comparisons. Genomics 3:207–216.

Brown, A.H.D., A.C. Matheson and K.G. Eldridge. 1975. Estimation of the mating system of *Eucalyptus obliqua* L'Hérit. by using allozyme polymorphisms. Aust. J. Bot. 23:931–949.

Brown, W.M., E.M. Prager, A. Wang and A.C. Wilson. 1982. Mitochondrial DNA sequences of primates: Tempo and mode of evolution. J. Mol. Evol. 18:225–239.

Buri, P. 1956. Gene frequency in small populations of mutant *Drosophila.* Evolution 10:367–402.

Cavalli-Sforza, L.L. and W.F. Bodmer. 1971. *The Genetics of Human Populations.* Freeman, San Francisco.

Cavalli-Sforza, L.L. and A.W.F. Edwards. 1967. Phylogenetic analysis: Models and estimation procedures. Am. J. Human Genet. 19:233–257.

Ceppellini, R., M. Siniscalco and C.A.B. Smith. 1955. The estimation of gene frequencies in a random-mating population. Ann. Human Genet. 20:97–115.

Chakraborty, R. 1986. Estimation of linkage disequilibrium from conditional haplotype data: Application to β-globin gene cluster in American Blacks. Genet. Epidemiol. 3:323–333.

Chakraborty, R., M. Shaw and W.J. Schull. 1974. Exclusion of probability: The current state of the art. Am. J. Human Genet. 26:477–488.

Chakravarti, A., C.C. Li and K.H. Buetow. 1984. Estimation of the marker gene frequency and linkage disequilibrium from conditional marker data. Am. J. Human Genet. 36:177–186.

Cheliak, W.M., K. Morgan, C. Strobeck, F.C.H. Yeh and B.P. Dancik. 1983. Estimation of mating system parameters in plant populations using the EM algorithm. Theor. Appl. Genet. 65:157–161.

Chimera, J.A., C.R. Harris and M. Litt. 1989. Population genetics of the highly polymorphic locus D16S7 and its use in paternity testing. Am. J. Human Genet. 45:926–931.

Christiansen, F.B. and O. Frydenberg. 1973. Selection component analysis of natural polymorphisms using population samples including mother-offspring combinations. Theor. Pop. Biol. 4:425–445.

Clegg, M.T. and R.W. Allard. 1973. The genetics of electrophoretic variants in *Avena.* II. The esterase E_1, E_2, E_4, E_5, E_6 and anodal peroxidase APX_4 loci in *A. fatua.* J. Hered. 64:2–6.

Clegg, M.T., A.L. Kahler and R.W. Allard. 1978. Estimation of life cycle components of selection in an experimental plant population. Genetics 89:765–792.

Cockerham, C.C. 1967. Group inbreeding and coancestry. Genetics 56:89–104.

Cockerham, C.C. 1969. Variance of gene frequencies. Evolution 23:72–84.

Cockerham, C.C. 1973. Analyses of gene frequencies. Genetics 74:679–700.

Cockerham, C.C. 1984. Drift and mutation with a finite number of allelic states. Proc. Natl. Acad. Sci. U.S.A. 81:530–534.

Cockerham, C.C. and B.S. Weir. 1986. Estimation of inbreeding parameters in stratified populations. Ann. Human Genet. 50:271–281.

Cockerham, C.C. and B.S. Weir. 1987. Correlations, descent measures: Drift with migration and mutation. Proc. Natl. Acad. Sci. U.S.A. 84:8512–8514.

Cochran, W.G. 1954. Some methods for strengthening the common χ^2 tests. Biometrics 10:417–451.

Dayhoff, M.O. 1979. Survey of new data and computer methods of analysis. Pp. 1–8 in *Atlas of Protein Sequence and Structure.* Volume 5, Supplement 3, 1978. National Biomedical Foundation, Washington, DC.

Dayhoff, M.O., R.M. Schwartz and B.C. Orcutt. 1979. A model of evolutionary change in proteins. Pp. 345–351 in *Atlas of Protein Sequence and Structure*, Volume 5, Supplement 3, 1978. National Biomedical Research Foundation, Washington, DC.

Degens, P.O. 1983. Hierarchical cluster methods as maximum likelihood estimators. Pp. 249–253 in *Numerical Taxonomy*, J. Felsenstein (Editor), Springer-Verlag, Berlin.

Dempster, A.P., N.M. Laird and D.B. Rubin. 1977. Maximum likelihood from incomplete data via the EM algorithm. J. Roy. Stat. Soc. B 39:1–38.

Dobzhansky, T. and H. Levene. 1951. Development of heterosis through natural selection in experimental populations of *Drosophila pseudoobscura.* Am. Nat. 85:247–263.

Dodds, K.G. 1986. *Resampling Methods in Genetics and the Effect of Family Structure in Genetic Data.* Institute of Statistics Mimeo Series 1684T, North Carolina State University, Raleigh, NC.

Donis-Keller, H., P. Green, C. Helms, S. Cartinhour, B. Weifenbach, K. Stephens, T.P. Keith, D.W Bowden, D.R. Smith, E.S. Lander, D. Botstein, G. Akots, K.S. Rediker, T. Gravius, V.A. Brown, M.B. Rising, C. Parker, J.A. Powers, D.E. Watt, E.R. Kaufman, A. Bricker, P. Phipps, H. Muller-Kahle, T.R. Fulton, S. Ng, J.W. Schumm, J.C. Braman, R.G. Knowlton, D.F. Barker, S.M. Crooks, S.E. Lincoln, M.J. Daly and J. Abrahamson. 1987. A genetic linkage map of the human genome. Cell 51:319-337.

DuMouchel, W.H. and W.W. Anderson. 1968. The analysis of selection in experimental populations. Genetics 58:435–449.

Edwards, A.W.F. 1972. *Likelihood.* Cambridge Univ. Press, Cambridge.

Edwards, A.W.F. 1986. Are Mendel's results really too close? Biol. Rev. 61:295–312.

Edwards, A.W.F. and L.L. Cavalli-Sforza. 1963. The reconstruction of evolution. Heredity 18:553.

Edwards, A.W.F. and L.L. Cavalli-Sforza. 1964. Recosntruction of evolutionary trees. Pp. 67–76 in *Phenetic and Phylogenetic Classification* (Systematics Association Publication No. 6), V.E. Heywood and J. McNeill (Editors). Systematics Association, London.

Efron, B. 1982. *The Jackknife, the Bootstrap and Other Resampling Plans.* CBMS-NSF Regional Conference Series in Applied Mathematics, Monograph 38. SIAM, Philadelphia.

Elston. R.C. 1896a. Probability and paternity testing. Am. J. Human Genet. 39:112–122.

Elston, R.C. 1986b. Some fallacious thinking about the paternity index: Comments. Am. J. Human Genet. 39:670–672.

Elston, R.C. and J. Stewart. 1971. A general method for the analysis of pedigree data. Human Hered. 21:523–542.

Estabrook, G.F., C.S. Johnson and F.R. McMorris. 1975. An idealized concept of the true cladistic character. Math. Biosci. 23:263–272.

Ewens, W.J. 1983. The role of models in the analysis of molecular genetic data, with particular reference to restriction fragment data. Pp. 45–73 in *Statistical Analysis of DNA Sequence Data,* B.S. Weir (Editor). Dekker, New York.

Felsenstein, J. 1978a. The number of evolutionary trees. Syst. Zool. 27:27–33.

Felsenstein, J. 1978b. Cases in which parsimony or compatibility methods will be positively misleading. Syst. Zool. 27:401–410.

Felsenstein, J. 1981. Evolutionary trees from DNA sequences: A maximum likelihood approach. J. Mol. Evol. 17:368–376.

Felsenstein, J. 1983. Methods for inferring phylogenies: A statistical view. Pp. 315–334 in *Numerical Taxonomy*, J. Felsenstein (Editor). Springer-Verlag, Berlin.

Felsenstein, J. 1985. Confidence limits on phylogenies: An approach using the bootstrap. Evolution 39:783–791.

Felsenstein, J. 1988. Phylogenies from molecular sequences: Inference and reliability. Annu. Rev. Genet. 22:521–565.

Finney, D.J., R. Latscha, B.M. Bennett and P. Hsu. 1963. *Tables for Testing Significance in a 2 × 2 Contingency Table.* Cambridge Univ. Press, Cambridge.

Fisher, R.A. 1925. *Statistical Methods for Research Workers*, 13th Edition, Hafner, New York.

Fisher, R.A. 1935. The logic of inductive inference. J. Roy. Stat. Soc. 98:39–54.

Fisher, R.A. 1936. Has Mendel's work been rediscovered? Annals of Science 1:115–137.

Fisher, R.A. and B. Balmakund. 1928. The estimation of linkage from the offspring of selfed heterozygotes. J. Genet. 20:79–92.

Fitch, W.M. 1983. Random sequences. J. Mol. Biol. 163:171–176.

Fitch, W.M. 1971. Toward defining the course of evolution: Minimum change for a specific tree topology. Syst. Zool. 20:406–416.

Fitch, W.M. and M. Margoliash. 1967. Construction of phylogenetic trees. Science 155:279–284.

Fu, Y.X. and J. Arnold. 1990. A table of exact sample sizes for use with Fisher's exact test for 2 × 2 tables. Biometrics (in press).

Fukami, K. and Y. Tateno. 1989. On the maximum likelihood method for estimating molecular trees: Uniqueness of the likelihood point. J. Mol. Evol. 28:460–464.

Fyfe, J.L. and N.T.J. Bailey. 1951. Plant breeding studies in leguminous forage crops. I. Natural cross-breeding in winter beans. J. Agric. Sci. 41:371–378.

Gjertson, D.W., M.R. Mickey, J. Hopfield, T. Takenouchi and P.I. Teraski. 1988. Am. J. Human Genet. 43:860–869.

Goodman, M.M. 1972. Distance analysis in biology. Syst. Zool. 21:174–286.

Goodman, M.M., C.W. Stuber, K. Newton and H.H. Weissinger. 1980. Linkage relationships of 19 enzyme loci in maize. Genetics 96:697–710.

Green, A.G., A.H.D. Brown and R.O. Oram. 1980. Determination of outcrossing rate in a breeding population of *Lupinus albus* L. (White lupin). Z. Pflanzenzüchtg.

84:181–191.

Gürtler, H. 1956. Principles of blood group statistical evaluation of paternity cases at the University Institute of Forensic Medicine, Copenhagen. Acta Med. Leg. Soc., Liège 9:83–93.

Haldane, J.B.S. 1919. The combination of linkage values and the calculation of distances between the loci of linked factors. J. Genet. 8:299–309.

Haldane, J.B.S. 1954. An exact test for randomness of mating. J. Genet. 52:631–635.

Haldane, J.B.S. 1956. The estimation of viabilities. J. Genet. 54:294–296.

Haldane, J.B.S. and C.A.B. Smith. 1947. A new estimate of the linkage between the genes for colour-blindness and haemophilia in man. Ann. Eugen. 14:10–31.

Hartl, D.L. and A.G. Clark. 1989. *Principles of Population Genetics,* Second Edition. Sinauer, Sunderland, MA.

Hasegawa, M., H. Kishino and T. Yano. 1985. Dating of the human-ape splitting by a molecular clock of mitochondrial DNA. J. Mol. Evol. 22:160–174.

Haynam, G.E., Z. Govindarajulu and F.C. Leone. 1970. Tables of the cumulative non-central chi-square distribution. Pp. 1–78 in *Selected Tables in Mathematical Statistics,* Volume 1, H.L. Harter and D.B. Owen (Editors). Am. Math. Society, Providence, RI.

Hernández, J.L. and B.S. Weir. 1989. A disequilibrium coefficient approach to Hardy-Weinberg testing. Biometrics 45:53–70.

Hill, W.G. and B.S. Weir. 1988. Variances and covariances of squared linkage disequilibria in finite populations. Theor. Pop. Biol. 33:54–78.

Hunt, L.T. and M.O. Dayhoff. 1982. Evolution of chromosomal proteins. Pp. 193–239 in *Macromolecular Sequences in Systematic and Evolutionary Biology*, M. Goodman (Editor). Plenum, New York.

Immer, F.R. 1930. Formulae and tables for calculating linkage intensities. Genetics 15:81–98.

Jeffreys, A.J., V. Wilson and S.L. Thein. 1985. Individual-specific "fingerprints" of human DNA. Nature (London) 316:76–79.

Jukes, T.H. and C.R. Cantor. 1969. Evolution in protein molecules. Pp. 21–123 in *Mammalian Protein Metabolism,* H.N. Munro (Editor). Academic Press, New York.

Kan, Y.W. and A.M. Dozy. 1978. Polymorphism of DNA sequence adjacent to the human β-globin structural gene: Relationship of sickle cell anemia. Proc. Natl. Acad. Sci. U.S.A. 75:5631–5635.

Karlin, S., G. Ghandour, F. Ost, S. Tavaré and L.J. Korn. 1983. New approaches for computer analysis of nucleic acid sequences. Proc. Natl. Acad. Sci. U.S.A. 80:5660–5664.

Katz, R.W. 1981. On some criteria for estimating the order of a Markov chain. Technometrics 23:243–249.

Kendall, M.G. and A. Stuart. 1961. *The Advanced Theory of Statistics.* Volume 2. *Inference and Relationship.* Hafner, New York.

Kim, J. and M.A. Burgman. 1988. Accuracy of phylogenetic-estimation methods under unequal evolutionary rates. Evolution 42:596–602.

Kimura, M. 1980. A simple method for estimating evolutionary rates of base substitutions through comparative studies of nucleotide sequences. J. Mol. Evol. 16:111–120.

King, M.-C. 1989. Genetic testing of identity and relationship. Am J. Human Genet. 44:179–181.

Knuth, D.E. 1981. *The Art of Computer Programming.* Volume 2. *Seminumerical Algorithms,* Second Edition. Addison-Wesley, Reading, MA.

Kosambi, D.D. 1944. The estimation of map distances from recombination values. Ann. Eugen. 12:172–175.

Kreitman, M. 1983. Nucleotide polymorphism at the alcohol dehydrogenase locus of *Drosophila melanogaster.* Nature (London) 304:412–417.

Kruskal, J.B. 1983. An overview of sequence comparison. Pp. 1–44 in *Time Warps, String Edits, and Macromolecules.* D. Sankoff and J.B. Kruskal, (Editors). Addison-Wesley, Reading, MA.

Lander, E.S. 1989. DNA fingerprinting on trial. Nature (London) 339:501–505.

Lange, K. and R.C. Elston. 1975. Extensions to pedigree analysis. I. Likelihood calculations for simple and complex pedigrees. Human Hered. 25:95–105.

Langley, C.H., Y.N. Tobari and K.I. Kojima. 1974. Linkage disequilibrium in natural populations of *Drosophila melanogaster.* Genetics 78:921–936.

Langley, C.H., E. Montgomery and W.F. Quattlebaum. 1982. Restriction map variation in the *Adh* region of *Drosophila.* Proc. Natl. Acad. Sci. U.S.A. 79:5631–5635.

Laurie-Ahlberg, C.C. and B.S. Weir. 1979. Allozyme variation and linkage disequilibrium in some laboratory populations of *Drosophila melanogas-ter.* Genetics 92:1295–1314.

Levine, L., M. Asmussen, O. Olvera, J.R. Powell, M.E. de la Rosa, V.M. Salceda, M.I. Gaso, J. Guman and W.W. Anderson. 1980. Population genetics of Mexican *Drosophila.* V. A high rate of multiple insemination in a natural population of *Drosophila pseudoobscura.* Am. Nat. 116:493–503.

Lewin, B. 1987. *Genes III.* Wiley, New York.

Lewis, P.A.W., A.S. Goodman and J.M. Miller. 1969. A pseudo-random number generator for the system/360. IBM Syst. J. 8:136–146.

Lewontin, R.C. 1974. *The Genetic Basis of Evolutionary Change.* Columbia Univ. Press, New York.

Lewontin, R.C. 1988. On measures of gametic disequilibrium. Genetics 120:849–852.

Lewontin, R.C. and C.C. Cockerham. 1959. The goodness-of-fit test for detecting natural selection in random mating populations. Evolution 13:561–564.

Li, C.C. 1988. Pseudo-random mating populations. In celebration of the 80th anniversary of the Hardy-Weinberg law. Genetics 119:731–737.

Li, C.C. and A. Chakravarti. 1988. An expository review of two methods of calculating the paternity probability. Am. J. Human Genet. 43:197–205.

Macpherson, J.N., B.S. Weir and A.J. Leigh-Brown. 1990. Extensive linkage disequilibrium in the achaete-scute complex of *Drosophila melanogaster.* (submitted)

Maizel, J.V. Jr. and R.P. Lenk. 1981. Enhanced graphic matrix analysis of nucleic acid and protein sequences. Proc. Natl. Acad. Sci. U.S.A. 78:7665- 7669

Majumder, P.P. and M. Nei. 1983. A note on the positive identification of paternity by using genetic markers. Human Hered. 33:29–35.

Margush, T. and F.R. McMorris. 1981. Consensus n-trees. Bull. Math. Biol. 43:239–244.

Mather, K. 1951. *The Measurement of Linkage in Heredity,* Second Edition. Methuen, London.

Meagher, R.B., M.D. McLean and J. Arnold. 1988. Recombination within a subclass of restriction fragment length polymorphisms may help link classical and molecular genetics. Genetics 120:809–818.

Meagher, T.R. 1986. Analysis of paternity within a natural population of *Chamaelirium luteum.* I. Identification of most-likely male parents. Am. Nat. 128:199-215.

Mendel, G. 1866. Versuche über Pflanzen-Hybriden. Verhandlungen des naturforschenden Vereines in Brünn 4,3-47. (English translation by Royal Horticultural Society of London, reprinted in Peters, J.A. 1959. *Classic Papers in Genetics.* Prentice-Hall, Englewood Cliffs, NJ.)

Morton, N.E. 1955. Sequential tests for the detection of linkage. Am. J. Human Genet. 7:277–318.

Morton, N. 1964. Genetic studies in northeastern Brazil. Cold Spring Harbor Symp. Quant. Biol. 29:69–79.

Mott, R.F., T.B.L. Kirkwood and R.N. Curnow. 1989. A test for the statistical significance of DNA sequence similarities for application in databank searches. CABIOS 5:123–131.

Mourant, A.E., A.C. Kopec and K. Domaniewska-Sobczak. 1976. *The Distribution of the Human Blood Groups and Other Polymorphisms*, Second Edition. Oxford Univ. Press, Oxford.

Mukai, T., R.A. Cardellino, T.K. Watanabe and J.F. Crow. 1974. The genetic variance for viability and its components in a local population of *Drosphila melanogaster.* Genetics 78:1195–1208.

Needleman, S.B. and C.D. Wunsch. 1970. A general method applicable to the search for similarities in the amino acid sequence of two proteins. J. Mol. Biol. 48:443–453.

Nei, M. 1972. Genetic distance between populations. Am. Nat. 106:283–292.

Nei, M. 1978. Estimation of average heterozygosity and genetic distance from a small number of individuals. Genetics 89:583–590.

Nei, M. 1987. *Molecular Evolutionary Genetics.* Columbia Univ. Press, New York.

Nei, M., F. Tajima and Y. Tateno. 1983. Accuracy of estimated phylogenetic trees from molecular data. II. Gene frequency data. J. Mol. Evol. 19:153–170.

Nussinov, R. 1984. Doublet frequencies in evolutionary distinct groups. Nucleic Acids Res. 12:1749–1763.

Nussinov, R. 1987. Nucleotide quartets in the vicinity of eukaryotic transcriptional initiation sites: Some DNA and chromatin structural implications. DNA 6:613–622.

Ott, J. 1974. Estimation of the recombination fraction in human pedigrees: Efficient computation of the likelihood for human linkage studies. Am. J. Human Genet. 34:630–649.

Ott, J. 1985. *Analysis of Human Genetic Linkage.* Johns Hopkins Press, Baltimore.

Pamilo, P. and M. Nei. 1988. Relationships between gene trees and species trees. Mol. Biol. Evol. 5:568–583.

Parker, R.C., R.M. Watson, and J. Vinograd. 1977. Mapping of closed circular DNAs by cleavage with restriction endonucleases and calibration by agarose gel electrophoresis. Proc. Natl. Acad. Sci. U.S.A. 74:851-855.

Pearson, W.R. 1982. Automatic construction of restriction site maps. Nucleic Acids Res. 10:217–227.

Piegorsch, W.W. 1983. The questions of fit in the Gregor Mendel controversy. Commun. Statist. Theor. Meth. 12:2289–2304.

Piegorsch, W.W. 1986. The Gregor Mendel controversy: Early issues of goodness-of-fit and recent issues of genetic linkage. Hist. Sci. 24:173–182.

Posten, H.O. 1989. An effective algorithm for the noncentral chi-squared distribution function. Amer. Statist. 43:261–263.

Prout, T. 1965. The estimation of fitness from genotypic frequencies. Evolution 19:546–551.

Queen, C.L. and L.J. Korn. 1980. Computer analysis of nucleic acids and proteins. Pp. 595–609 in *Methods in Enzymology.* Volume 65, L. Grossman and K. Moldave (Editors). Academic Press, New York.

Quenouille, M. 1956. Notes on bias in estimation. Biometrika 43:253–260.

Race, R.R., R. Sanger, S.D. Lawler and D. Bertinshaw. 1949. The inheritance of the MNS blood groups: A second series of families. Heredity 3:205–213.

Rao, D.C., N.E. Morton, J. Lindsten, M. Hultén and S. Yee. 1977. A mapping function for man. Human Hered. 27:99–104.

Reynolds, J. 1981. *Genetic Distance and Coancestry.* Institute of Statistics Mimeo Series 1341, North Carolina State University, Raleigh, NC.

Reynolds, J., B.S. Weir and C.C. Cockerham. 1983. Estimation of the coancestry coefficient: Basis for a short-term genetic distance. Genetics 105:767–779.

Rogers, J.S. 1972. Merasures of genetic similarity and genetic distance. In *Studies in Genetics VII.* University of Texas Publication 7213, Austin.

Rohlf, F.J. and R.R. Sokal. 1981. *Statistical Tables,* Second Edition, Freeman, San Francisco.

Saitou N. and T. Imanishi. 1989. Relative efficiencies of the Fitch-Margoliash, maximum-parsimony, maximum-likelihood, minimum-evolution and neigh-bor-joining methods of phylogenetic tree construction in obtaining the correct tree. Mol. Biol. Evol. 6:514–525.

Sankoff, D., C. Morel and R.J. Cedergren. 1973. Evolution of 5S RNA and the non-randomness of base replacement. Nature New Biol. 245:232–234.

Santibánez-Koref, M. and J.G. Reich. 1986. Dinucleotide frequencies in different reading frame positions of coding mammalian DNA sequences. Biomed. Biochim. Acta 45:737–748.

Sarich, V.M. and A.C. Wilson. 1973. Generation time and genomic evolution in primates. Science 179:1144–1147.

Schaffer, H.E. 1983. Determination of DNA fragment size from gel electrophoresis mobility. Pp. 1–14 in *Statistical Analysis of DNA Sequence Data.* B.S. Weir (Editor). Dekker, New York.

Schrage, L. 1979. A more portable Fortran random number generator. ACM Trans. Math. Software 5:132–138.

Schroeder, J.L. and F.B. Blattner. 1978. Least-squares method for restriction mapping. Gene 4:167–174.

Schwartz, R.M. and M.O. Dayhoff. 1979. Matrices for detecting distant relationships. Pp. 353-358 in *Atlas of Protein Sequence and Structure,* Volume 5, Supplement 3, 1978. National Biomedical Research Foundation, Washington, DC.

Shaw, D.V., A.L. Kahler and R.W. Allard. 1981. A multilocus estimator of mating system parameters in plant populations. Proc. Natl. Acad. Sci. U.S.A. 78:1298–1302.

Shen, S., J.L. Slightom and O. Smithies. 1981. A history of the human fetal globin gene duplication. Cell 26:191–203.

Smith, C.A.B. 1957. Counting methods in genetical statistics. Ann. Human Genet. 21:254–276.

Smith, C.A.B. 1977. A note on genetic distance. Ann. Human Genet. 21:254–276.

Smith, T.F., M.S. Waterman and J.R. Sadler. 1983. Statistical characterization of nucleic acid sequence functional domains. Nucleic Acids Res. 11:2205–2220.

Sneath, P.H.A. and R.R. Sokal. 1973. *Numerical Taxonomy.* Freeman, San Francisco.

Sokal, R.R. and F.J. Rohlf. 1981. *Biometry,* Second Edition. Freeman, San Francisco.

Steel, R.G.D. and J.H. Torrie. 1980. *Principles and Procedures of Statistics,* Second Edition. McGraw-Hill, New York.

Sytsma, K.J. and B.A. Schaal. 1985. Genetic variation, differentiation, and evolution in a species of complex tropical shrubs based on isozymic data. Evolution 39:582–593.

Tateno, Y., M. Nei and F. Tajima. 1982. Accuracy of estimated phylogenetic trees from molecular data. J. Mol. Evol. 18:387–404.

Tavaré, S. and B.W. Giddings. 1989. Some statistical aspects of the primary structure of nucleotide sequences. Pp. 117–132 in *Mathematical Methods for DNA Sequences,* M.S. Waterman (Editor). CRC Press, Boca Raton, FL.

Thomson, G. and M.P. Baur. 1984. Third order linkage disequilibrium. Tissue Antigens 24:250–255.

Tukey, J. 1958. Bias and confidence in not quite large samples. Ann. Math. Stat. 29:614.

Vithayasai, C. 1973. Exact critical values of the Hardy-Weinberg test statistic for two alleles. Commun. Stat. 1:229–242.

Vogel, F. and A.G. Motulsky. 1986. *Human Genetics,* Second Edition. Springer-Verlag, New York.

Weir, B.S. 1979. Inferences about linkage disequilibrium. Biometrics 25:235–254.

Weir, B.S. 1989a. Sampling properties of gene diversity. Pp. 23–42 in *Plant Population Genetics, Breeding and Genetic Resources,* A.H.D. Brown, M.T. Clegg, A.L. Kahler and B.S. Weir (Editors). Sinauer, Sunderland, MA.

Weir, B.S. 1989b. Locating the cystic fibrosis gene on the basis of linkage disequilibrium with markers? Pp. 81–86 in *Multipoint Mapping and Linkage Based upon Affected Pedigree Members: Genetic Analysis Workshop 6.* R.C. Elston, M.A. Spence, S.E. Hodge and J.W. MacCluer (Editors), Liss, New York.

Weir, B.S., R.W. Allard and A.L. Kahler. 1972. Analysis of complex allozyme polymorphisms in a barley population. Genetics 72:505–523.

Weir, B.S., R.W. Allard and A.L. Kahler. 1974. Further analysis of complex allozyme polymorphisms in a barley population. Genetics 78:911–919.

Weir, B.S. and C.J. Basten. 1990. Sampling strategies for DNA sequence distances. Biometrics (in press).

Weir, B.S. and L.D. Brooks. 1986. Disequilibrium on human chromosome 11p. Genet. Epidemiol. Suppl. 1:177–183

Weir, B.S. and C.C. Cockerham. 1978. Testing hypotheses about linkage disequilibrium with multiple alleles. Genetics 88:633–642.

Weir, B.S. and C.C. Cockerham. 1979. Estimation of linkage disequilibrium in randomly mating populations. Heredity 42:105–111.

Weir, B.S. and C.C. Cockerham. 1984. Estimating F-statistics for the analysis of population structure. Evolution 38:1358–1370.

Weir, B.S. and C.C. Cockerham. 1989. Complete characterization of disequilibrium at two loci. Pp. 86–110 in *Mathematical Evolutionary Theory,* M.E. Feldman (Editor). Princeton Univ. Press, Princeton.

Weir, B.S. and G.B. Golding. 1990. Gametic linkage disequilibrium at two, three and four loci. (submitted)

Weir, B.S., J. Reynolds and K.G. Dodds. 1990. The variance of sample heterozygosity. Theor. Pop. Biol. (in press).

Weir, B.S. and S.R. Wilson. 1986. Log-linear models for linked loci. Biometrics 42:665–670.

Wellso, S.G., D.J. Howard, J.L. Adams and J. Arnold. 1988. Electrophoretic monomorphism in six biotypes and two populations of the Hessian Fly (*Diptera: Cecidomyiidae*). Ann. Entomol. Soc. Am. 81:50–53.

Williams, C.J. and S. Evarts. 1989. The estimation of concurrent paternity probabilities in natural populations. Theor. Pop. Biol. 35:90–112.

Williams, C.J., W.W. Anderson and J. Arnold. 1990. Generalized linear modeling methods for selection component experiments. Theor. Pop. Biol. (in press)

Workman, P.L. and S.K. Jain. 1966. Zygotic selection under mixed random mating and self-fertilization: theory and problems of estimation. Genetics 54:159–171.

Wright, S. 1951. The genetical structure of populations. Ann. Eugen. 15:323–354.

Wu, C.-I. and W.-H. Li. 1985. Evidence for higher rates of nucleotide substitution in rodents than in man. Proc. Natl. Acad. Sci. U.S.A. 82:1741–1745.

Yasuda, N. and M. Kimura. 1968. A gene-counting method of maximum likelihood for estimating gene frequencies in ABO and ABO-like systems. Ann. Human Genet. 31:409–420.

Yates, F. 1934. Contingency tables involving small numbers and the X^2 test. J. Roy. Stat. Soc. Suppl. 1:217–235.

Yates, F. 1984. Tests of significance for 2 × 2 contingency tables. J. Roy. Stat. Soc. A 147:426–463.

Yeh, F.C. and K. Morgan. 1987. Mating system and multilocus associations in a natural population of *Pseudotsuga menziesii* (Mirb.) Franco. Theor. Appl. Genet. 73:799–808.

Zuker, M. 1989. The use of dynamic programming algorithms in RNA secondary structure prediction. Pp. 159–184 in *Mathematical Methods for DNA Sequences,* M. Waterman (Editor). CRC Press, Boca Raton, FL.

Author Index

Subject Index